Build Your Own Electric Vehicle

About the Authors

Seth Leitman (Briarcliff Manor, New York) is currently President and Managing Member of Green Living Guy©, which talks about organic, natural, and sustainable products for business and home use (from energy-efficient bulbs to electric vehicle conversion referrals). Previously, he worked for the New York Power Authority and the New York State Energy Research and Development Authority, where he helped develop, market, and manage electric and hybrid vehicle programs serving New York State and the New York metropolitan area. For green living news, follow Seth on Twitter @Seth_Leitman; for electric vehicle conversion and electric transportation news, @BuildYourOwnEV.

Bob Brant was the author of the first edition of this book, published in 1993, and some might say ahead of his time in his passion to convert to electric. While there have obviously been updates and technological advances since then, many of the concepts in the first edition are still in use today. Bob grew up in New York City, got a BSEE, and worked on NASA projects such as the Apollo program, the Lunar Excursion Module, and the Earth Resources Technology Satellite. He then went on to get an MSEE and MBA, and worked for a company that worked on the Lunar Rover. Bob was always fascinated with every electric vehicle breakthrough, was convinced of the electric vehicle's personal and environmental benefits, and was curious why stronger steps had not been taken to make electric vehicles a reality.

Build Your Own Electric Vehicle

Seth Leitman
Bob Brant (Deceased)

Third Edition

New York   Chicago   San Francisco
Lisbon   London   Madrid   Mexico City
Milan   New Delhi   San Juan
Seoul   Singapore   Sydney   Toronto

Cataloging-in-Publication Data is on file with the Library of Congress

McGraw-Hill Education books are available at special quantity discounts to use as premiums and sales promotions, or for use in corporate training programs. To contact a representative, please e-mail us at bulksales@mcgraw-hill.com.

Build Your Own Electric Vehicle, Third Edition

1 2 3 4 5 6 7 8 9 0   DOC/DOC   1 9 8 7 6 5 4 3

ISBN 978-0-07-177056-9
MHID   0-07-177056-9

 The pages within this book were printed on acid-free paper containing 100% postconsumer fiber.

Sponsoring Editor
Judy Bass

Editorial Supervisor
Stephen M. Smith

Production Supervisor
Pamela A. Pelton

Acquisitions Coordinator
Bridget L. Thoreson

Project Manager
Patricia Wallenburg, TypeWriting

Copy Editor
James Madru

Proofreader
Claire Splan

Indexer
Judy Davis

Art Director, Cover
Jeff Weeks

Composition
TypeWriting

Contents

Preface xiii
Acknowledgments xix

1   Why Electric Vehicles Are for Today! **1**
Build That Car! 4
What Is an Electric Vehicle? 8
    Electric Motors 8
    Batteries 9
    Controllers 9
Have You Driven an EV Lately? 10
    Electric Vehicles Offer a "Total Experience" 10
    Electric Vehicles Are Fun to Drive 10
    Electric Vehicles Make a Difference by Standing Out 10
    Electric Vehicles Save Money 12
    Electric Vehicles Are Customizable 12
    Safety First 13
    Electric Vehicles Save the Environment 15
Electric Vehicle Myths (Dispelling the Rumors) 15
    Myth #1: Electric Vehicles Can't Go Fast Enough 16
    Myth #2: Electric Vehicles Have Limited Range 16
    Myth #3: Electric Vehicles Are Not Convenient 17
    Myth #4: Electric Vehicles Are Expensive 18
The Disadvantages of Electric Cars Have Been
    Reduced or Eliminated 20
Time to Purchase/Build Your Own Brand-New Electric Car! 20
Get Your EV Built! 25

2   Electric Vehicles Save the Environment and Energy **27**
Why Do Electric Vehicles Save the Environment? 27
Save the Environment and Save Some Money Too! 28
Petroleum Will Not Last Forever 30
Clean Energy Is the Future 30
Fuel-Efficient Vehicles 32
Time Is Running Out! 32
U.S. Transportation Depends on Oil 33
    Increasing Long-Term Oil Costs 34
    What's Better for the Environment: Raising the Gas Tax
      or Fuel Efficiency Standards? 34
    Summary 41

3   Electric Vehicle History **53**
1900s 53

Timeline of Vehicle History   54
The Timeline of Electric Cars   54
    Up to 1915   55
    Huff and Puff—Steam First, Then Electric, Then Oil,
        and Then Electric Again   55
The Golden Age of Internal Combustion Engines   56
A World Awash in Oil After World War II   56
Twilight of the Oil Gods   57
Electric Vehicles   59
    Forget Oil! Electric Cars Have the Need for Speed!   59
    National Electric Drag Racing Association   60
    The Need for Distance   61
    Casey Mynott Builds for Speed   61
The Need for an Association   61
    Worldwide   61
    North America   62
    Europe   63
Lithium Ion Just Starts the EV Market   63
The Need for Events, Cars, Books, and Movies   63
    The 1990s Until Today   63
    Regulation in California   64
    9/11, Oil, and Our New Understanding of EVs   66
    GM's Awakening—The Volt   68
    EVs for the Twenty-First Century   70
    Near-Future Trends for Electric Drives   78
Electric Mobility Still Is an Academic and International Move   80
    TurnE   80
    EVE (Italy)   81
    Tesla Is Building Model X Electric Cars and They Are
        Selling Like Hot Cakes   81
Summary   83

4   The Best Electric Vehicle for You   85
EV Purchase Decisions   85
    Conversions Can Save You Money and Time   85
    Buying or Leasing a Ready-to-Run EV Saves
        You a Lot of Time   86
    EVs Have Some Big-Name Backers   91
    Other Cool Electric Cars   92
    The Netherlands Is Building Its Own EV   95
    Mitsubishi i-MiEV   96
    REVA Is Loved in the United Kingdom and India.
        It's Selling!   97
EV Conversion Shops   97
    Buying a Ready-to-Run EV from an Independent
        Manufacturer   97

Converting a Vehicle 98
    Converting Existing Vehicles 99
    Converting Existing SUVs and Van-Type Cars 100
    EV Conversion Decisions 105
The Procedure 110
Electric Car Motors II: The AC Versus DC Debate 111
    Where AC Electric Motors Aren't the Best Fit 112
    How Much Is This Going to Cost? 112
    Analysis 114
Conclusion 114

5 **Chassis and Design** **117**
Comprehensive Testing Under Way—and There's More on the Way ... 118
Choose the Best Chassis for Your EV 118
    Know Your Options! 120
Optimize Your EV 121
    Conventions and Formulas 121
It Ain't Heavy, It's My EV 124
    Remove All Unessential Weight 124
    Weight and Acceleration 125
    Weight and Climbing 126
    Weight Affects Speed 126
    Weight Affects Range 128
    Remove the Weight but Keep Your Balance 128
    Remember the 30 Percent Rule 129
Streamline Your EV Thinking 130
    Aerodynamic Drag Force Defined 130
    Choose the Lowest Coefficient of Drag 130
    Relative Wind Contributes to Aerodynamic Drag 133
    Aerodynamic Drag Force Data You Can Use 134
    Shape Rear Airflow 134
    Shape Wheel Well and Underbody Airflow 134
    Block and/or Shape Front Airflow 135
Roll with the Road 135
    Rolling Resistance Defined 136
    Pay Attention to Your Tires 136
    Rolling Resistance Force Data You Can Use 137
Less Is More with Drive Trains 138
    Drive Trains 138
Difference in Motor Versus Engine Specifications 140
Going Through the Gears 144
    Automatic Versus Manual Transmission 145
    Use a Used Transmission 145
    Heavy Versus Light Drive Trains and Fluids 145
Design Your EV 146
    Horsepower, Torque, and Current 147

Calculation Overview 148
Torque-Required Worksheet 149
Torque-Available Worksheet 149
Torque-Required and Torque-Available Graph 149
Buy Your EV Chassis 154
EV Conversions 154
The Other Side of Conversion 155
How to Get the Best Deal 155
All-Over Aerodynamic Aesthetics 160

6   **Electric Motors** **163**
Why an Electric Motor? 164
Horsepower 164
DC Electric Motors 165
Magnetism and Electricity 166
Conductors and Magnetic Fields 166
Ampere's Law or the Motor Rule 167
Electromagnets and Motors 167
DC Motors in General 168
DC Motors in the Real World 169
Armature 169
Commutator 169
Field Poles 169
Series Motors 169
Brushes 170
Motor Case, Frame, or Yoke 170
DC Motor Types 171
Series DC Motors 172
Shunt Motors 173
Compound DC Motors 175
Permanent-Magnet DC Motors 176
Brushless DC Motors 177
Universal DC Motors 177
AC Electric Motors 178
Transformers 179
AC Induction Motors 180
Single-Phase AC Induction Motors 180
Polyphase or Split-Phase AC Induction Motors ... 181
Wound-Rotor Induction Motors 182
Tomorrow's Best EV Motor Solution 185
AC Motors: HiPer Drive by ElectroAutomotive 185
Motor Ratings 187
Tuning 188
Keep It Simple 188
Conclusion 188

7   **The Controller** **193**
    Controller Overview 193
        Solid-State Controllers 195
        Electronic Controllers 195
        AC Controllers 196
    Controller Choice 196
        An Off-the-Shelf Curtis PWM DC Motor Controller 196
        AC Controllers 198
    Today's Best Controller Solutions 198
        Zilla Controller (One of the Best DC Controllers
            for Conversions) 199
    EV Controllers Help to Dispel All Myths About EVs Today 203
    AC Propulsion, Inc., to the Rescue—Today 204
        AC-150 Gen 2 System: A Reliable Proven Performer
            for Passenger Car Needs 204
        AC-150 Gen 3 System: For Next-Generation EVs 205
    Tesla Controllers Use AC Propulsion Technologies 214
    Summary 215

8   **Batteries** **217**
    Metals, Salts, and Ions: Nature's Sad Love Story 218
    Building a Battery 220
    The Circuit 221
        Amperes Versus Volts 221
    Battery Anatomy 222
    Shapes and Sizes of Lithium-Ion Batteries 223
        Battery Formats 223
    Ampere-Hours and Voltage 225
    Wiring in Series or Parallel 225
    C-Rate 226
    Flat Discharge Curve 226
    The Diamond in the Discard Pile: $LiFePO_4$ 227
        The Scandal 227
        That Beautiful LiFePO4 Cathode 228
        Molecular $LiFePO_4$ 228
    Tinkering with the $LiFePO_4$ Recipe 230
    Stumbling onto Something Good 230
    Going with the Grain 231
    Impedance and Fast Charging 231
    Energy Density Versus Power Density 232
        Power Density 232
    Mother Nature's Battery Management System 233
    Thermal Runaway Explained 234
    Best Practices for Your $LiFePO_4$ EV Batteries 235
        Balancing 236

Battery Management 236
Cooling 237
Warm Enough? 237
Charging 237
Battery Box Placement 237
Battery Wiring 238
Safety Tip—First Startup 239
Summary 239

9   The Charger and Electrical System **241**
Charger Overview 241
Battery Discharging and Charging Cycle 242
What You Can Learn from a Battery Cycle-Life Test 242
Battery Discharging Cycle 243
Battery Charging Cycle 243
The Ideal Battery Charger 244
Charging Between 0 and 20 Percent 246
The Real-World Battery Charger 246
ChargePoint America Program 247
The Manzita Micro PFC-20 247
The Zivan NG3 248
Other Battery-Charging Solutions 250
ECOtality Partners with Regency Centers at
19 Locations Nationwide 251
The EV Project 254
Rapid Charging 254
Induction Charging 255
Nissan and Nichicon Launched the Leaf to Home
Power-Supply System with EV Power Station in Japan ... 255
The EV Power Station: Specifications 257
Replacement Battery Packs 258
Beyond Tomorrow 259
Your EV's Electrical System 266
High-Voltage, High-Current Power System 266
Low-Voltage, Low-Current Instrumentation System 269
DC-to-DC Converter 275
Wiring It All Together 275
Wire and Connectors 275
Connections 276
Routing 276
Grounding 276
Checking 276
Summary of the Parts 277

10   Electric Vehicle Builds and Conversions **281**
Conversion Overview 281
Before Building or Converting 284

Arrange for Help 284
Arrange for Space 285
Arrange for Tools 285
Arrange for Purchases and Deliveries 285
Building or Converting 286
Chassis 288
Mechanical 292
Mounting and Testing Your Electric Motor 298
Fabricating Battery Mounts 300
Additional Mechanical Components 302
Cleanup from Mechanical to Electrical Stage 302
Electrical 302
High-Current System 303
Low-Voltage System 308
Junction Box 312
Charger System 312
Batteries 320
Battery Installation 320
Battery Wiring 322
12-V Accessory Battery 328
After Conversion 328
System Checkout on Blocks 329
Further Improved Cooling 330
Improving Heating Too 330
Neighborhood Trial Run 332
Paint, Polish, and Sign 332
Onward and Upward 332
Put Yourself in the Picture 335
Summary 338

11   How We Can Maximize Our Electric Vehicle Enjoyment 341
Registration and Insurance Overview 341
Getting Registered 342
Getting Insured 342
Safety Footnote 342
Driving and Maintenance Overview 343
Driving Your EV 343
Running Out of Power 344
Regular Driving 344
Caring for Your EV 345
Battery Care 345
Tire Care 346
Lubricants 346
Checking Connections 348
Safety Information 349
Before Working on Any Electric Motor 349

      Motors 350
      Emergency Kit 351
Recommendations for EV Mass Deployment 352
EV Update 354
Conclusion 357

Notes 363

Index 369

Preface

The electric vehicle's time has come. That is what I said in the second edition of this book, and I will say it again today and for the future when electric cars can go 300 miles for a reasonable price. People in Germany are getting amazing range and price. Why not worldwide?

The electric vehicle (EV) movement has broadened to multiple levels of the public debate. *Who Killed the Electric Car?* and *Revenge of the Electric Car* prove that electric transportation is not going away. We have more hybrids on the road—plug-in hybrids are the next wave—and all-electric is becoming vogue with the car companies, the U.S. government, and celebrities.

Well, EVs solve a lot of problems quickly. EVs bypass high energy prices. EVs cost pennies to charge. EVs have zero tailpipe emissions.

While they charge up on electricity from power plants, they also can charge on electricity from solar, wind, or any other renewable resource. Also, if you compare emissions from power plants for every car on the road versus gasoline emissions, electric cars are always—*always*—cleaner. In addition, as power plants get cleaner and begin to reduce emissions, EVs will only get cleaner.

EVs also help to develop the economy. We all know that we need to increase the number of electric cars, yet we have more electric transport on the roads than ever before. Hybrid electrics, plug-in hybrids, and low-speed electric vehicles all expand electric transportation. We as a country—no, we as the world—are increasing our involvement in this industry. From China to India, to Great Britain, to France, and back here in the United States, electric transportation will and can only create a new industry that will increase our manufacturing sector's ability to build clean, efficient cars.

This can only increase domestic jobs in the United States and help our economy. We have been depending on fossil fuels from countries that predominantly do not support the best financial interests of the United States. It is a national security issue for all of us to be sending over billions of dollars to countries that are politically unstable and/or antagonistic toward Western nations. Another way to ask the question is: Should we be sending more money to Iran and Venezuela, or should we keep it in our own pockets? This is why I believe in a pollution-free, oil-free form of transportation. My first company's tag line use to be, "Pollution-Free, Oil-Free, It's Good to Be Free." This is the mantra I would like to provide for this book! When you drive an EV, that is how you feel—*free*.

Who I Am

My name is Seth Leitman, and I had always been known in "green circles" as *that green living guy*. I am the guy who answers questions "people are afraid to ask" about going green. I guess that's why they call me the *green living guy*.

Author and Editor

The books on green living that I've written or edited are:

Build Your Own Electric Vehicle by Seth Leitman and Bob Brant, Second Edition

Build Your Own Plug-In Hybrid Electric Vehicle by Seth Leitman

Build Your Own Electric Motorcycle by Carl Vogel

Green Lighting by Brian Clark Howard, William Brinsky, and Seth Leitman

Solar Power for Your Home by David Findley

Renewable Energies for Your Home by Russel Gehrke

Do-It-Yourself Home Energy Audits by David Findley

Build Your Own Small Wind Power System by Kevin Shea and Brian Clark Howard

and more green living books to follow.

These green living guides focus on implementing environmentally friendly technologies and making them work for you. Currently, I am developing a video series about each of these books.

I am managing member for Eco, Going Green Sustainability Consulting Services, which is part of my ETS Energy Store, LLC. It is intended to help people, companies, and government to go green and live greener and provide more going-green opportunities. I also have a division that helps clients with LEED certification, solar power, wind, geothermal, energy efficiency, lighting, indoor air quality, and more. We get things done! Whether it is asking a third party about their audits, lighting change-outs, or whatever they are working on, Go Green, Save Green, and Learn to Love Green Living (saving cash, that is)!

Also, I have formed a company under the ETS Energy Store, LLC called *Green Living Guy Productions*. Its purpose is to provide social media, editorial, video, and movies to sustainable companies or corporations trying to go green through the consulting services. Bottom line: Ask me to consult you, I'll promote you.

I am also a reporter, writer, blogger, interviewer, videographer, eco-consultant, and sustainability expert under the ETS Energy Store, LLC, which owns GreenLivingGuy .com. The Green Living Guy has written for or is currently writing for Findthebest.com, Greenopia.com, GreenTowns.com, EcoSnobberySucks.com, The Daily Green, *Huffington Post*, *Modern Hippie Magazine*, *Inhabitat*, *Planet Green*, *New York House Magazine*, the local Patch.com, MotherEarthNews.com, and other green living websites, newspapers, and magazines. I've been interviewed for the local Journal News to the BBC.

I am even now a local expert for WPIX Channel 11! Now I am getting into film, production, and acting. This should be fun!

How I Have Helped New York and the World to Go Green

I worked for the New York State Energy Research and Development Authority (NYSERDA) on marketing and managing alternative-fueled vehicle (AFV) and electric vehicle (EV) programs throughout the State of New York. At NYSERDA, I was the project manager for the Clean Fuel Bus Program that funded over $100 million to the incremental cost of AFV buses.

I learned about green buildings and indoor air quality in schools and how it affects children and asthma rates, having an indirect impact on environmental justice. Then I knew that school buses and the school's own energy consumption affect the bottom line, indoor air quality for the students, and pollution from power plants.

In 2000, I worked for the New York Power Authority (NYPA), where we created the NYPA/TH!NK Clean Commute Program, which leased 100 EVs in the New York metropolitan area. This program was covered by *USA Today*, the Associated Press, Reuters, *The New York Times*, CNN, *Good Morning America*, the *Today Show*, and others. While at NYPA, I also worked on the donation program for low-speed vehicles.

I also worked with the two large low-speed EV companies in the United States. They were Global Electric Motor Cars (GEM) and the TH!NK Neighbor (previously from Ford Motor Company) all across New York State. From Brooklyn to western New York, we placed thousands of these cars to spread them around for people to see and appreciate electric cars (in one shape or form).

While at NYPA, I had the pleasure to work with New York City transportation staff on electric vehicles and watched as well as helped a little the Lighting Department in the Energy Services Group of NYPA. Ninety-nine percent of the time I did electric cars but I had a little part in the change-out of all the traffic lights from incandescent to light-emitting diodes (LEDs). That's what gave me the impetus to write the book *Green Lighting*. The group even showed me CFL technology. The Energy Services Group works every day to provide inexpensive, reliable, clean power for many customers in the State of New York.

NYPA even keeps the lights on and the backup power for New York City too. In addition, whenever you ride a subway or look at a street lamp, that power is brought to you by NYPA for the City of New York and the MTA.

They have renewable energy, clean transport, and energy efficiency services that are bar none for agencies statewide, localities, and the Power for Jobs program.

Radio Host

I'm even hosting a really fun and funny blog talk radio show at Green Living Guy.

Published Government and Electric Car Reports

I've published reports with the Electric Power Research Institute, the U.S. Department of Energy, and NYSERDA on the Clean Commute Program and Green Schools. In addition, when I did my master's degree in comparative international development, I had the advisor to the president of the World Bank and former U.S. Congressman Matthew McHugh attend my graduate school, the Rockefeller College of Public Affairs and Policy.

So this is why I am considered the green living guy. I'm not perfect. I am just a guy. However, if we can make everyone green gurus, then that is some good green living.

Just like Bob Brant, I am a New Yorker who rode the electric-powered subway trains. In fact, when I worked for NYPA, which powers those subways, I gained a new appreciation for electric transportation every time I took the train into and around New York City.

My interest in renewable energy and energy efficiency, however, began in graduate school at the Rockefeller College of Public Affairs and Policy in Albany, New York, where I received a master of public administration degree. I concentrated on comparative international development, which focused on the World Bank and the International Monetary Fund. After I read about the World Bank funding inefficient and environmentally destructive energy projects, such as coal-burning power plants in China and dams in Brazil that had the potential to destroy the Amazon, I decided to take my understanding to another level. For my master's thesis, I interviewed members of the World Bank, the International Monetary Fund, and the Bretton Woods Institution.

I was fortunate enough to be able to ask direct questions to project managers who oversaw billions that went to China to build coal-burning power plants. I asked them how the bank could fund an environmentally destructive energy project when there were no traps or technologies to recapture the emissions and use that energy. The answers were not good, but since I researched the bank, attention to environmental issues has expanded by leaps and bounds, and the bank is starting to work toward economic and environmental efficiencies. While it still has a lot of work to do, it's clear that progress is possible.

This passion to understand how organizations could create positive environmental, energy-efficient, and economic development programs (all in one) led me to work for NYSERDA. While there, I was the lead project manager for the USDOE Clean Cities Program (a grass roots–based program that developed AFV projects across the country) for five of the seven Clean Cities in New York. I was also the lead project manager for the Clean Fuel Bus Program for the Clean Air/Clean Water Bond Act, which provided over $100 million in incremental cost funding for transit operators to purchase alternative-fueled buses. When I funded programs and realized the benefits of electric cars or hybrid electric buses versus their counterparts, I was transformed. I saw that electric transportation was the way to go.

I fell so in love with electric transportation that I went to work for the NYPA Electric Transportation Group. In total, I helped to bring over 3,500 alternative-fueled vehicles and buses into New York. I was the market development and policy specialist for NYPA, the nation's largest publicly owned utility. I worked on the development, marketing, and management of electric and hybrid vehicle programs serving the New York metropolitan area. I developed programs that expanded the NYPA fleet from 150 to over 700 vehicles while enhancing public awareness of these programs.

I was the lead manager of the NYPA/TH!NK Clean Commute Program, which under my leadership expanded from 3 to 100 vehicles. This program was the largest public-private partnership of its time. I secured and managed a $6.5 million budget funded by the federal government, project partners, participants (EV drivers), and Ford Motor Company. I developed an incentive program that offered commuters up-front parking at train stations with electric charging stations. We provided insurance rebates and reduced train fares. To date, this was the largest EV station car program in the world. Figures 1 and 2 show station parking and chargers for cars leased in Chappaqua, New York. This was one of the most successful train stations (excluding Huntington and Hicksville in Long Island, New York) for the program. Figure 1 is a great overhead shot of the cars lined up in the areas we set up right next to the train station platform. Figure 2 shows one of the charging stations up close using an AVCON charger with an overhead light. The station connector cables were designed to be like a regular gas station (thanks to Bart Chezar, former manager of the Electric Transportation Group;

FIGURE 1 An electric car dream. TH!NK City electric cars at Chappaqua train station in the NYPA/ TH!NK Clean Commute Program™. (*Town of New Castle.*)

Sam Marcovicci, who was the electric charger specialist for NYPA at the time; and ETEC out in Arizona). On a related note, TH!NK is reemerging into the international automotive marketplace and is starting to open offices in the United States. The TH!NK is coming back!

I also led and worked with multiple state and local agencies to place over 1,000 GEM and TH!NK Neighbor low-speed vehicles for their respective donation programs to meet zero-emission vehicle credits for New York State.

FIGURE 2 TH!NK City cars charging with AVCON chargers at Chappaqua train station. (*Town of New Castle.*)

My intention for this book can be summed up in a conversation I had with Bill Ford, Jr., in Florida. I handed him a copy of *Build Your Own Electric Vehicle* and told him that I converted a Ford Ranger in it. I also told him that I am working on a third edition for future car manufacturers and for Ford and all the other car companies, car conversion companies, and people who just want to buy or lease a Ford Focus Electric, or a Tesla Model S, or a Nissan Leaf. This book does the tie-ins to research institutes by showing people how to convert their cars to electric. In addition, this version has many more pictures to add to the amazing text.

I also want to tell people why they should build their own electric car. Now look, I am not telling you that if you don't want to convert one yourself you shouldn't check out this book. I want you to realize that this book is for you too. You will see all the car companies that are in the electric car game and what conversion car companies are making.

In addition, for the builders or mechanics, the intent is to create a useful guide to get you started, to encourage you to contact additional sources in your own "try-before-buy" quest, to point you in the direction of the people who have already done it (i.e., EV associations, consultants, builders, suppliers, and integrators), to familiarize you with EV components, and finally, to go through the process of actually building/converting your own EV. My intention is never to make a final statement, only to whet your appetite for EV possibilities. I hope that you have as much fun reading about the issues, sources, parts, and building process as I have had.

Another timely point to add concerns the price of oil and its relation to this book. When Bob Brant first wrote this book, he determined that $100 for a barrel of oil was the worst-case scenario. When the second edition was being produced, oil was trading at $142 a barrel, and I saw my local gas station prices at $4.50 per gallon for gas and $5.03 per gallon for diesel. My friends in the financial markets were telling me that $170 to $200 a barrel for oil can happen in the future. Therefore, since the price of gas is a moving target, I left all mathematical equations assuming $4.50 per gallon. You might not be at this price today, but there is nothing saying that it can't get back to that level and above in the future.

Finally, while completing this third edition, Hurricane Sandy destroyed the New York metropolitan area. We saw gas lines that reminded people of the gas lines in the late 1970s. Electric cars can help us even in times of a disaster since we would need no oil.

—*Seth Leitman*

Acknowledgments

To paraphrase an actor who just won an Emmy, "There are so many people to thank." However, I want to first dedicate this book to my family. They have watched my involvement in electric cars expand over the years and, with global warming and green becoming the new black, they really appreciate what I am doing. More importantly, I am in this movement so that my children can have a better earth to live in, since climate change and global warming is a reality.

There are also so many people to thank. I want to thank Carl Vogel and Steve Clunn. Carl started Vogelbilt Corporation. He developed an electric motorcycle with the assistance of the first edition of this book. He has added his comments on the chapters in this book and has always been a great friend to me. More importantly, he wrote a book called *Build Your Own Electric Motorcycle*—check it out!

Steve owns a company in Florida called Grass Roots EV and now Green Shed Conversions. We have been talking about car conversions for years. I have always supported him and believed he was doing great things. Now, he has so many cars to convert that he has a waiting list. That is so great for him and a great sign for the conversion industry.

Thanks to Paul Liddle for his continued movement of the all-electric Porsche. His conversions inspire people to do better!

Thanks also go to Ron Freund, Chairman of the Electric Auto Association. When I told him about this book he immediately said, "Look no further." He wasn't interested in anything but making sure the book was a success. Not for personal glory, but for the electric car movement. "They don't build them like that anymore," a friend once told me and he is right. This organization represents all the clubs across the country that promote electric cars and they need all of our support. There are some great, honorable people involved in these clubs with a love and desire to do the right thing: to make sure electric cars flourish. To me, that deserves our appreciation and thanks.

I'd also like to thank Chelsea Sexton. We all know her from the movie *Who Killed the Electric Car?* When I asked her for her assistance, she was more than willing. More than that, there was no ego or posturing. I say this because she came to help out on the book with a smile, decency, and greatness. Chelsea founded Plug-In America, which is working to make sure plug-in hybrid electric cars become a reality. We'll talk about plug-in hybrids later. What I like is that the organization does their work in a bipartisan manner and they do it for all the right reasons. Even Felix Kramer from Plug-In America once told me that he always tells people that the best type of car is an electric car. They deserve to receive kudos too.

Don Francis from Southern Company has known me since my time with the State of New York and when I worked in the Electric Transportation Group of the New York Power Authority. He has seen my involvement expand over the years and has even helped out with this book with a simple phone call. He is a great friend and colleague.

Credit must also be extended to the many other electric vehicle enthusiasts who made this book possible, such as Lee Hart, Bill Moore from EVWorld, and Remy Chevalier from Electrifying Times and Rock the Reactors; they are some of the experts of the electric vehicle industry. I also want to thank Lynne Mason from electric-cars-are-for-girls.com for contributing so much to this book. Check out her website and support her efforts. A big thank you goes to Ken Koch. He was part of the first edition of this book and some thought he was not working on EV conversions. Well, his conversions are in this new edition. KTA Services (his previous company) is still in business too. Talk about staying power and support for the electric vehicle movement! Also, thanks go to Christian von Hösslin from TurnE Conversions, Ruediger Hild from Germany, Ford Motor Company, Tesla Motors, Toyota Motor Corporation, Tom Gage formerly from AC Propulsion, Rick Woodbury from Commuter Cars Corporation, Ian from Zero Emission Vehicles Australia, and Chris Isles. Chris was one of the clean commuters in Chappaqua, and this past year he and I connected again. His kids even started babysitting my kids. He said he wished the program was back. I just hope this book helps create more programs and electric vehicles are used more regularly.

I also want to thank Bob Brant for writing the first edition of this masterpiece, as well as his wife, Bonnie Brant.

Most importantly, I want to thank Judy Bass from McGraw-Hill Education. She is such a sweet, loving person who believed in me. Also, a shout out to the newest crew at McGraw-Hill Education, Bridget Thoreson. She is the best and the future for this company.

I cannot forget Patty Wallenburg, my compositor extrodinare. If you need a book, call her!

I have to thank Michael Vincent, Director of Publication Operations for Commodities and Commercial Markets at The McGraw-Hill Companies. He has coached my son in lacrosse, and his friendly support helped me during the times I worked on my books. He didn't edit anything but he loved what I was doing.

A shout out, too, to my friend Steve Monaco from high school. You have always worked to keep me centered, like gifting me *The Tao of Pooh* on the night of the party for the second edition of this book.

I dedicate this book to my beautiful wife, Jessica, and our beautiful sons, Tyler and Cameron. While sometimes things have been difficult for a person starting a company after the realities of living the Northeast version of the movie *Who Killed the Electric Car?*, Jessica always supported me, loved me, wished for my happiness, and was a rock during the tough times. My sons Tyler and Cameron are two great boys who only inspire me with their happiness, love, and intellect. As I see them grow up in the world we live in today, I am glad that this book can help to do its part for their future.

I also want to thank my Aunt Carol, Uncle Dick, Mom and Dad, Uncle Bob, Aunt Pam, and all the cousins. I love you all! You've been there and that's all one can ask from anyone.

Finally, a big thank you to all of my supporters and collaborators: MSI Lighting, for LED bulbs; Brian Clark Howard, Remy Chevalier, and Jim Motavalli, the masters of disasters; Josh Dorfman, Summer Rayne Oakes, Remy Chevalier, Phu Styles, Jane Graves, Lori Hamilton, Taryn Hipwell, Ruben Estrada, Evette Rios, John Metric, and everyone at NEDRA; Christopher Gravanga, Ileana Garcia Jacalow (the best photographer in the world), Bradford Rand and his posse (which even includes the likes of Rachael Ray and John Cusimano, her husband, who I interviewed a few months ago regarding his rock band, The Cringe), and

Amanda Farrar, for working on the *Green Travel Girl* series for Green Living Guy Productions; Aysia Wright, Darren Moore, Tamara Henry, Matt Petersen and Diane Simon of Global Green, Jessica Zappatero, Bianca Bucaram, the Environmental Media Association, and all the firms I work with to get the message of going green out to the masses.

I also must thank the Rocky Mountain Institute, the Natural Resources Defense Council, Oceana, the Sierra Club, and all the great organizations that have added electric vehicle support to their agendas since 2008. It is great to see and I appreciate their contributions to this book.

Build Your Own
Electric Vehicle

CHAPTER 1

Why Electric Vehicles Are for Today!

Thank you for building your own electric vehicles!
—Seth Leitman to Elon Musk, CEO of Tesla Motors
(as he handed Elon a copy of the Second Edition
of *Build Your Own Electric Vehicle*)

As I asked in the first and second editions, "Why should anyone buy, convert, or build an electric vehicle today?"

How about zero tailpipe emissions and 100 percent torque at 0 rev/min? Also, electric vehicles are available in all price ranges. When I worked for the State of New York, I always used to say that electric cars were almost maintenance-free: They never require oil changes, new spark plugs, or any other regular repairs.

When a person would ask me to list all the great things about electric cars, they would start by saying, "Really, are electric cars that efficient?" I would then say, "Well, not quite—you need to change the washer fluid, brake fluid, and coolants, but you get the point. No oil, no gas, no brake replacements, or anything!"

Electric vehicles (EVs) are part of everyday society: EVs include subway trains, electric cars, forklifts, and airport tugs. They are a Tesla Roadster, a Nissan Leaf, and a Ford Focus Electric.

What do you need an EV to be: big, small, powerful, fast, ultra-efficient? Design to meet that need: the THINK City, Toyota RAV4, Fisker, Tesla Roadster Sport, Tesla Model S or new Model X, Nissan Leaf, Ford Focus Electric, Tango, or any EV conversion that you have seen on the street. Man oh man, what will you create when you start with a clean sheet of paper. Closer to home and the subject of this book, do you want an EV car, sport utility vehicle (SUV), pickup truck, or delivery van? When you build your own electric vehicle or even have someone build one for you, you decide. You are in the driver's seat!

Since car companies continue producing SUVs that cannot meet federal fuel standards or reduce emissions that are harmful to our environment, think about some of the statistics that come from the U.S. Department of Energy (USDOE) and various notable sources: The USDOE states that more than half the oil we use every day is imported. This level of dependence on imports (55 percent) is the highest in our history. The USDOE even goes on to say that this dependence on foreign oil will increase as we use up domestic resources. Also, as a national security issue, we should all be concerned that the vast majority of the world's oil reserves are concentrated in the Middle East (65 to 75 percent) and controlled by the members of the OPEC oil cartel. We cannot be controlled by a fuel. We must control our fuel costs. The only way to realistically reduce our energy costs on transportation is to go all-electric![1]

Further, the USDOE goes on to state that 133 million Americans live in areas that failed at least one National Ambient Air Quality Standard. Transportation vehicles produce 25 to 75 percent of the key chemicals that pollute the air, causing smog and health problems.

All new cars must meet federal emissions standards. But, as vehicles get older, the amount of pollution they produce increases. In addition, when you get a new car, 20 percent (yes, only about 20 percent) of the energy you get from gas actually moves the car and operates the accessories, air conditioning, power steering, satellite radio, or that amazing sound system you just had installed.

Yes! Twenty percent of the fuel is used. The rest goes out into air!

So if you're getting angry or thinking that you have no ability to make changes, you are wrong. Just start driving an electric car. It's just that simple, not rocket science. Well, the rocket science can be left to the engineers, converters, and manufacturers. Your job is to get your EV built!

Here are some reasons why:

1. Although they are only at a relatively embryonic stage in terms of market penetration, EVs represent the most environmentally friendly vehicle in terms of fuel because they have absolutely no tailpipe emissions. In addition, the energy generated to power an EV is 97 percent cleaner in terms of noxious pollutants.

2. Another advantage of electric motors is their ability to provide power at almost any engine speed. Only about 20 percent of the energy in gasoline gets converted to the useful purpose of moving the wheels of an internal combustion vehicle, whereas 75 percent or more of the energy from a battery reaches the wheels of an EV.

3. One of the big arguments made by car companies against EVs is that they are powered by power plants that primarily consume coal. Less than 2 percent of U.S. electricity is generated from oil, so using electricity as a transportation fuel would greatly reduce our dependence on imported petroleum.[2] Yes, there are coal-burning power plants, but most of the power plants in this country are getting cleaner and more efficient by the day with smokestack scrubbers or the use of natural gas versus coal. And better than that, there is a revitalization of renewable energies such as solar, wind, geothermal, and tidal power that will help to propel our vehicles to be even cleaner than expected. Zero tailpipe and plant emissions is the future, and the future is here today![3]

4. Even assuming that the electricity to power the EV is not produced by rooftop solar panels or natural gas (let's assume that it comes 100 percent from coal), it is *still* much cleaner than gasoline produced from petroleum![4] The best part is that these facts still stand today.

5. The major concerns facing the EV industry are range, top speed, and cost. As we see more power tools and cell phones using lithium-based technologies,

we will see the cost of batteries for EVs come down too! The best part is that we can then expect a 300-mile-range electric car with an almost immediate charge time and pickup of 0 to 60 mi/h in the 3.4 seconds all my "car guys" demand. The only way EVs are going to make a big difference in people's lives is if they can do everything a gas-powered car can do and more. They have to look great (almost be an extension of the person buying the car), and they have to be safe.

EV conversions use frames currently approved by the National Traffic and Highway Administration (NHTSA) to which vehicle identification numbers (VINs) can be attached, and this is why this book has stood the test of time. When you go to the local Department of Motor Vehicles (DMV) or insurance company, just give them the VIN, and everything will be all right! When a car company gives you the title to a car or an EV, the same thing happens because it has been approved by those who could make your life a living hell to get it on the road!

When you add electric motors and controllers (basically, the brains) reaching 1,000 to 2,000 amps, high-end car batteries, and the light weight of a Porsche 911 chassis or an old sports car, electric cars can provide respectable performance. They are fun to drive and virtually silent, and they coast very easily when you let off the accelerator.[5] In other words, you can convert an old Porsche 911 to go over 100 mi/h with a 50-mile range using lead-acid batteries alone! With lithium-ion technology, you can get the car to go 300 miles, and the cost is still less than that of some brand-new SUVs or expensive sports cars.

In an effort to move the market toward electric cars, some people are trying other alternatives, including:

- Converting hybrid cars to either grid-connected or plug-in hybrids (www .calcars.org)
- Buying hybrid electric cars in droves
- Buying the Nissan Leaf, Chevy Volt, Ford Focus Electric, Tesla Roadster Model S or Model X, and all the other cars expected to come online in 2012 or 2013
- Purchasing low-speed electric vehicles, such as the GEM car
- Driving the last remaining car company–built electric cars, such as the EV1, the Toyota RAV4 EV, or the TH!NK City
- Taking an old junked or unused car and converting it to an all-electric car

It is amazing to see, but we are closer than we have ever been to getting electric cars in all brands across the board in my lifetime. When I first authored this book, I never thought that would be possible. After the release of the second edition of this book and the immediate ramp-up of electric cars, hybrids, plug-in hybrids, and more conversion companies than I have ever seen before, we are moving toward a massive electric car market by the car companies.

My point is that you can get an EV today. You can also take any vehicle you want and convert it to an EV. You can also encourage the fix-it person down the street to help you with your conversion so that more mechanics across the country are building electric cars.

Build That Car!

EVs are not difficult for mechanics or car lovers to build, and they are easy to convert. For car companies, it is easy because they are already working with a frame for starters. Right? There are so many reasons to convince you of the need to go electric:

- The cost of a gallon of gas
- Higher asthma rates
- The need to reduce our reliance on imported oil
- The prospect of owning a car that is cost-effective, fun, and longer lasting than most cars on the road today

Once again: How about the fact that you can convert a vehicle to electricity today? Right now! And it would cost you less than some new cars on the market.

Another part of building an EV must be brought into focus. When a person goes to a dealership and buys a car, that car was *built*. Thus, when a car company *builds an electric car*, it is the same thing as *building your own EV*. Therefore, for those of you out there who are not mechanics or future car manufacturers, have no fear: Someone will build *you* your own EV. When you put a deposit on a new Tesla Model X, Tesla then builds you your own electric car. This is why I thanked Elon Musk from Tesla Motors for building his own electric cars. He is building them for *you*!

There are so many television programs showing people tricking out their cars, adding better engines, or doing just about anything to make them go fast and be safe. EV conversions do all that and more. They allow the next generation to have a safer world without relying on foreign sources of oil, and they provide our kids with really clean and cool cars to drive.

In very practical terms, the 2001 Porsche and Mazda EV conversions shown in Figures 1-1 and 1-2 (which you'll learn how to convert in Chapter 11) go 80 mi/h, get 60 or 300 miles on a charge, and use conventional lead-acid batteries or lithium-ion or nanophosphate batteries, and using off-the-shelf components and newer technologies, we can all be building our own EVs.

In this chapter you learn what an EV is and explore the change in consciousness responsible for the upsurge in interest surrounding EVs. You'll discover the truths and untruths on which EV myths are based. You'll also learn about the advantages of EVs and why their benefits—assisted by technological improvements—will continue to increase in the future.

To really appreciate an EV, it's best to start with a look at the internal combustion engine vehicle. The difference between an EV and a regular car is a study in contrasts. People's continued fascination with the internal combustion engine vehicle is an enigma. The internal combustion engine is a device that inherently tries to destroy itself: Numerous explosions drive its pistons up and down to turn a shaft. For example, a drive shaft rotating at 6,000 rev/min (rpm) produces 100 explosions every second. These explosions, in turn, require a massive vessel to contain them—typically a cast-iron cylinder block.

Additional systems are necessary:

- A cooling system to keep the engine temperature within a safe operating range
- An exhaust system to remove the heated exhaust products safely

FIGURE 1-1   A converted electric Porsche.

FIGURE 1-2   Joe Porcelli and Dave Kimmins of Operation Z's Nissan.

- An ignition system to initiate the combustion at the right moment
- A fueling system to introduce the proper mixture of air and gas for combustion
- A lubricating system to reduce wear on high-temperature, rapidly moving parts
- A starting system to get the whole cycle going

It's complicated to keep all these systems working together. This complexity means that more things can go wrong (more frequent repairs and higher repair costs). Figure 1-3 summarizes the internal combustion engine vehicle systems.

Today's internal combustion engine is more evolved than ever. There are some car companies that believe that they can get more out of the gas car. Yet all the technologies they use were developed through EVs. Therefore, inevitably, those at car companies who believe that the gas engine is the next new frontier should be reminded it is EVs that are the next new frontier.

However, we still have a carbon-based combustion process that creates heat and pollution. Everything about the internal combustion engine is toxic, and it is still one of the least efficient mechanical devices on the planet. Unlike lighting a single match, the use of hundreds of millions (soon to be billions) of internal combustion engine vehicles threatens to destroy all life on earth. You'll read about environmental problems caused by internal combustion engine vehicles in Chapter 2.

While an internal combustion engine has hundreds of moving parts, an electric motor or even a controller has only one moving part or no moving parts. This is one of the reasons why EVs are so efficient. To make an EV out of a gas-powered car, pickup, or SUV you are driving now, all you need to do is take out the internal combustion engine along with all related ignition, cooling, fueling, and exhaust system parts and add an electric motor, batteries, and a controller.

Hey, it doesn't get any simpler than this for mechanics out there! Even car manufacturers find EVs easier to build! Figure 1-4 shows all there is to it: Batteries and a charger are your "fueling" system, an electric motor and controller are your "electrical" system, and the "drive" system was as before (although today's advanced EV designs don't even need the transmission and drive shaft).

This infamous simple diagram of an EV looks like a simple diagram of a portable electric shaver from Norelco or another company: a battery, a motor, and a controller or switch that adjusts the flow of electricity to the motor to control its speed. That's it. Nothing comes out of your electric shaver, and nothing comes out of your EV. EVs are simple (therefore highly reliable), have lifetimes measured in millions of miles, need no periodic maintenance (i.e., filters, etc.), and cost significantly less per mile to operate. They are highly flexible as well, using electrical energy readily available anywhere as input fuel.

In addition to all these benefits, if you buy, build, or convert your EV from a gas-powered vehicle chassis as suggested in this book, you perform a double service for the environment: You remove one polluting car from the road and add one nonpolluting EV to service.

You've had a quick tour and side-by-side comparison of EVs and internal combustion engine vehicles. Now let's take a closer look at EVs.

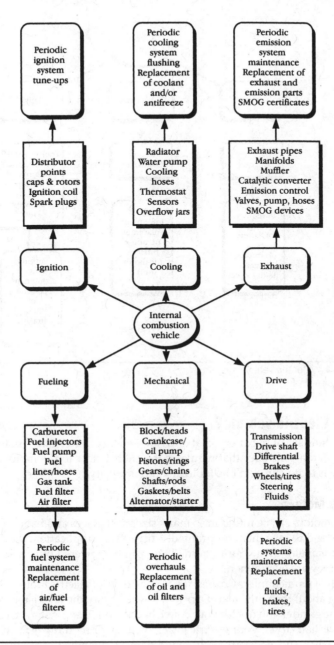

FIGURE 1-3   Internal combustion engine systems.

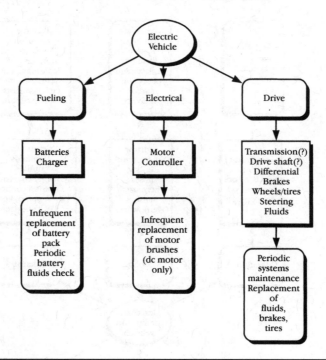

FIGURE 1-4 Electric vehicle systems.

What Is an Electric Vehicle?

An EV consists of a battery that provides energy, an electric motor that drives the wheels, and a controller that regulates the energy flow to the motor. Figure 1-5 shows all there is to it—but don't be fooled by its simplicity.

Electric Motors

Electric motors can be found in so many sizes and places and have so many varied uses that we tend to take them for granted. Universal in application, they can be as big as a house or smaller than your fingernail and can be powered by any source of electricity. In fact, they are everywhere!

Each of us encounters dozens, if not hundreds, of electric motors daily without even thinking about them: the alarm clock that wakes you; the television you turn on for the news; the grinder you use for your coffee beans; your electric shaver, electric toothbrush, or electric hair dryer; your electric juicer, blender, or food mixer; your vacuum cleaner; your washer and dryer; the automatic gate or door in the subway; the elevator or escalator at work; your computer, cell phone, iPhone, or BlackBerry; your fax machine, copier, and scanner; and your fan, heater, or air conditioner. In addition, you might use an electric garage door opener, program your TiVO/DVR, or use an electric power tool on a project. And the list goes on and on! Get it?

In gas-powered automobiles, in addition to your all-important electric starter motor, you typically find electric motors in the passenger compartment heating/cooling

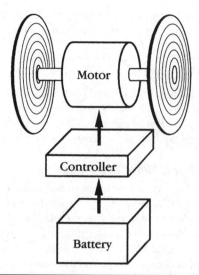

FIGURE 1-5 Simple block diagram of an EV.

system; radiator fan; windshield wipers; electric seats, windows, door locks, and trunk latch; outside rear-view mirrors, outside radio antenna, and more.

Batteries

No matter where you go, you cannot get away from batteries either. Rechargeable batteries can be connected to a charger or source of electrical power to build them up to capacity. There are different types of batteries. There are rechargeable lead-acid batteries, nickel–metal hydride batteries, and lithium-ion batteries (as some examples), and they can be used in your EV to manage the recharging process invisibly via an under-the-hood generator or alternator that recharges the battery while you are driving.

Another great thing about EVs is the promise of lithium-ion, nanophosphate, lithium-air (yes, that is coming), or other battery technologies coming rapidly and cost-effectively to market. Such technologies are moving rapidly into the marketplace and dropping in price. Over the next few years we can expect further drops in price, making EV conversions more affordable. Soon enough, the standard will be lithium-ion batteries in any conversion kit.

Controllers

Controllers have become much more intelligent, and AC Propulsion really is leading the way on this. The same technology that reduced computers from room-sized to desk-sized allows you to exercise precise control over an electric motor. Regardless of the voltage source, current needs, or motor type, today's controllers—built with reliable solid-state electric components—can be designed to meet virtually any need and can easily be made compact to fit conveniently under the hood of your car.

Back in the early 1990s when the first edition of this book was published, EVs resembled your battery-operated electric shaver, portable power tool, or kitchen appliance. Today's and tomorrow's EV more closely resemble your portable laptop computer, Smartphone, or flat-screen TV in terms of both sophistication and capabilities.

Have You Driven an EV Lately?

Besides all the discussion of EVs and the California Air Resources Board (CARB) mandates to incentivize EVs, hybrids, and fuel cells that are replicated in several other states, very few people have bought, built, or converted an EV because they want to save the planet. Here are some of the reasons why people *do* get into EVs.

Electric Vehicles Offer a "Total Experience"

Word of mouth and personal experience make a difference. The documentary movies *Who Killed the Electric Car?* and *Revenge of the Electric Car!* have also helped to spread the word. The cumulative effect of numerous people attending EV symposiums, rallies, and Electric Automobile Association meetings and movies all over the world—and experiencing first-hand what it's like to ride or drive an EV—has gradually done the job. Almost universally, people enjoy their consciousness-raising EV experience, are impressed by it, and tell a friend. This is the real reason for the resurgence in interest in EVs.

Electric Vehicles Are Fun to Drive

Imagine starting a car and hearing nothing! The only way you can tell that the car is on is by looking at the battery/fuel gauge on the dashboard. This is only the first surprise of many when you get into an EV. Recently I was interviewed by my local TV station, Channel 11, WPIX, and the interviewer was amazed at the silence and pickup of a hybrid electric car. Wait until I get him into an all-electric car!

One story I love to tell concerns when I used to work for the New York Power Authority (NYPA) and do ride-and-drives for the public. I always used to say that once you got in an EV, you were changed forever. It's true! Every single time a person got out of the EV, he or she would have a smile on his or her face and a sense of real excitement, and then the inevitable first question would pop up: "Where can I get one?"

EVs are first and foremost practical—but they are also fun to own and drive. Owners say that they become downright addictive. Tooling around in breezy EV silence gives you all the pleasure without the noise. As I liked to put it to my friends: *You can really hear your stereo.*

Electric Vehicles Make a Difference by Standing Out

EVs are a great way to drive and make a real contribution to the country. By driving an oil-free, gasoline-free car, you reduce our country's reliance on imported oil; that will make you friends.

Whether you've owned or even driven an EV1 from GM (what a ride!), a Nissan Leaf, a Tesla Roadster Sport, a Ford Focus Electric, a Fisker, a Model S or X, a Toyota RAV4 (old or the new ones coming out with a Tesla drive system), a TH!NK City, a really old Solectria Force, a converted Porsche 914, or a converted Ford Ranger pickup truck, your EV is a sexy, quiet, technologically spiffy show-stopper.

Believe me, if the words *electric car* or *electric vehicle* appear prominently on the outside of your car, pickup, or van, you will not want for instant friends at any stoplight, shopping center, or gas station or just playing stop and go on the highway. One time I was driving a Toyota RAV4 EV on the West Side Highway in Manhattan with the window down. All different types of people were intrigued:

- "Is that an electric car?"
- "When can I get one?"
- "How far does it go?"
- "Is it as good as people say?"
- "That is one phat ride!"

Or you can park it in a conspicuous spot, lift the hood, and wait for the first passerby to ask questions, as Figure 1-6 suggests.

There's a level of respect you receive, a pride in riding in the car, and a feeling of leading the pack in those experiences, at shows, and during demonstrations. One suggestion too: Have plenty of literature always on hand so that you can keep your EV discussions to under five minutes in length. On the other hand, if you just want to meet people, make the letters on your sign real big, and you will never want for company.

The first owners of anything new always have an aura of prestige and mystique about them. You will be instantly crowned in your own neighborhood. You are driving what others have only talked about. While hopefully everyone will own one in the

FIGURE 1-6 "Hey, where's the engine?" (Inside the Electric DeLorean at the NY International Auto Show 2010.)

future, you are driving an EV *today*. When TV sets were introduced in the 1950s, whole neighborhoods crowded into the first houses with the first tiny black-and-white screens. Expect the same with your EV project. This will show that you can reduce your carbon footprint with an EV, and it won't cost hundreds of thousands of dollars.

Electric Vehicles Save Money

All this emotional stuff is nice, but let's talk out-of-pocket dollars. Ask any EV conversion owner, and he or she will tell you that it transports him or her where he or she wants to go, is very reliable, and saves him or her money. Let's examine operating, purchase, and lifetime ownership costs and summarize their savings.

Operating Costs

EVs only consume electricity. In between charge-ups, there are no other things to do except get some washer fluid or clean the inside of the car. These figures are covered in more detail later, but an EV conversion that will be discussed in Chapter 10 has an average operating cost of about 7.3 cents per mile versus 27 cents per mile for its gasoline-powered equivalent—almost three times the cost of the electric vehicle.

Purchase Costs

Commercially manufactured EVs are not as prohibitive as they used to be, but a conversion is about 99 percent of the time cheaper than an EV from a car company.

You can convert an existing internal combustion engine vehicle to an EV. You remove the internal combustion engine and all systems that go with it and add an electric motor, controller, and batteries. If you start with a used gas-powered vehicle chassis, you can save even more (with the advantage of having the drive-train components already broken in, as later chapters point out). (This is true even if you buy a Porsche or other high-end car because if you were planning to buy a Porsche in the first place, you already would have been willing to spend more money than the average consumer.)

To this must be added the cost of your internal combustion engine vehicle chassis.

If you start with a brand-new vehicle, this could mean $10,000 or more (less any credit for the removed internal combustion engine components). A good, used chassis might cost you just $2,000 to $3,000 (or less if you take advantage of special situations, as mentioned in Chapter 5). So your total purchase costs are in the $8,000 to $25,000 ballpark. This is substantially less than already converted EVs! Obviously, you can do better if you buy carefully and scrounge for parts. Equally obviously, you also can spend more if you elect to have someone else do the conversion labor, decide you must have a brand-new Ferrari Testarossa chassis, or elect to build a Kevlar-bodied roadster with titanium frame from scratch.

Electric Vehicles Are Customizable

EVs are modularly upgradable. See a better motor and controller? Bolt them on. Find some more efficient batteries? Take out the old ones and in with the new! Then you don't need to get an entirely new vehicle; you can adapt new technology incrementally as it becomes available.

EVs are easily modified to meet special needs. Even when EVs are manufactured in volume, the exact model needed by everyone will not be made because it would be prohibitively expensive to do so. But specialist shops that add heaters for those living in the north, air conditioners for those living in the south, and both for those living in the heartland will spring up. Chapter 4 will introduce you to *some* of the conversion specialists who operate today.

Safety First

EVs are safer for you and everyone around you. EVs are a boon for safety-minded individuals. EVs are called *zero emission vehicles* (ZEVs) because they emit nothing, whether they are moving or stopped. In fact, when stopped, EV motors are not running and use no energy at all.

As my colleague Brian Clark Howard wrote on the *Daily Green*:

> It's been true for a while that advancing battery technology is a critical part of the development of cleaner cars. As Tony Posawetz, vehicle line director for the Chevy Volt, said at the recent Popular Mechanics Breakthrough Awards, "We will electrify the future. Batteries have gotten better, but consumers have gotten more demanding."
>
> While some experts have projected that the batteries of the near future may be literally assembled from dirt, it's true that today's devices have some amount of toxicity to them and some environmental concerns. Still, widely circulated concerns over the impact of batteries are overblown.
>
> The *Daily Green* recently received an e-mail from Kathryn Mayer, a reader who said she is considering buying a 2011 hybrid Hyundai Sonata. However, she had second thoughts after a friend asked her: What would happen to the battery at the end of its life? The friend's argument was that because hybrids (and electric cars) have bigger batteries than conventional gas cars, shouldn't that mean the batteries are more toxic? Shouldn't that mean the environmental benefits of driving a hybrid are canceled out?
>
> Well, there are a number of things Kathryn's friend should know:
>
> 1. *Hybrid and EV batteries come with long warranties.* In the case of the Chevy Volt, that's eight years, and in the case of the 2011 Hyundai Sonata, the warranty is 10 years. In contrast, the warranty on an ACDelco conventional car battery is three years or 36,000 miles, whichever comes first.
>
>   What this means is that you would likely go through about three conventional car batteries for every hybrid or electric battery. That takes us to the next point.
> 2. *Conventional car batteries are more toxic.* Standard car batteries are based on lead-acid chemistry, and lead is toxic stuff that can leak into groundwater, where it can lead to developmental disabilities and other illnesses. This means that conventional car batteries are actually more toxic, not less, than hybrid and EV batteries, which are typically based on nickel–metal hydride (NiMH) and lithium-ion chemistries, respectively (although the 2011 Sonata uses a lithium battery). Although these materials are not totally benign, they are less dangerous than lead.
>
>   (True, a number of clean vehicles, including the Nissan Leaf, also have a conventional battery as well as a primary battery, for starting and a few other functions. But we think it's still instructive to point out that the technology we have been using for years is actually dirtier than newer advanced batteries.)
> 3. *Car companies are working on battery recycling infrastructure.* Remember, hybrid and EV batteries last a long time, and consumer uptake of EVs in particular thus far has been slow. In the words of Larry Dominique, vice president of product

planning at Nissan, "We have some time to figure this out." Dominique told the panel at the *Popular Mechanics* conference that the industry is committed to supporting a responsible disposal and recycling infrastructure for spent batteries. As of now, it is illegal in many states to toss any lithium-ion batteries in the regular trash, and a recycling industry is gearing up. Lithium is fairly valuable, as are some of the other materials involved, and there is economic incentive to reuse the components.

This is in direct contrast to gas-powered vehicles that not only consume fuel but also do their best polluting when stopped and idling in traffic. EVs are obviously the ideal solution for minimizing pollution and energy waste on congested stop-and-go commuting highways all over the world, but this section is about saving yourself: As an EV owner, you are not going to be choking on your own exhaust fumes. EVs are easily and infinitely adaptable. Want more acceleration? Put in a bigger electric motor. Want greater range? Want more speed?

When you buy, convert, or build an EV today, all these choices and more are yours to make because there are no standards and few restrictions. The primary restrictions regard safety (you want to be covered in this area anyway) and are taken care of by using an existing gas-powered automobile chassis that has already been safety qualified. Other safety standards to be used when buying, mounting, using, and servicing your EV conversion components are discussed later in this book.

It gets better, as Bob Brant explains: "EVs carry no combustible fuels, 20,000-volt spark plug ignition circuits, hot exhaust manifolds, catalytic converters, or hot radiators on board."

Those who enjoy

- Engine compartment fires (caused by the ignition system or hot manifolds—as seen by the side of the road)
- Hot radiator coolant explosions (caused by improper radiator cap removal—sadly, many of us have experienced this firsthand), or
- Needless chores associated with gas-powered vehicle ownership

should stick with their gas-powered vehicles. Contrast the numerous periodic gas-powered engine vehicle activities related to the systems shown in Figure 1-3 with the far simpler EV system requirement shown in Figure 1-4.

In the "yuck" but not really dangerous category, EV owners don't have to mess with oil (no dark, slippery spots on the garage floor), antifreeze (no lighter, slippery spots on the garage floor), or filters (the kind you hold in a rag far away from your body because they are filthy or gunky). How about the acid part in lead-acid batteries or the lithium? You'll learn about battery details in Chapter 8, but most EV batteries are recyclable, so those concerns of the past get washed away.

On another safety issue, while EVs do not emit noise pollution, there has been concern about hybrid vehicles being unsafe for seeing-impaired pedestrians because the engines don't make noise. However, the Baltimore-based National Federation of the Blind presented written testimony to the U.S. Congress asking for a minimum sound standard for hybrids to be included in the state's emissions regulations. As the president of the group, Marc Maurer, stated, he's not interested in returning to gas-guzzling vehicles; the group just wants fuel-efficient hybrids to have some type of warning noise.

"I don't want to pick that way of going, but I don't want to get run over by a quiet car, either," Maurer said.

Manufacturers are aware of the problem but have made no pledges yet. Toyota is studying the issue internally, said Bill Kwong, a spokesman for Toyota Motor Sales USA.

"One of the many benefits of the Prius, besides excellent fuel economy and low emissions, is quiet performance. Not only does it not pollute the air, it doesn't create noise pollution," Kwong said. "We are studying the issue and trying to find that delicate balance."

The Association of International Auto Manufacturers, Inc., a trade group, is also studying the problem, along with a committee established by the Society of Automotive Engineers. These groups are considering "the possibility of setting a minimum noise level standard for hybrid vehicles," said Mike Camissa, the safety director for the manufacturers' association."[6]

Well, the best part now is there is a Smartphone app that can give your car any sound you want. It's true!

Electric Vehicles Save the Environment

EV ownership is visible proof of your commitment to help clean up the environment. Chapter 2 will cover in detail the environmental benefits of this choice. EVs produce no tailpipe emissions of any kind to harm the air, and virtually everything in them is recyclable. Plus, every EV conversion represents one less polluting internal combustion vehicle on the road. EVs are not only the most modern and efficient form of transportation, but they also help to reduce our carbon footprint today.

I have worked with hundreds of early adopters of all-electric cars over the years, and electric cars recoup the most, but at present, the car companies build hybrid electric cars which are not as cost-effective as they really could be, and this is an issue for the Toyota Prius to expand on year after year. This happens to all new technologies. People buy EVs, hybrids, or plug-in hybrid electric vehicles (PHEVs) because it's more about doing the right thing and making a public statement than the product itself. There is no red or blue about buying a hybrid or an EV. It's just the right thing to do!

Once car companies offer the free market EVs to compete with every gas-powered brand, people will want them. They can only pretty much get hybrids at present, except for the Tesla, Nissan Leaf, and Ford Focus, which for now have limited market participation. The free market usually does not prevail on EV. On average, the oil industry subsidies are $20 billion annually versus the $16 billion spent on electric transportation in the United States.[7]

Prices for lithium-ion battery technology from A123 Systems and controllers such as the AC Propulsion system are coming down because we are building EVs and building them with U.S. technologies and becoming the world leader in this new transportation era.

Electric Vehicle Myths (Dispelling the Rumors)

There are four widely circulated myths/rumors about EVs that are not true. Because the reality in each case is the 180-degree opposite of the myth, you should know about them.

Myth #1: Electric Vehicles Can't Go Fast Enough

The reality is that EVs can go as fast as you want—just choose the electric vehicle model (or design or build one) with the speed capability you want. One example of how fast they can accelerate occurred when I was driving a TH!NK City (really small City EV) in New York City. I was at a traffic light next to a Ford cab (how appropriate, since Ford owned TH!NK at the time). The cabbie wanted to see how fast the EV could go, so I said, "I know it can beat you." (Please note that all of this was done well within the legal speed limits on the Manhattan Street!) He said, "You're crazy!" The light turned green, and I hit the accelerator. The look on the cabbie's face was worth a million dollars. He was more than surprised at the torque and acceleration. People on the street were screaming, "Go, go, go." I blew him away. We met up at the next traffic light, and he said, "Where can I get one?" Enough said.

A second true story occurred back in 2009 when I was able to test drive a Tesla Roadster Sport electric car on the streets of Manhattan with a good friend of mine. While we drove the car around town, my friend was impressed (besides the fact that we were in a $200,000 vehicle) that it was quiet and seemed easy to handle. Then we got onto an avenue where there was little to no traffic. I hit the accelerator and went from 0 to 50 mi/h in about 3 seconds. Then we continued on for a good 10 seconds and came to a complete stop. We missed the parking lot, so I accelerated again and hit the brakes right in front of the parking garage, where the rep from Tesla Motors was waiting for us. My friend said, "My dad use to tell me electric cars were nothing but glorified golf carts. *If these are electric cars, I want one for everyone in my family!*"

Myth #2: Electric Vehicles Have Limited Range

Nothing could be further from the truth, but unfortunately, this myth has been widely accepted. The reality is that EVs can go as far as most people need. Remember, this book advocates an EV conversion only as your second vehicle. While lithium-ion batteries will expand your range dramatically, and there are some people traveling cross-country in EVs, EVs are not yet your best choice for a massive road trip at this time.

But what is the range? The federal government reports that the average daily commuter trip distance for all modes of vehicle travel (i.e., auto, truck, and bus) is 10 miles, and this figure hasn't changed appreciably in 20 years of data gathering. An earlier study showed that 98 percent of all vehicle trips are under 50 miles per day; most people do all their driving locally and take only a few long trips. Trips of 100 miles and longer account for only 17 percent of total miles.

As *EV World* reported in 2011:

> Electric cars are just beginning to hit U.S. highways, but if the most recent Nielsen Energy Survey is any indication, they should receive a huge welcome. An overwhelming 85 percent of American consumers say they would be interested in purchasing a plug-in electric vehicle, either right away (3 percent), when their current car needs replacement (57 percent), or when the technology is proven and becomes mainstream (25 percent).

One consumer who won't be waiting is MXenergy CEO Jeffrey Mayer. Mayer, head of one of the nation's leading independent energy providers, was one of the first in line to order his "victory red" Chevy Volt. "The Volt and other plug-in vehicles getting ready to hit showroom floors represent a huge step forward for the sustainability movement," said Mayer. "The large-scale production of electric cars is something many people feared might remain just a dream. I for one can't wait to be a part of that dream becoming a reality."

While U.S. drivers seem ready to embrace the EV, they also have a strong preference as to the type of vehicle they prefer. Fifty-eight percent of customers prefer plug-in hybrid electric vehicles (PHEVs) such as the Chevy Volt to electric-only models such as the Nissan Leaf. PHEVs have a much greater range than their electric-only counterparts. According to E Source, younger drivers are more willing to go the electric-only route with their vehicle choice. Regardless of which version people prefer, Mayer says that the greening of America's highways is exciting to see.

> People realize we have to dramatically reduce our dependence on foreign oil and do all we can to lessen our environmental impact as well. The introduction of electric vehicles into the marketplace is a transformational moment for the automotive industry and the sustainability movement.

Even the first and second editions of this book noted that back in 1990s, General Motors' own surveys (taken from a sampling of drivers in Boston, Los Angeles, and Houston) indicated the same as today:

- Most people don't drive very far.
- More than 40 percent of all trips are under 5 miles.
- Only 8 percent of all trips are more than 25 miles.
- Nearly 85 percent of drivers drive fewer than 75 miles per day.

Virtually any of today's 120-volt EV conversions will go 75 miles—using readily available off-the-shelf components—if you keep the weight under 3,000 pounds.

This means an EV can meet more than 85 percent of average driving needs. If you're commuting to work—a place that presumably has an electric outlet available—you can nearly double your range by recharging during your working hours. In addition, if range is really important, optimize your EV for it. It's that simple.

Myth #3: Electric Vehicles Are Not Convenient

The myth that EVs are not effective as a real form of transportation or that they are not convenient is really silly. Car companies and others have complained that there is not enough recharging infrastructure across the country or that you cannot charge an EV anywhere you would like, as with fueling a car. A popular question is, "Suppose you're driving and you are not near your home to charge up or you run out of electricity; what do you do?" Well, my favorite answer is, "I would do the same thing I'd do if I ran out of gas—call AAA or a tow truck."

Why would I have reported on my blog that ABB, the leading power and automation technology group, announced recently the acquisition of Epyon BV, an early leader in EV charging infrastructure solutions focusing on direct-current (DC) fast-charging stations and network charger software? The acquisition is in line with ABB's strategy to expand its global offering of EV infrastructure solutions in late 2011.

"This acquisition gives ABB access to competitive products, key network management software, and a robust maintenance service business model, which ideally complements our own offering," said Ulrich Spiesshofer, head of ABB's Discrete Automation and Motion Division. "ABB's brand recognition and strong global presence will accelerate the growth of a combined Epyon-ABB offering and provide access to key customers and partners," said Hans Streng, Epyon's CEO, who led the company over the last year and who will stay as an experienced industrial leader of the combined business. "Epyon's existing business is complemented by ABB's strong power electronics platform, global manufacturing footprint as well as its supply, marketing and service network."

Furthermore, and according to ABB, the growing number of EVs is driving a global market opportunity for charging solutions, including supporting technologies to equip the electrical grid with more sophisticated monitoring systems and software. EV charging-station unit sales are expected to multiply rapidly over the next five years and reach 1.6 million units globally by 2015, according to Pike Research.[8]

The reality is that EVs are extremely convenient. Recharging is as convenient as your nearest electric outlet, especially for conversion cars using 110-volt charging outlets. Here are some other reasons:

- You can get electricity anywhere you can get gas—there are no gas stations without electricity.

- You can get electricity from many other places—there are few homes and virtually no businesses in the United States without electricity. All these are potential sources for you to recharge your EV.

- As far as being stuck in the middle of nowhere goes, other than taking extended trips in western U.S. deserts (and even these are filling up rapidly), there are only a few places you can drive 75 miles without seeing an electric outlet in the contiguous United States. Europe and Japan have no such places.

- Plug-in-anywhere recharging capability is an overwhelming EV advantage. No question it's an advantage when your EV is parked in your home's garage, carport, or driveway. If you live in an apartment and can work out a charging arrangement, it's an even better idea: A very simple device can be rigged to signal you if anyone ever tries to steal your car.

How much more convenient could EVs be? There are very few places you can drive in the civilized world where you can't recharge in a pinch, and your only other concern is to add water once in a while. Electricity exists virtually everywhere; you just have to figure out how to tap into it. If your EV has an onboard charger, extension cord, and plug(s) available, charging is no more difficult than going to your neighbor's house to borrow a cup of sugar. Except, of course, you probably want to leave a cash tip in this case.

Myth #4: Electric Vehicles Are Expensive

While perhaps true of EVs that are manufactured in low volume today—and partially true of professionally done conversion units—it's not true of the do-it yourself EV conversions, and even today, a Nissan Leaf is starting to get really competitive in pricing. The reality, as we saw earlier in this chapter, is that EVs cost the same to buy (you're not going to spend any more for an EV than you would have budgeted anyway for your second gas-powered vehicle), the same to maintain, and far less per mile to operate. In the long term, future volume production and technology improvements will only make the cost benefits favor EVs even more.

Although the Volt and Leaf Have Stratospheric Sticker Prices, Both Have Pretty Low Ownership Costs

by Jessica L. Anderson, Associate Editor, Kiplinger's Personal Finance

Sales of hybrid and electric cars jumped nearly 40 percent in the first quarter of the year. The primary reason: soaring gas prices. More than one-third of consumers now say fuel economy is the most important factor in their next vehicle purchase, according to a recent study by Maritz Research, a market-research firm.

Setting aside their environmental cred, are hybrids, diesels and electric vehicles actually wallet-friendly? Prices on hybrids run about $3,300 higher on average than stickers on their gas-engine siblings. The diesel difference is about $2,800 more. And the two EVs on the market—the Chevrolet Volt and Nissan Leaf—are each more than $18,000 pricier than their closest gas-engine match. But the long-term ownership costs are the real measure of whether buying green is worth it.

Running the Numbers

Using five-year ownership costs from Vincentric, an automotive data firm, we compared 2011 hybrid, diesel and electric vehicles with their closest gasoline-engine counterparts. In most cases, that's the same model with a different power train; when a hybrid (like the Prius) had no counterpart, we chose the closest match from the carmaker's lineup.

The numbers assume you drive 15,000 miles a year and that regular gasoline costs $3.64 a gallon, premium is $3.91, and diesel is $3.97—the average prices nationwide in early summer—with a 3.5 percent annual increase for each fuel. In addition to fuel costs, depreciation, maintenance and repairs, the math also includes finance costs for a five-year loan after a 15 percent down payment, insurance and taxes.

The Volt and Leaf are both eligible for the $7,500 tax credit for electric vehicles, and that, too, is factored in.

Best Deals

Pump prices have a lot to do with making green cars a good value. Two years ago, when gas prices were hovering close to $2 a gallon, few hybrids and diesels earned back their premium price with savings at the pump. But with gas prices now closer to $4, more buyers will save green by buying green.

Perhaps the biggest surprise is that although the Volt and Leaf have stratospheric sticker prices—nearly double those of the gas-engine Chevrolet Cruze LTZ and Nissan Versa S hatchback—both have pretty low five-year ownership costs. The Volt's costs come within $1,600 of the Cruze's and the Leaf is only $800 more than the Versa over five years. (Run your own comparisons of these models and many more.[9]

In addition, Toyota recently reported to me:

The 2012 RAV4 EV will definitely be sold to the general public. We anticipate robust public interest in the RAV4 EV and are keen to inform consumers that their future vehicle options include a battery electric Toyota.

Toyota is the only manufacturer bringing two battery EVs to the market in 2012—the RAV4 EV and the Scion iQ EV. While the RAV4 EV will be available to the public, for starters the Scion iQ EV will be marketed to fleet and car-sharing programs only.

In addition, back in 2007, in India, *Archana Mohan/Ahmedabad* reported on March 8, 2007[10]:

The Bangalore-based Reva Electric Car has created a record: Over 700 of its cars are running on the streets of London, the highest number of electric vehicles functioning in any city across the world. The company, which last year became the highest selling on-road electric vehicle globally, is all set to touch the 1,000 car mark in London following its upcoming consignments in the next four to six months.

Currently, the company has around 1,800 cars on road, of which 1,000 cars are in India, while the others are in London and other parts of the UK.

The Disadvantages of Electric Cars Have Been Reduced or Eliminated

Well, there had to be a downside. If any one of the following factors is important to you, you might be better served by taking an alternate course of action.

Extended trips, as already mentioned, were something that could not be done across the United States. Now the EV is becoming more and more a great choice for nationwide travel. There are some places that are still not realistic, but on Route 66, Cracker Barrel locations and Best Buy stores are building the charging infrastructure to go long distances. Until this is completed, you'd better bring a cord.

However, I am not telling you not to take an airplane and rent an electric car when you get there, right? An electric train, hybrid electric (soon all-electric) buses, and rental EVs are here to stay.

Time to Purchase/Build Your Own Brand-New Electric Car!

Regardless of your decision to buy, build, or convert an EV, it is going to take you time to do it—but certainly less than it used to. There is a growing network of new and used EV dealers and conversion shops. However, the highest grade controllers and motors do take time to produce, and this might slow down your conversion process. However, as demand for these products increases, the supply will increase too, and the time it takes for these products to be built decrease. (Check the online sources section at the end of this book.) But plan on taking a few weeks to a few months to arrive at the EV of your choice.

EV resale (if you should decide to sell your EV) will take longer—for the same reason. While a reasonably ready market exists via the Electric Auto Association chapter and national newsletters, it is still going to take you longer and be less convenient than going down to a local automobile dealer.

What you are seeing today in EVs is just the tip of the iceberg. It is guaranteed that future improvements will make them faster, longer-ranged, and even more efficient.

There are five prevailing reasons that guarantee that EVs will always be with us in the future. The only one not previously discussed is technological change. All the available technology has just about been squeezed out of internal combustion engine vehicles, and they are going to be even more environmentally squeezed in the future. This will hit each buyer right in the pocketbook. Incremental gains will not come inexpensively. Internal combustion engines are nearly at the end of their technological lifetime. Almost all improvements in meeting today's higher corporate average fuel economy (CAFE) requirements have been achieved via improving electronics technology. CAFE is the sales weighted-average fuel economy expressed in miles per gallon (mi/gal) of a manufacturer's fleet of passenger cars or light trucks with a gross vehicle weight rating (GVWR) of 8,500 pounds or less manufactured for sale in the United States for any given model year. *Fuel economy* is defined as the average mileage traveled by an automobile per gallon of gasoline (or equivalent amount of other fuel) consumed as measured in accordance with the testing and evaluation protocol set forth by the Environmental Protection Agency (EPA).[11]

By 2020, CAFE standards mandate that all new cars built will have a fuel economy of 55 mi/gal, as approved by the U.S. Congress. Since the fuel efficiency of an EV is more than that, an EV will always be the best approach.

In addition, once lithium battery technology becomes the standard, EVs will be able to go 300 to 600 miles (depending on the technology). Unquestionably, the future looks bright for EVs because the best is yet to come. The four remaining reasons were covered in this chapter and are summarized in Figure 1-7:

- EVs are fun to drive and own.
- EVs are cost-efficient.
- EVs are performance-efficient.
- EVs are environmentally efficient.

Any one of these reasons is compelling by itself; the benefits of all of them taken together are overwhelming.

Also as I wrote in 2011:

Within six months, drivers will be recharging their electric vehicles, like the Nissan Leaf and Chevrolet Volt, along Washington's Electric Highways. The Washington State Department of Transportation (WSDOT) selected Monrovia, Calif.-based AeroVironment (NASDAQ: AVAV) to transform Interstate 5 and US 2 into the premier interstates of the 21st century, serving the latest generation of electric cars.

AeroVironment rose to the top during a competitive contract award process in which six companies submitted proposals to electrify I-5 and US 2 on a budget of $1 million.

AeroVironment will manufacture, supply, install and operate a network of nine fast-charging stations for electric vehicles. Stations will be located every 40 to 60 miles along stretches of I-5 between the Canadian border and Everett and between Olympia and the Oregon border, as well as along US 2 between Everett and Leavenworth.

"A network of charging stations linking Washington to Oregon and Canada will make electric vehicles more attractive to consumers and businesses, and transportation better for the environment," said Paula Hammond, Washington secretary of trans-

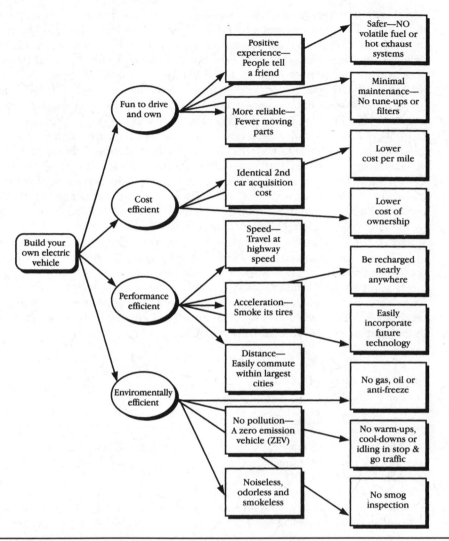

FIGURE 1-7   Four reasons why EVs will always be with us in the future.

portation. She said AeroVironment provides the technology needed for efficient and sustainable transportation, a green economy and new jobs in a growing field.

The fast-charging stations will be operational by Nov. 30, and will power an electric vehicle from zero to fully charged in less than 30 minutes. Each station also will include a Level 2 "medium-speed" charging station, which will cost less for users and take up to four to six hours for a full charge. The stations will be located at private retail locations such as shopping malls, fueling stations and travel centers with easy access to the highway

In the area between Everett and Olympia, additional charging stations will be installed through a federal program, the EV Project, administrated by the U.S.

Department of Energy. Combined, the two projects will connect Washington drivers along the entire 276 miles of I-5 between Canada and Oregon.

"As the leading hydroelectric power producer in the United States, Washington state is ideally suited to support emissions-free electric vehicles," said Mike Bissonette, AeroVironment's senior vice president and general manager of Efficient Energy Systems. "Electrifying I-5 is not just a vision about a cleaner future, it's about American jobs and building the economy today. We're working with Washington to turn that vision into reality."

While AeroVironment helps power the electric-vehicle charging network in Washington, the company will simultaneously electrify I-5 in Oregon through a similar project managed by the Oregon Department of Transportation. Both states' projects will complement the EV Project, which will install thousands of home and public electric-vehicle charging stations in six states, including Washington, Oregon and California.

Washington's electric-highway infrastructure is a key component of a future West Coast Green Highway. When complete, it will extend a seamless network of recharging stations along all 1,350 miles of I-5 from Canada to Mexico, serving more than 2 million electric vehicles that market analysts say will be sold in Washington, Oregon and California in the next decade.[12]

The People Want Green Electric Cars

As I wrote in 2011 on my website[13]:

Growth of alternative power train vehicles sales will be limited by consumer concerns about costs as well as functionality, according to the J.D. Power and Associates 2011 U.S. Green Automotive Study released today. Despite a rapid increase in the number of alternative power train vehicle models projected for the next several years, automakers will be fighting over the relatively few consumers who are willing to drive green.

The inaugural study examines attitudes of U.S. consumers toward four primary alternative power train technologies: hybrid electric vehicles, clean diesel engines, plug-in hybrid electric vehicles, and battery electric vehicles. The study gauges consumer consideration rates of these power train types for their next new vehicle purchase and explores specific perceived benefits and concerns that factor into the decision-making process.

For Most Consumers, Cost Matters More than the Environment

While consumers often cite saving money on fuel as the primary benefit of owning an alternative power train vehicle, the reality for many is that the initial cost of these vehicles is too high, even as fuel prices in the United States approach record levels. Reduced expenditure on fuel is the predominant benefit cited by considerers for each of the primary alternative power train technologies examined in the study. Although the environmental benefits of these vehicles are recognized, they are mentioned far less frequently than saving money on fuel. For example, 75 percent of consumers who indicate they would consider a hybrid electric vehicle cite lower fuel costs as a main benefit. In contrast, only 50 percent cite "better for the environment" as a main benefit of these vehicles.

Consumers who are not considering an alternative power train vehicle also recognize the fuel-cost savings these vehicles can offer. However, they cite significant perceived or actual impediments to ownership in addition to purchase price, including driving range, increased maintenance costs, and compromised vehicle performance. These consumers are far more likely to switch into a more fuel-efficient vehicle powered by a traditional internal combustion engine than an alternative powertrain vehicle.

"Alternative powertrains face an array of challenges as they attempt to gain widespread acceptance in the market," said Mike VanNieuwkuyk, executive director of global vehicle research at J.D. Power and Associates. "It is the financial issues that most often resonate with consumers, whether it is the higher price of the vehicle itself, the cost to fuel or charge the vehicle, or the fear of higher maintenance costs. The bottom line is that most consumers want to be green, but not if there is a significant personal cost to them."

According to VanNieuwkuyk, concern about the purchase price of alternative power train vehicles—particularly for hybrid electric vehicles—has become even more of an issue in 2011. At the end of 2010, tax credits from the Energy Policy Act of 2005 were phased out.

"Hybrid electric vehicles have been available in the automotive market for more than 10 years, and consumer awareness and understanding of them have grown during that time," said VanNieuwkuyk. "As concerns about the functionality and performance of hybrid vehicles have abated, vehicle price has become more prevalent as the primary purchase impediment. Without a tax credit to offset the price premium, consumers must absorb all of this additional cost. Furthermore, aggressive government subsidies are unlikely to be sustainable over the long term. Ultimately, the true cost of the technology needs to come down substantially."

Although there are also significant price premiums for battery electric vehicles, functional concerns are more likely to limit consideration rates for this power train. Driving range and the availability of charging sites away from home are the two concerns cited most often by those not considering this power train. This "range anxiety" contributes to the lowest consideration levels of the primary alternative power train technologies.

For clean diesel engines, fuel prices and availability—factors largely out of the control of vehicle manufacturers—have long been impediments to acceptance of the technology. Furthermore, negative perceptions of older diesel-powered vehicles continue to affect perceptions of clean diesel vehicles, as concerns about emissions and exhaust odor are mentioned frequently.

"Advocates of clean diesel engines tend to be some of the most vocal among consumers who tout the benefits of their chosen technology," said VanNieuwkuyk. "However, this consumer group is relatively small. Clean diesel technology continues to struggle not only against concerns about cost and perceived fuel availability, but also against the lingering perception that diesel is 'dirty.'"

Implications for Automakers

Overall, the study reveals interest in alternative power train vehicles among a majority of consumers, with perceptions of green vehicles being largely positive. However, converting this interest into actual sales will require concerted efforts to improve the technology and infrastructure and reduce the cost to consumers.

By the end of 2016, J.D. Power and Associates expects there to be 159 hybrid and electric vehicle models available for purchase in the U.S. market. This is a significant increase from only 31 hybrid and electric models in 2009. Despite this, according to VanNieuwkuyk, automakers, along with government entities and others, have considerable work to do in educating consumers as to the true costs and benefits of these technologies. Only through promotion and education will significant numbers of U.S. consumers become sufficiently comfortable with both the financial investment and, in some cases, lifestyle changes required to make the leap from traditional vehicles to alternative power train vehicles.

The 2011 U.S. Green Automotive Study combines information and insight from J.D. Power's primary consumer research, social media intelligence, forecasting, and transactional sales data. The primary research includes a study of more than 4,000 consumers who indicate that they will be in the market for a new vehicle within the next one to five years. The study was fielded in February 2011.

Get Your EV Built!

So the future is here! Build an electric car or get someone to build it for you! Do not despair or feel like electric is not for you. It is and will become more a part of the transport sector of our economy.

CHAPTER 2

Electric Vehicles Save the Environment and Energy

The World Needs Electric Vehicles Now!

Besides the fact that consumers have been consistently interested in electric cars (despite popular reporting by car companies), there is a newly sparked (no pun intended) excitement in either plug-in hybrid electric cars (which are more electric car than hybrid) and all-electric cars (e.g., Tesla, TH!NK City, RAV4, Reva, THINK, Ford Focus, and Nissan Leaf). This only means great things for the planet. Specifically, zero tailpipe emissions and greater air quality in our major metropolitan cities. And with this is a significant reduction in overall energy use.

Why Do Electric Vehicles Save the Environment?

Electric vehicles (EVs) are zero-emission vehicles (ZEVs). They do not emit toxic compounds into our atmosphere. Even the power plants that generate the power for EVs are held to a higher standard (meaning a lower level of toxic emissions) than the requirements for gasoline-powered vehicles. Everything going into and coming out of an internal combustion vehicle, on the other hand, is toxic, and it's still among the least efficient mechanical devices on the planet.

Besides, the internal combustion engine has an ineffective and destructive path plus the legacy of environmental problems (summarized in Figure 2-1). Now multiply that by hundreds of millions/billions of vehicles. The greatest of the internal combustion engine vehicle's problems include the following:

- Dependence on foreign oil (environmental and national security risk)
- Greenhouse effect (atmospheric heating)
- Toxic air pollution
- Wasted heat generated by inefficiency

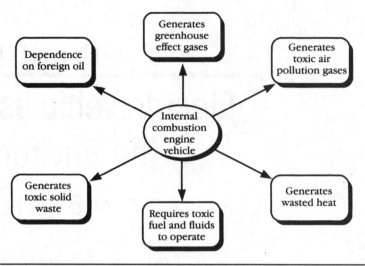

FIGURE 2-1   Internal combustion engine vehicles create many problems.

Save the Environment and Save Some Money Too!

Because EVs use less energy than gasoline-powered vehicles, their effect on the environment is much less than vehicles powered by fossil fuels. Because EVs are more efficient than gasoline-powered vehicles, they cost less to run.

EVs have existed for more than a hundred years (predating internal combustion engine vehicles). The aerospace-derived technology to improve them has existed for decades. Unquestionably, EVs will be the de facto transportation mode of choice for years to come if life on our planet is to continue to exist in its modern form with the conveniences we count on today. Figure 2-2 shows the reasons why.

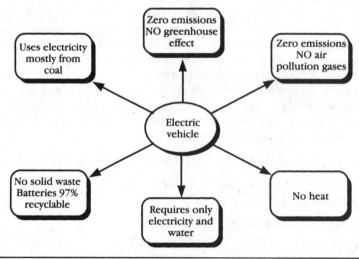

FIGURE 2-2   EVs create no environmental problems.

In direct contrast to the problems created by internal combustion engine vehicles, EVs:

- Use electricity (today, electricity comes mostly from coal).
- Are zero-emission vehicles; they emit no pollutants.
- Generate little toxic waste (lead-acid batteries are 98 percent recyclable).
- Require no toxic input fluids such as water and wash fluid (only occasional watering of batteries if you choose cheaper lead-acid batteries).
- Are highly efficient (motors and controllers 90 percent, batteries 75 to 80 percent).
- Benefit electric utilities (a market for electricity sales).

Even the cost advantage goes to the EV because it converts about 70 percent of the charging energy into motor energy, whereas a typical gas-powered vehicle converts only about 20 percent. An automatic transmission represents another significant loss, as do auxiliaries such as charging your Smartphones, radio or satellite radio, power steering, and air conditioning. (Around town, air conditioning can consume 40 percent or more of available power.) Imagine greater security for our service members abroad. Imagine freedom from the rising prices of the gas pump. Imagine freedom from massive oil spills like the one in 2011 in the Gulf of Mexico or the one in 2012 in Yellowstone National Park. Imagine freedom from auto tailpipe emissions. Imagine greater economic freedom through reduced national debt and thousands of new, clean tech jobs. Imagine freedom from oil altogether. Look at what our oil dependence is subjecting Americans to: The U.S. armed forces spend up to $83 billion annually protecting vulnerable infrastructure and patrolling oil transit routes. In 2012, U.S. Navy Secretary Roy Mabus said that "out of every 24 fuel convoys we use [in Afghanistan], a soldier or marine is killed or wounded guarding that convoy. That's a high price to pay for fuel."

Our nation sends up to a third of a trillion dollars overseas each year to purchase foreign oil, often produced by countries that are unstable or unfriendly to American interests. Foreign oil purchases are also responsible for about 50 percent of the U.S. trade deficit.

The statement we issued on freedom from oil includes companies and organizations with leaders who have served in our Armed Services—such as Brian Patnoe, vice president of fleet sales at CODA Automotive, a Los Angeles, CA–based EV and battery company. "I'm proud to be working for a company committed to supporting oil independence and the emerging EV supply chain," said Patnoe. "As a former Marine, it's also exciting for me to see successful business opportunities that support a prosperous oil-free future—from CODA's own parts manufacturers and assembly-line workers to electrical workers installing EV chargers and the customers purchasing a whole new type of vehicle."

New EVs are on the market in select cities nationwide, and by next year, they will be available in nearly every state. Now we need to put the right policies, infrastructure, and programs in place to support a cleaner and safer shift in the way we power our vehicles. The EV sign-on statement spells out some of the types of policies that will allow us to become EV-ready ASAP:

1. Expand national, regional, and local efforts that help to attract greater concentrations of EVs in communities across the country.

2. Remove unnecessary bureaucratic and market obstacles to vehicle electrification nationwide through a variety of policies that:

- Bolster nationwide installation of and access to basic charging infrastructure, both at people's homes and in public places
- Incentivize the purchase of EVs and EV charging equipment and streamline the permitting application process for EV charging equipment
- Educate the public about the benefits of EVs and the costs, opportunities, and logistical considerations involved with EV charging infrastructure
- Ensure appropriate training for workers installing EV charging equipment and for first responders
- Encourage utilities to provide attractive rates and programs for EV owners and increase off-peak charging
- Assist in the deployment of clean energy, efficiency, and energy management technologies jointly with vehicle charging
- Accelerate advanced battery cost reduction by boosting EV use in fleets, in second use, and in stationary applications

3. Ensure U.S. leadership in manufacturing of electric drive vehicles, batteries, and components.

EVs will help us to achieve freedom from oil, and they also mean less air pollution. Emissions from EVs are at least 30 percent lower than those from traditional vehicles—and that's on today's electric grid. As we clean up our grid and rely more and more on renewable sources of power, EVs get even cleaner over time.[1]

Petroleum Will Not Last Forever

Petroleum is a limited resource and a vexing source of price spikes, geopolitical instability, and environmental disasters of epic proportions. Petroleum currently fuels 95 percent of the U.S. transportation sector, a sector that demands nearly 28 percent of total energy usage. Globally, demand for personal transportation is increasing, whereas reserves are decreasing. Not only is petroleum a diminishing resource, but it is also a significant source of greenhouse gas emissions.

Fortunately, reducing the use of oil for transportation can quickly increase independence and reduce emissions. Tesla vehicles are seminal in developing a cleaner, more independent transportation paradigm.

Clean Energy Is the Future

As Tesla Motors explains in the environmental section of its website:

> Reducing oil consumption in transportation sectors means increasing our global use of electricity [I say better make it renewable electricity]. As transportation increases its use of the electricity grid, the demand for energy will rise. This demand can be met by bringing online increasingly efficient power plants and renewables.

Many countries have committed to decreasing dependence on fossil fuels by increasing their renewable energy portfolios. The United States Federal Government has pledged to reduce greenhouse gas emissions by 28 percent by 2020. In Europe, renewable energy should account for 20 percent of energy consumption by 2020. Germany alone has more than quadrupled the amount of renewable energy feeding its grid in less than a decade.

Some may argue that EVs simply replace one polluter (petroleum) with another (coal). To use the U.S. electricity grid as an example, Tesla vehicles cut in half the carbon dioxide emissions of its petroleum-burning rivals despite the fact that about half the grid is powered by coal. We can do better. Over the past five years, U.S. electricity generation from renewables has increased, while coal has decreased. As this shift happens, EVs become even cleaner. In California, renewables (13 percent) contribute more to the grid than coal (12 percent).

Bob Batson from Electric Vehicles of America recently reviewed those costs and added:

For the conversion: 100,000 Btu (1 therm) = 29.3 kWh = 39.3 hp/h (horsepower per hour). Thus, for the same price, you have 47 hp/h for the electric and 21 hp/h for the vehicle powered by California gasoline. In actual use you can have a RAV4 EV costing $0.04 per mile and a gasoline-powered RAV4 costing $0.16 per mile just for fuel.

The economics also have changed recently owing to the high cost of gasoline. The G-Van byVehma International is the first commercially available production van. The G-Van selling price is $50,000 compared with the $16,000 internal combustion engine (ICE) vehicle.For comparison, a pickup truck conversion similar to my own has been added. Its principal advantage is that it was converted at about 50,000 miles, when many of the components start to require replacement and when engine efficiency starts to decrease.

Detroit Edison had done extensive testing of the G-Van; it has found that maintenance has decreased from 2.5 to 1.4 hours per month due to the learning curve. In the United Kingdom, there is significant experience in EVs for retail delivery vehicles. The EV cost per mile is only 60 percent of the cost of an internal combustion engine (ICE) vehicle.

In the late 1970s, the U.S. Postal Service in California used EVs as postal vehicles. At that time, the cost was 2.5 cents per mile for an EV versus 5.6 cents per mile for a gasoline-powered jeep. The principal advantage of the EV in this type of stop-and-go service is that an EV does not use any energy when it is stopped. The jeep got only 6.5 miles per gallon of gasoline.

We can also look at a more developed industry—that of lift trucks. Initial cost for an electric lift truck is $10,000 more because of the batteries and charger. However, there is substantial saving in fuel and maintenance costs.

- Fuel cost:
  - EV: $1.25/day
  - ICE: $7–9/day

Electric trucks have 20 percent less driving time but 25 percent greater service life; annual maintenance cost is only $1,700. The maintenance cost for an ICE truck ranges from \$3,000 to \$7,500 depending on whether it operates for one or two shifts per day. EVs are fully capable of different types of service, although the present range is limited to 50 to 80 miles daily.[2]

Fuel-Efficient Vehicles

Reducing greenhouse gas emissions would lower the economic and human health risks associated with a changing climate. The primary greenhouse gas is carbon dioxide (CO_2), and according to the Environmental Protection Agency (EPA), about 20 percent of total U.S. emissions of CO_2 are from passenger vehicles (cars and light trucks).[3] Those emissions are directly related to the amount of gasoline a vehicle uses, which, in turn, depends on the number of miles the vehicle is driven and on its fuel economy.

Car companies are now making an effort to develop or provide more fuel-efficient cars. The market wants them, large organizations have responded to the market, and the car companies that are doing well (i.e., Toyota and Honda) are producing fuel-efficient and hybrid cars. Now Ford, GM, and the other car companies are either producing or developing these type of cars as well. This is to be expected in business.

Much has been said about creating a new type of car that can get 35 mi/gal. In other countries where fuel is very expensive, such cars already exist. The cars are much smaller and have smaller engines. What has not been said is the great extent to which drivers control the range of these vehicles. In countries where fuel is expensive, drivers tend to drive at slower speeds. Driving twice as fast requires four times the energy to overcome aerodynamic losses. To go from 50 to 100 mi/h increases the rate fuel or electrical energy is used almost by a factor of eight. (Since you get there in half the time, total energy used is increased by a factor of four.) It is the driver's right foot more than anything else that controls mile per gallon or miles per kilowatt-hour for a particular vehicle.

Even if you do not have an electric car, plan on converting a car, or plan on buying a hybrid electric car, one thing to take away from this book is that driving more efficiently will reduce your carbon footprint.

Time Is Running Out!

Time is running out on cheap petroleum fuels. Recent research states that we have little more than 40 years of steadily decreasing supplies of that type of energy while concurrently its price grows prohibitively high. Then it's no more cheap oil. Then what?

This is the question being asked and answered by Howard Johnson, an engineer and author of the book, *Energy, Convenient Solutions: How Americans Can Solve the Energy Crisis in Just Ten Years*[4]:

> We are running out of oil, period. That being said, it's time to get down to the business of seriously developing alternatives. It is paramount that we develop realistic solutions to the energy crisis from among the multitude of products and systems that are in use, under development, or even latent ideas in the minds of America's creative genius. We must collect and examine descriptions of fuels and energy systems—past, present, and future—and the many possible and practical ways to replace fossil fuels with renewable fuels or energy systems.

According to Johnson, "new combinations of old and emerging technologies promise amazing new ways to generate, distribute, and use energy of many kinds. The large variety of proven technologies and systems is astounding. There is no single right answer. The astonishing thing is the variety. In the long run, some will flourish while others will fall by the wayside."

He thinks the United States should select new energy solutions based on some or all of the following criteria. "It matters not to a driver what powers his vehicle when he presses down on the accelerator pedal." Johnson added:

> Any power system that provides adequate mobile power economically when that pedal is pressed will satisfy his needs. All of the new systems could replace fossil fuels as the prime energy source for our nation and even the world. We need to develop an entirely new and more efficient means of generating, transporting, storing and using energy that will generate profitable domestic business, good jobs and stop the hemorrhaging of billions of American dollars to nations that hate us.

U.S. Transportation Depends on Oil

According to the Pacific Northwest National Laboratory for the U.S. Department of Energy, more than $172 billion dollars of government money was spent on new energy technology between 1961 and 2008, with the bulk of it being used during the 1970s. In the 1980s, the spending accounted for only 1 percent of all federal investment.

Although small amounts of natural gas and electricity are used, the U.S. transportation sector is almost entirely dependent on oil. A brief look at Figure 2-3 will demonstrate the facts. It doesn't take a rocket scientist to figure out that this situation is both a strategic and economic problem for us.

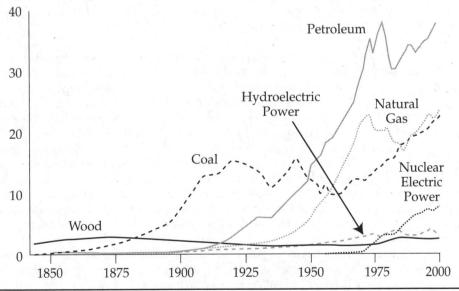

Figure 2-3   U.S. energy consumption by source from 1775 to 2001. (*U.S. Department of Energy.*)

Forty percent of our energy comes from petroleum, 23 percent from coal, and 23 percent from natural gas. The remaining 14 percent comes from nuclear power, hydroelectric, and renewables. As Bob Brant stated, "Our entire economy is obviously dependent on oil."

The United States consumes 20.8 million barrels of petroleum a day, of which 9 million barrels is gasoline. Automobiles are the single largest consumer of oil, consuming 40 percent, and they are also the source of 20 percent of the nation's greenhouse gas emissions.

The United States has about 22 billion barrels of oil reserves while consuming about 7.6 billion barrels per year. Problems associated with oil supply include volatile oil prices, increasing world and domestic demand, and falling domestic production.

While it's our own fault for letting it happen, the Organization of Petroleum Exporting Countries (OPEC) price hikes have had a disastrous impact on our economy, our transportation system, and our standard of living. The Arab oil crisis of 1973 and subsequent crises were not pleasant experiences. After each crisis, the United States vowed to become less dependent on foreign oil producers—yet exactly the opposite has happened.

Increasing Long-Term Oil Costs

There is a fixed amount of oil/petroleum in the ground around the world, and there isn't going to be any more. We're going to run out of oil at some point. Before that happens, it's going to get expensive.

What's Better for the Environment:
Raising the Gas Tax or Fuel Efficiency Standards?

The Obama administration recently announced that it was gradually raising fuel efficiency standards to 54.5 mi/gal by 2025. This is good news for the environment, but it's not the best *possible* news for the environment, argues Eric Morris at Freakonomics. That would be raising the federal gas tax:

> This raises the true problem with CAFE [the fuel economy standard]. It misses out on a potentially key part of the solution to reducing fuel use: driving less. In fact, ironically, increased CAFE standards will have a perverse and unwelcome effect; better fuel economy will increase the fixed cost of driving (i.e., vehicle prices) but will actually reduce the marginal cost (i.e., fuel expenditures). To a degree, less thirsty cars will actually cause people to increase the number of miles they drive.
>
> With increased gas taxes, on the other hand, less driving will be part of the consumer's toolkit. Some who absolutely need vehicles with poor fuel economy will have the option of avoiding the tax by driving less instead. As long as their fuel use goes down, why not give them that choice? Greater economic efficiency would result. In fact, the Congressional Budget Office (CBO) ran the numbers in 2004 and found that cutting fuel use through taxes was considerably cheaper in the long run than raising CAFE.
>
> Yet the CBO reported in 2008 that human activities are producing increasingly large quantities of greenhouse gases, particularly CO_2, and their accumulation in the atmosphere is expected to affect the climate throughout the world.

Charging a price for CO_2 emissions would raise the price of gasoline, but that increase—and the resulting decrease in vehicle emissions—would be relatively small. Most of the reduction in CO_2 emissions would occur in other sectors.

The initial impact on vehicle emissions would be particularly small: People could drive less and at slower speeds, and some could switch to public transit, but in the short run, they would have few other alternatives. Over time, consumers could respond to higher gasoline prices by buying more fuel-efficient vehicles and reducing their commuting distance when an opportunity arises. Substantial increases in gasoline prices in recent years have triggered measurable responses of both types. But a CO_2 price high enough to induce sizable reductions from other sources of emissions would have only a small effect on vehicle emissions of CO_2.

Recent changes to the automobile fuel economy standards—greatly increasing their stringency—will result in a substantial decline in vehicle emissions whether gasoline prices increase or not.[5]

Morris makes some strong points here—without even mentioning the fact that a gas tax increase, unlike higher fuel economy standards, will also generate revenue for the highway trust fund. (If anything, raising CAFE will *hurt* gas tax revenues because fuel-efficient cars don't have to be filled up as often.) The gas tax is also better for congestion, and, consequently, the accident rate, and we still get the benefit of a cleaner fleet over time because high gas costs will create a demand for fuel-efficient cars.

Let's examine the CBO report from 2004,[6] cited by Morris, that directly compares the two approaches. The authors estimate that raising CAFE to roughly 31 mpg would reduce gas consumption 10 percent after 15 years, by which time the entire U.S. fleet would be replaced. This change would come at a cost of $3.6 billion a year to the economy (largely through higher car costs).

To achieve the same 10 percent reduction through a gas tax, it would have to be raised 46 cents a gallon. This would cost the economy $2.9 billion a year (mainly through reduced gas sales, *not* through the tax itself). In the first 14 years of the comparison—before the full fleet is replaced—the advantage in both cost and gas consumption goes clearly to raising the gas tax.

CAFE Versus Gas Tax in First 14 Years

It's important to note that the authors only figure on a fuel economy of roughly 31 mi/gal, well below the 2025 mandate of 54.5 mi/gal. Still, some energy economists are convinced about this idea, including Michael Giberson at Knowledge Problem. He's ready to repeal the new CAFE laws right now in favor of a higher gas tax. He states:[7]

I'd like to propose the following deal: Repeal CAFE, raise the gasoline tax in stages over the next several years, and offset the revenue increases with reductions in other federal taxes. That is "Repeal" with a capital R. . . . Repeal CAFE and raise the gasoline tax instead. No net increase in federal taxes, and we toss out a cumbersome, bureaucratic, inefficient regulatory system that has been burdening automakers and auto consumers for years.

Their argument is as intriguing as it is moot. The CAFE raise is in the books. The gas tax, meanwhile, might disappear entirely. The *Washington Post* reports[8] that some officials believe the congressional debate over transportation funding is heading for a stalemate even worse than the one recently endured by the Federal Aviation Administration. And if that process is any guide, then drivers should not expect any cost reprieve at the tank if collection of the federal gas tax is suspended; rather, they should expect gas companies to charge them the same price and simply pocket the difference, like the airlines did.[9]

How did we get into this situation? Today we are already past that amount. No one can accurately predict what fuel prices will be this summer or next year or whether there will be a shortage or abundance of supply. Everyone agrees that this is a bad situation. We need to take real steps to correct the problem. Since none of us has the luxury of skipping going to work, if gasoline suddenly cost $4.50 a gallon in the United States or a price for a barrel of oil costs more than $150, many creative solutions would come forward—for example, EVs.

How Electric Vehicles Can Help

Maintaining stringent toxic air pollution emission levels, along with conforming to increasingly higher mandated corporate average fuel economy levels, puts an enormous burden on gas-powered vehicle technology and on your pocketbook. Auto manufacturers have to work their technical staff's overtime to accomplish these feats, and the costs will be passed on to new buyers.

Pollution-control equipment is a problem each gas-powered vehicle owner has to revisit every year: smog checks, emission certificates, and replaced valves, pumps, filters, and parts—all extra cost-of-ownership expenses and inconveniences.

How can EVs reduce toxic air pollution emissions? Easy. All EVs are by definition zero-emission vehicles (ZEVs): They emit nothing. This is why California and other states have mandated EVs to solve their air pollution problems. To quote Quanlu Wang, Mark A. DeLuchi, and Dan Sperling, who studied the subject extensively:

> The unequivocal conclusion of this paper is that in California and the United States the substitution of electric vehicles for gasoline-powered vehicles will dramatically reduce carbon dioxide, hydrocarbons and to a lesser extent, nitrogen oxide emissions.[10]

The earlier comments regarding power-plant emissions associated with electricity production for EV use are equally applicable to air pollution as they are to greenhouse gas production. In addition, shifting the burden to coal-powered electricity-generating plants for EV electricity production has these effects:

- It focuses on smokestack scrubber and other mandated controls on stationary sites that are far more controllable than gas-powered vehicle tailpipes.

- It shifts automotive emissions from congested, populated areas to remote, less populated areas, where many coal-fired power plants are located.

- It shifts automotive emissions to nighttime (when most EVs will be recharged), when fewer people are likely to be exposed and emissions are less likely to react in the atmosphere with sunlight to produce smog and other by-product pollutants.

EVs generate no emissions whatsoever and reduce our reliance on imported oil. Frankly, until you get an appreciable number of EVs on the road (hundreds of thousands to millions), they will not have an impact on emissions from electricity-generating plants.

Toxic Solid Waste Pollution

Almost everything going into and coming out of an internal combustion engine is toxic. In addition to the gas-powered vehicle's greenhouse gas and toxic air pollution outputs, consider its liquid waste (i.e., fuel spills, oil, antifreeze, grease, etc.) and solid waste by-products (i.e., oil/air/fuel filters, mufflers, catalytic converters, emission-control system parts, radiators, pumps, spark plugs, etc.). This does not bode well for our environment, our landfills, or anything else—especially when multiplied by hundreds of millions of vehicles.

How can the EV help? The only waste elements of an EV are its batteries. For example, lead-acid batteries—the kind commonly available today—are 99.99 percent recyclable. In processing many tons per day, almost every ounce is accounted for. This means that 99.99 percent of all such batteries and the products that go into them (i.e., the sulfuric acid, the lead, and even the plastic of their cases) is recoverable.

Study Finds Lithium-Ion Batteries Cause Less Environmental Damage

Recent research undertaken at the Norwegian University of Science and Technology (NTNU) and published in the current issue of *Environmental Science and Technology*, a scientific journal, provides details on the three major types of batteries that are currently in use, their efficiency, and the quantity of environmental impact during their manufacturing process.

The research finds that on a per-storage basis, the nickel–metal hydride (NiMH) battery affects the environment worst, closely accompanied by the nickel-cobalt-manganese lithium-ion (NCM) and the iron-phosphate lithium-ion (LFP) batteries. The researchers have observed 11 types of ecological impacts, such as greenhouse gas emissions, ecotoxicity in freshwater, human toxicity, and freshwater eutrophication, excluding the ozone-depletion potency. They also have found global warming discharges with higher-level life cycles than reported previously during the research.

The researchers found that the two lithium-ion batteries performed better than the NiMH battery owing to their two to three times increased energy-storage capability in their lifetime and higher level of usage-period efficiency in comparison with NiMH batteries. The NCM and LFP lithium-ion batteries were observed to contain less nickel and no trace of rare earth metals. The LFP batteries were found to offer more ecological benefits than NCM batteries owing to their increased usage life period and use of metals that do not affect the environment.

All the three types of batteries needed the same level of energy during their production process, thus releasing the same levels of greenhouse gases. The manufacture of polytetrafluoroethylene, which is used as a dispersant/binder in the electrode paste in the batteries, was found to be the cause for over 97 percent of the ozone-depletion qualities of the batteries. The two lithium-ion batteries released at least 14 to 15 percent increased levels of CO_2 discharges during the production process owing to the release of halogenated methane discharges. The cell containers, module packages, materials used for separation, and the electrolyte used in the packing of the batteries were found to cause less than 10 percent ecodamage in all categories. The researchers concluded that LFP batteries have a lesser environmental impact owing to their increased life expectancy.[11]

Toxic Input Fluids Pollution

Remember, almost everything going into and coming out of an internal combustion engine is toxic. The fuel and oil that you put into an internal combustion engine, the fuel vapors at the pump (and those associated with extracting, refining, transporting, and storing fuel), and the antifreeze you use in its cooling system are all toxic and/or carcinogenic, as a quick study of the pump and container labels will point out. On the output side, when burning coal, oil, gas, or any fossil fuel, you create more problems either by the amount of CO_2 produced or by the type of other toxic emissions produced.

Everything you pour into an internal combustion engine is toxic, but some chemicals are especially nasty. In addition to more than 200 compounds on its initial hazardous list, the Clean Air Act of 1990 amendments stated: "[This] . . . study shall focus on those categories of emissions that pose the greatest risk to human health or about which significant uncertainties remain, including emissions of benzene, formaldehyde and 1,3-butadiene."

Fouling the environment, as in the *Exxon Valdez* oil spill disaster of the 1990s or the BP spill in the Gulf Coast several years ago is one thing. Poisoning your own drinking water is another. Those enormous holes in the ground near neighborhood gas stations everywhere (as they rush to be compliant with federal regulations regarding acceptable levels of gasoline storage tank leakage) make the point. So does the recall of millions of bottles of Perrier drinking water, in which only tiny levels of benzene contamination were found.

Maintaining stringent toxic air pollution emission levels, along with conforming to increasingly higher average miles per gallon levels, is not going to do it with an internal combustion engine. It's just based on the fact that an internal combustion engine is only 20 percent efficient to begin with versus about 90 to 95 percent efficiency on electric AC and DC motors.

EVs generate no emissions whatsoever and reduce our reliance on imported oil or oil at all. However, once a significant percentage of cars become electric or even plug-in hybrid electric, which means hundreds of thousands to millions of cars, the EVs we are talking about for now don't even have an impact on emissions from power plants.

According to *Science Daily* and Treehugger.com, there was a study by Rice University's Baker Institute for Public Policy that finally said, duh, if the United States wants to significantly reduce its oil consumption, it should prioritize the fast adoption of electric cars over "the proposed national renewable portfolio standard."

This makes a lot of sense because a renewable portfolio standard would mostly have an impact on electricity production and not liquid fuels, but what the study fails to mention is that we'll need both approaches to truly start moving toward greener transportation.

The people from Rice University further stated:

In fact, mandating that 30 percent of all vehicles be electric by 2050 would both reduce U.S. oil use by 2.5 million barrels a day beyond the 3 million barrels-per-day savings already expected from new corporate average fuel efficiency standards, and also cut emissions by 7 percent, while the proposed national renewable portfolio standard (RPS) would cut them by only 4 percent over the same time.

How can the EV help?

In 2001, the latest phase of fuel economy standards announced by President Obama will result in total savings of 3.5 million barrels of oil a day by 2030, helping to break

America's long-standing dependence on foreign oil. Representative Ed Markey (D-Mass.), ranking member of the Natural Resources Committee, senior member of the Energy and Commerce Committee, and House coauthor of the legislative standards that triggered the announcement, praised the president for aggressively implementing the goals of the legislation:

> These fuel-efficiency standards strike at the heart of the oil cartel that has held America in a stranglehold for decades, constricting our economic growth, threatening our national security and poisoning our planet. . . . This is a groundbreaking moment in our energy history that will finally help shatter the bonds of our dependence on Middle East oil.

The 54.5 mi/gal standard was enabled by the Energy Independence and Security Act of 2007, which included fuel economy standards coauthored by Representative Markey and championed by then-Speaker Nancy Pelosi (D-Calif.). That law included Markey language that said the standard must be at least 35 mi/gal by 2020 and that the "maximum feasible standard" must be set every year. The bill was signed by President George W. Bush in December 2007.

The fuel economy legislation, combined with the 2007 Supreme Court decision of *Massachusetts v. the U.S. Environmental Protection Agency*, which affirmed the agency's authority to reduce greenhouse gas pollution from automobiles, paved the way for the announcement.

That bill also included provisions to require the deployment of advanced biofuels. Combined, the fuel efficiency and biofuels provisions will save 5.1 million barrels of oil per day by 2030. The United States currently imports 4.6 million barrels of oil per day from OPEC.

Representative Markey first offered his fuel economy amendment in 2001, following years of Republican legislative riders that prevented fuel economy increases from being adopted. He brought his legislation up for a vote in successive sessions of Congress, finally succeeding in his attempts in the 2007 energy bill.[12]

Waste Heat Owing to Inefficiency

Although its present form represents its highest evolution to date, the gas-powered internal combustion engine is classified among the least efficient mechanical devices on the planet. The internal combustion engine is close to 20 percent efficient. The efficiency of an advanced DC electric motor runs between 80 and 90 percent, sometimes lower.

In gas-powered vehicles, only 20 percent of the energy of combustion becomes mechanical energy; the rest becomes heat lost in the engine system. Of the mechanical energy generated,

- One-third overcomes aerodynamic drag (energy ends up as heat in the air).

- One-third overcomes rolling friction (energy ends up as heated tires).

- One-third powers acceleration (energy ends up as heat in the brakes).

In contrast to the hundreds of moving parts of internal combustion engines, the electric motor has just one. This is why such motors are so efficient. Today's EV motor efficiencies typically are 90 percent or more. The same applies to today's solid-state controllers (with no moving parts), and today's lead-acid batteries come in at 75 percent or more efficiency.

Combine all these and you have all-electric vehicle efficiency far greater than anything possible with an internal combustion engine vehicle.

Electric Utilities Love Electric Vehicles

Even the most wildly optimistic EV projections show only a few million EVs in use by early in the twenty-first century. At somewhere around that level, EVs will begin making a dent in the strategic oil, greenhouse gas, and air-quality problems. Until we reach the 10 to 20 million or more EV population level, though, we're not going to require additional electricity-generating capacity. This is due to the magic of *load leveling*.

Load leveling means that if EVs are used during the day and recharged at night, they perform a great service for their local electrical utility, whose demand curves almost universally look like that shown in Figure 2-4.

How electricity is generated varies widely from one geographic region to another and even from city to city in a U.S. region. In 2007, the net electricity mix generated by electrical utilities was 48.6 percent coal, 19.4 percent nuclear fission, 21.5 percent natural gas, 5.8 percent hydropower, and 1.6 and 2.5 percent for geothermal, solar, and wind, with other miscellaneous sources providing the balance.[13]

Electric utility plants producing electricity at the lowest cost (e.g., coal and hydro) are used to supply base-load demands, whereas peak demands are met by less economical generation facilities (e.g., gas and oil).[14] By recharging their EVs in the evening hours (valley periods), owners receive the benefit of an off-peak (typically lower) electrical rate. By raising the valleys and bringing up its base-load demand, the electric utility is able to use its existing plant capacity more efficiently. This is a tremendous near-term economic benefit to our electric utilities because it represents a new market for electricity sales with no additional associated capital asset expense.

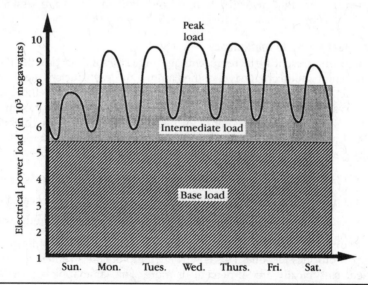

FIGURE 2-4 Weekly peak power-demand curve for a large utility operating with a weekly load factor of about 80 percent.

Summary

EV ownership is the best first step you can take to help save the planet. But there is still more you can do. Do your homework. Write your senator or congressperson. Voice your opinion. Get involved with the issues. But don't settle for an answer that says we'll study it and get back to you. Settle only for action—Who is going to do what by when and why? I leave you with a restatement of the problem, a possible framework for a solution, and some additional food for thought.

The Legacy of the Internal Combustion Engine Is Environmental Problems

Internal combustion engine technology and fuel should be priced to reflect its true social cost, not just its economic cost, because of the environmental problems it creates:

- Our dependence on foreign oil and the subsequent security-risk problem
- The greenhouse gas problem
- The air-quality problems of our cities
- The toxic waste problem
- The toxic input fluids problem
- The inefficiency problem

Our gasoline's cost should reflect our cost to defend foreign oil fields, reverse the greenhouse gas effect, and solve the air-quality issues. Best of all, our gasoline's cost should include substantial funding to research solar energy generation (and other renewable sources) and EV technologies—the two most environmentally beneficial and technologically promising gifts we can give to our future generations.

A Proactive Solution

People living in the United States have been extremely fortunate for most of our nation's history. However, and more now than ever, we have issues with clean air, our natural resources, instable governments, expensive energy costs, and maintaining a convenient and true standard of living second to no other country on the planet. But nothing guarantees that our future generations will enjoy the same birthright. In fact, if we fold our hands behind our backs and walk away from today's environmental problems, we guarantee that our children and children's children will not enjoy the same standard of living that we do. For the sake of our children, we cannot walk away. We *must* do something. We must attack the problem straight on, pull it up by its roots, and replace it with a solution.

Figure 2-5 suggests a possible approach. We need to look at the results we want in the mid–twenty-first century and work backward—on both the supply and demand sides—to see what we must start doing today. Clearly, it's time for a sweeping change, but we all have to want it and work toward it for it to happen. No one has to be hurt by the change if they become part of the change. Automakers can make more efficient vehicles. Suppliers can provide new parts in place of the old. The petrochemical industry can alter its mix to supply less crude as oil and gas and more as feedstock material used in making vehicles, homes, roads, and millions of other useful items. Long before any of these things happen, you can do your part by building your own EV. Today!

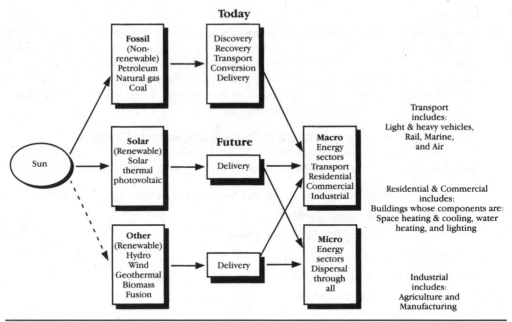

FIGURE 2-5   Model of balanced future energy usage made possible by working from future desired goal back to today.

After the BP oil spill in the Gulf of Mexico, international ocean conservation group Oceana presented a game plan that would eliminate the need for offshore drilling in the Gulf within the next 10 years. Using data from the nonpartisan U.S. Energy Information Administration and the National Renewable Energy Laboratory, among others, Oceana offered pragmatic recommendations that, if followed, would eliminate the need for drilling by 2020 and alleviate the need to import oil from the Persian Gulf by 2023, a mere 12 years from today. The plan also outlines an approach to eliminating imports altogether by 2030 (Figure 2-6).

"Rather than being held hostage by the status quo, Oceana devised a practical plan to end America's oil habit. We can make major changes within the next 12 years," said Oceana senior campaign director, Jacqueline Savitz. "Our plan focuses on four sectors: heating, transportation, shipping and power generation, and substitutes clean energy for current uses of oil. With conservative projections of renewable energy generation potential, we show that there are safer ways to meet our nation's energy needs," Savitz added.

On the anniversary of the *Deepwater Horizon* offshore drilling disaster, Oceana's plan provides a way to prevent additional, catastrophic oil spills that foul our waters and kill valuable marine life. This solution also can serve as a catalyst for a clean energy technological revolution, which will stimulate our economy and create jobs.

Cut Oil Use in Shipping

By slowing down by just 10 percent, large marine vessels can cut their fuel use by 23 percent, according to the International Maritime Organization. A number of other

FIGURE 2-6 The type of power needed for electric cars. (*Oceana.*)

actions to increase efficiency could be taken as well. Assuming a simple 10 percent slowing and a corresponding savings of 23 percent on fuel, the need for 105,000 barrels of oil per day could be eliminated.

Stop Using Oil for Electricity Generation

While we use relatively little oil to generate electricity, ending the use of oil in electricity generation would alleviate the need for 200,000 barrels of oil per day. This is 11 percent of the amount of oil the United States is projected to extract from the Gulf of Mexico in 2020. To replace electricity generated by oil, we would need 11.7 GW of offshore wind power.

Shift Residential and Commercial Heating from Oil to Electricity

Shifting just 25 percent of homes and businesses that still use oil for heat to heat generated by electricity not only would increase their efficiency, but it also would alleviate the need for about 228,000 barrels of oil daily. This savings is roughly 13 percent of the amount of oil we are projected to extract from Gulf of Mexico in 2020.

Electrify the Transportation Fleet

EVs and cellulostic ethanol in gas-powered cars could reduce oil consumption by 6.5 million barrels per day by 2030. About 41 percent of this reduction would be due to increased adoption of EVs, which could account for 40 percent of the light-duty vehicle fleet by 2030. Another 17 percent of the projected reduction would result from increased production of cellulosic ethanol. The balance of the reduction, roughly 42 percent, could occur as algal biofuel production is brought to market.

These are just some of the Oceana recommendations for reducing our reliance on oil. More specific recommendations can be found in the full Oceana "2020 Vision Plan," which is posted at www.stopthedrill.org.

Recent changes in the Middle East have brought U.S. dependence on foreign oil into sharp focus for business and political leaders. Many U.S. politicians and energy experts suggest that increased domestic production would help to break U.S. dependency on oil imports. However, increased production would not have that effect, according to economists. Oceana recommends a better alternative that doesn't endanger our environment or health—eliminating the need for oil imports in the first place.

The U.S. Energy Information Administration projects that oil imports will decrease slightly from current levels to about 9 million barrels per day in 2030, representing roughly 50 percent of the nation's oil consumption. Of these 9 million barrels of imported oil, about 1.6 million barrels will come from the Persian Gulf. Further advances in the previously described energy sectors would allow us to reduce all oil imports by 80 percent by 2030 and entirely eliminate Persian Gulf imports as early as 2023. Continued reductions beyond 2030 in these same sectors would allow the nation to achieve a crucial goal—the elimination of all oil imports.

Why Are Gasoline Prices So High?

Gasoline and diesel prices are high as a result of tight crude oil supplies and increased global demand for transportation fuels.

- OPEC production decisions will continue to influence the oil market situation.
- Low surplus production capacity of 2 to 3 million barrels per day, concentrated in Saudi Arabia, weakens the market's ability to respond to supply disruptions.
- Oil prices are likely to remain high at least through 2009.
- Many uncertainties could alter the outlook and create volatility in global oil markets.

Why Are Prices Getting So High?

- Production increases are expected in OPEC and non-OPEC countries—China, Nigeria, Angola, Saudi Arabia, Azerbaijan, Brazil, Kazakhstan, and U.S. Gulf of Mexico.
- The high oil price environment of the past several years has provided sufficient incentives to spur increased growth in unconventional supplies.
  - Canadian oil sands—this won't happen at realistic prices: $75 a barrel just preprofit
  - Brazilian and U.S. biofuels
  - Qatar gas to liquids
- Significant projected growth is expected in global natural gas liquids.

  However, here are all the things that can screw up the analysis shown!

Yet Green Cars Make "Cents"

Although the Volt and Leaf have stratospheric sticker prices, both have pretty low ownership costs. Sales of hybrid and electric cars jumped nearly 40 percent in the first quarter of the year. The primary reason—soaring gas prices. More than one-third of consumers now say that fuel economy is the most important factor in their next vehicle purchase, according to a recent study by Maritz Research, a market-research firm.

Setting aside their "environmental credibility," are hybrids, diesels, and EVs actually wallet-friendly? Prices on hybrids run about $3,300 higher, on average, than stickers on their gas-powered siblings. The diesel difference is about $2,800 more. And the two EVs on the market—the Chevrolet Volt and Nissan Leaf—are each more than $18,000 pricier than their closest gas-powered match. But the long-term ownership costs are the real measure of whether buying green is worth it.

Running the Numbers

Using five-year ownership costs from Vincentric, an automotive data firm, we compared 2011 hybrid, diesel, and EVs with their closest gas-powered counterparts. In most cases, this is the same model with a different power train; when a hybrid (such as a Prius) had no counterpart, we chose the closest match from the carmaker's lineup.

The numbers assume that you drive 15,000 miles a year and that regular gasoline costs $3.64 a gallon, premium $3.91, and diesel is $3.97—the average prices nationwide in early summer—with a 3.5 percent annual increase for each fuel. In addition to fuel costs, depreciation, maintenance, and repairs, the math also includes finance costs for a five-year loan after a 15 percent down payment, insurance, and taxes. The Volt and Leaf are both eligible for the $7,500 tax credit for EVs, and that, too, is factored in.

Best Deals

Pump prices have a lot to do with making green cars a good value. Two years ago, when gas prices were hovering close to $2 a gallon, few hybrids and diesels earned back their premium price with savings at the pump. But with gas prices now closer to $4 a gallon, more buyers will save green by buying green.

Perhaps the biggest surprise is that although the Volt and Leaf have stratospheric sticker prices—nearly double those of the gas-powered Chevrolet Cruze LTZ and Nissan Versa S hatchback—both have pretty low five-year ownership costs. The Volt's costs come within $1,600 of the Cruze's and the Leaf is only $800 more than the Versa over five years. (Run your own comparisons of these models and many more.)[15]

Debunking the Myth of EVs and Smokestacks

by Chip Gribben, Electric Vehicle Association of Greater Washington, DC (EVA/DC)

Introduction

As ozone levels in the U.S. remain at unhealthy levels, researchers and government officials continue to study alternatives to reduce air pollution from gasoline-powered cars.

1. Among the alternatives are ultra-low emission vehicles (ULEVs) and zero-emission vehicles (ZEVs). ULEVs are equipped with emission controls that release only 45 pounds of carbon monoxide per 12,000 miles.

2. ZEVs produce no tailpipe emissions at all. ZEVs include vehicles powered by electricity, flywheels, hydrogen fuel cells, and other zero emission energy sources. Although some

ZEVs are still in the experimental stage, electric vehicles (EVs) are available today. In fact, more EVs roamed the nation's roads in the early 1900s than gas-powered cars did.

Unlike a gasoline car that is powered by an internal combustion engine (ICE), an EV uses electricity stored in batteries to power one or more electric motors. When the batteries need recharging, you simply "plug in" from the convenience of your home. EVs have no tailpipe or evaporative emissions

3. Because they have no fuel, combustion, or exhaust systems, in fact, EVs are virtually maintenance free because they never need oil changes, air filters, tune-ups, mufflers, timing belts, or emission tests. One of the most common issues surrounding EVs today is their status as ZEVs. Critics proclaim that EVs are simply "elsewhere emission vehicles" because they transfer emissions from the tailpipe to the smokestack. Although there are emissions associated with coal and oil-fired power plants, smokestack emissions associated with charging EVs are extremely low.

4. In fact, EVs can charge from zero-emission sources such as nuclear, hydroelectric, solar, and wind power.

The purpose of this section is to prove that EVs recharging from today's power plants are substantially cleaner than even the most efficient ICE vehicles. The myth that EVs are "elsewhere emission vehicles" will be put to the test with facts that clearly show EVs and power plants are cleaner, more efficient and more reliable than the infrastructure that supports ICE vehicles.

The Effects of the ICE Age

The golden age of the automobile has lasted more than 50 years; however, the golden haze caused by our love affair with the ICE car will have long-lasting effects. Despite stringent standards to improve tailpipe emissions, the number of vehicles and miles traveled are increasing every year. Scientists predict that our increased reliance on the automobile could increase pollution levels 40 percent by the year 2010.

In California, where the automobile is considered a necessity, ICE vehicles account for 90 percent of the carbon monoxide, 77 percent of nitrous oxides, and 55 percent of reactive organic gases. In addition, greenhouse gases such as carbon dioxide are expected to increase approximately 33 percent by the year 2010.

Continual exposure to these pollutants can cause a variety of symptoms and aggravate existing medical conditions. The elderly and the young are more susceptible to the risks. Children in the Los Angeles area have 10 to 15 percent less lung capacity than children in cleaner cities such as Houston, Texas.

The following list describes the potential health risks associated with these emissions.

- *Carbon monoxide (CO).* An odorless and colorless gas that is highly poisonous. CO can reduce the blood's ability to carry oxygen and can aggravate lung and heart disease. Exposure to high concentrations can cause headaches, fatigue and dizziness.

- *Sulfur oxides (SO_x) and sulfur dioxide (SO_2).* When combined with water vapor in the air, SO_2 is the main contributor of acid rain. Gasoline typically contains 0.03 percent sulfur.

- *Nitrogen oxides (NO_x) and nitrogen dioxide (NO_2).* These chemicals are the yellowish brown haze seen over dirty cities. When combined with oxygen from the atmosphere, NO becomes NO_2, a poisonous gas that can damage lung tissue.

- *Hydrocarbons (HCs).* This is a group of pollutants containing hydrogen and carbon. HCs can react to form ozone. Some HCs are carcinogenic, and others can irritate mucous membranes. HCs include

  - Volatile organic compounds (VOCs)

  - Volatile organic gases (VOGs)

  - Reactive organic gases (ROGs)

  - Reactive organic compounds (ROCs)

  - Nonmethane hydrocarbons (NMHCs)

  - Nonmethane organic gases (NMOGs)

- *Ozone (O_3).* This is the white haze or smog seen over many cities. Ozone is formed in the lower atmosphere when NMOGs and NO_x react with heat and sunlight. Ozone can irritate the respiratory system, decrease lung function and aggravate chronic lung disease such as asthma. Ozone gases have contributed to smog levels as high as 80 parts per billion an average of 84.3 days per year since 1982 in Baltimore, Maryland. Federal safety standards state that the risk level is 120 parts per billion when exposed to smog for an hour. However, recent studies suggest that exposure to 80 parts per billion is enough to cause lung inflammation that can lead to permanent scarring.

- *Carbon dioxide (CO_2).* CO_2 is a naturally occurring gas in the atmosphere and is a necessary ingredient of the ecosystem. However, in large quantities, it can allow more light to enter the atmosphere than can escape. The excess heat from the trapped light can lead to the *greenhouse effect* and global warming.

Clearing the Air About Power Plant Emissions

EVs have the unique advantage of using electricity generated from a variety of fuels and renewable resources. The overall mix of power plants in the United States is 55 percent coal, 9 percent natural gas, and 4 percent oil. The other 32 percent include nuclear power and renewable energy sources such as hydroelectric, solar, wind, and geothermal..

Many EV critics point out that charging thousands of EVs from aging coal plants will increase greenhouse gases such as CO_2 significantly. Although half the country uses coal-fired plants, EVs recharging from these facilities are predicted to produce less CO_2 than ICE vehicles. According to the World Resources Institute, EVs recharging from coal-fired plants will reduce CO_2 emissions in this country from 17 to 22 percent.

Reductions in pollutants such as HCs, CO, NO_x, SO_2, and particulates vary according to a region's power-plant mix. If EVs were introduced on a global scale, urban pollution would improve significantly. See Table 2-1. In France, where most of the power comes from nuclear energy, emissions produced to charge EVs would be cut across the board. Countries such as the United States and the United Kingdom use a mix of coal- and oil-fired facilities that produce an elevated level of SO_2 and particulates. However, levels of HCs, CO and NO_x would decrease significantly.

Although half the electricity generated in the United States comes from coal-fired plants, larger regions of the country such as California and the Northeast are turning toward cleaner fuels such as natural gas. In California, where over half the state's pollution comes from ICE vehicles, the overall mix of power plants is one of the cleanest in the country. See Table 2-2. Power plants burning cleaner fuels, such as natural gas, account for a major share of the state's electricity. In fact, natural gas facilities in California emit 40 times less NO_x than exist-

	HCs	CO	NO$_x$	SO$_2$	Particulates
France	–99	–99	–91	–58	–59
Germany	–98	–99	–66	+96	–96
Japan	–99	–99	–66	–40	+10
United Kingdom	–98	–99	–34	+407	+165
United States	–96	–99	–67	+203	+122
California	–96	–97	–75	–24	+15

TABLE 2-1   Electric Vehicles Reduce Pollution

Power Plant	Percent
Natural Gas	33
Hydroelectric	20
Coal	16
Nuclear	15
Solar and Wind	6
Geothermal	6

TABLE 2-2   Power Plant Mix in California

ing coal-fired plants in the Northeast. Renewable sources such as hydro, solar, wind, and geothermal produce a respectable share of the electricity generated in California.

Taking advantage of California's abundance of sunlight, several utilities are using solar charge ports to charge EVs. Charge ports are facilities that have an array of solar panels placed strategically on the roof of the structure. The solar panels convert sunlight into electricity, which is distributed to vehicles or to the adjacent building's power supply. On cloudy days, the building supplies the electricity to charge the EVs. Charge ports are in operation in several cities in California, including Diamond Bar, Azusa, and Santa Monica.

Because California has a mix of cleaner fuels and renewable sources, several studies have concluded that improvements in air quality can be achieved easily by "plugging in" to EVs. The California Air Resources Board (CARB) estimates that EV's operating in the Los Angeles Basin would produce 98 percent fewer hydrocarbons, 89 percent fewer oxides of nitrogen, and 99 percent less carbon monoxide than ICE vehicles.

In a study conducted by the Los Angeles Department of Water and Power, EVs are significantly cleaner over the course of 100,000 miles than ICE cars. The electricity-generation process produces less than 100 pounds of pollutants for EVs compared with 3,000 pounds for ICE vehicles. See Table 2-3.

Engine Type	CO	ROG	NO$_x$	Total
Gasoline	2574	262	172	3008 lbs
Diesel	216	73	246	835 lbs
Electric	9	5	61	75 lbs

TABLE 2-3   Pounds of Emissions Produced per 100,000 Miles

CO_2 emissions are also significantly lower. Over the course of 100,000 miles, CO_2 emissions from EVs are projected to be 10 tons versus 35 tons for ICE vehicles. Many EV critics remain skeptical of such findings because California's mix of power plants is relatively clean compared with that in the rest of the country. However, in Arizona, where 67 percent of power plants are coal-fired, a study concluded that EVs would reduce greenhouse gases such as CO_2 by 71 percent.

Similar comparisons to those in California and Arizona can be found in the northeastern part of the country, where the majority of power plants are coal-fired. A study conducted by the Union of Concerned Scientists found that EVs in the Northeast would reduce CO emissions by 99.8 percent, VOCs by 90 percent, NO_x by 80 percent, and CO_2 by as much as 60 percent.

According to a Northeast States for Coordinated Air Use Management (NESCAUM) study, EVs result in significant reductions of CO, greenhouse gases, and ground-level ozone in the region with magnitudes cleaner than even the cleanest ULEV.

In the future, EVs in the Northeast will reap the benefits of switching to cleaner fuels such as natural gas. In the next 15 years, aging coal plants will be replaced by modern natural gas–fired plants. This improvement alone will reduce power-plant emissions significantly.

Several northeastern states are also exploring renewable sources such as solar energy to generate electricity for EVs. The EVermont Project is using a successful solar-powered system to charge a mail delivery truck used at the General Services Center in Middlesex, Vermont. A solar array was installed and wired into the system's power grid. The solar array generates electricity during the day, and the truck charges at night. Overall, the solar panels put out more power than the truck uses on its daily rounds.

The Efficiency Advantage of EVs and Power Plants

EVs recharging from fossil-fueled power plants have unique efficiency advantages over ICE vehicles. As a system, EVs and power plants are twice as efficient as ICE vehicles and the system that refines gasoline. See Table 2-4. Although there are losses associated with generating electricity from fossil-based fuels, EVs are significantly more efficient in converting their energy into mechanical power.

Since EVs operate more efficiently than their ICE-powered counterparts, overall fuel economy is higher. However, making a direct comparison between the fuel efficiencies of both vehicles is difficult. By applying a common unit of energy, such as the British thermal unit (Btu), we can get a fair comparison between the two.

For the following example, we will compare the fuel efficiencies of a 1995 Acura 3.2 TL and GM's new electric vehicle—the EV1. See Table 2-5. Both vehicles cost about $34,000 and can accelerate from 0 to 60 mi/h in 8.5 seconds.

Even though the GM EV1 has 43 percent fewer Btus after electricity generation, it can be driven almost 350 miles farther because the vehicle is more efficient than the Acura. In fact, the GM EV1 has the gasoline equivalency of 59 mi/gal even after factoring in losses from electricity generation and charging!

	EVs and Power Plants	ICE and Fuel Refining
Processing	39% (Electricity Generation)	92% (Fuel Refining)
Transmission Lines	95%	–
Charging	88%	–
Vehicle Efficiency	88%	15%
Overall Efficiency	28%	14%

TABLE 2-4   Operating Efficiency Comparison Between EVs and ICE Vehicles

Electric-Powered GM		EV1 Gasoline-Powered Acura	
Start with	1 million Btus	Start with	1 million Btus
Energy left after generation (39% efficiency)	390,000 Btus	Energy left after refining (92% efficiency)	920,000 Btus
Energy left after charging losses (88% efficiency)	343,200 Btus	Energy left after transportation (95% efficiency)	874,000 Btus
Btus per Kilowatt-hour	3412 Btus	Btus per gallon of gasoline	115,400 Btus
Electricity Available	100.6 kWhr	Gallons available	7.6 gallons
Energy Efficiency	.19 kWhr/mile	Fuel economy	24 mpg
Miles per million Btus	529.5 miles	Miles per million Btus	182.5 miles
Equivalent mpg	59 mpg	Equivalent mpg	24 mpg

TABLE 2-5   Fuel Efficiency Comparison Between EVs and ICE Vehicles

Scrubbing Out Power-Plant Emissions

We've discussed how the system of power plants and EVs can improve air quality, improve operating efficiencies, and save fuel, but just how efficient are power-plant emissions controls?

Controlling emissions from several hundred power plants is much easier than controlling the emissions from 187 million ICE vehicles. In fact, electrical utilities go through considerable efforts to monitor and remove emissions from their facilities. Teams of engineers carefully maintain the plants at peak operating efficiency. State-of-the-art equipment such as scrubbers are installed to remove emissions. Electrostatic precipitators (ESPs) between the boilers and smokestacks remove up to 99.75 percent of the ash emitted by power plants. Coal-fired plants in Texas using ESPs remove up to 13.4 million tons of ash each year, releasing only 3,000 tons into the atmosphere. The amount released falls below U.S. EPA regulations for ash emissions.

Over the next seven years, electrical utilities in the Northeast are committed to reducing NO_x emissions by 55 to 70 percent. When one power plant upgrades its emission controls, thousands of EVs immediately reap the benefits from this improvement.

Catalytic Clunkers

Upgrading and maintaining emissions for ICE vehicles is a different story. According to Drew Kodjak, a lawyer from NESCAUM, ICE vehicles pollute more over time, whereas power plants tend to pollute less over time. Over the course of its lifetime, a gasoline car will spew out 60 times more CO, 30 times more VOCs, and twice as much CO_2 as electric power plants.

The U.S. EPA estimates that tailpipe emissions increase 25 percent for every 10,000 miles traveled. As gasoline cars age, their engines, catalytic converters, and other emission-control devices become less efficient. The cleanest a gasoline car ever will be is the day it rolls off the assembly line.

The deterioration of emission-control systems on ICE vehicles can increase emissions up to 90 percent. To deal with increased emissions, state governments have adopted emission inspection programs with varied degrees of success. Many of these programs have been delayed owing to public concern for the cost of repairing emission components. In Maryland, drivers can receive a waiver if they document attempts to repair their ICE cars even though the cars continue to fail emission tests.

Newer cars entering the market are not necessarily the cleanest either. The hottest vehicles on the market today are sport utility vehicles (SUVs), which now account for 40 percent of all new car sales. These gas guzzlers are driving up this country's demand for imported oil, decreasing overall fuel efficiency, and increasing emissions.

Today's Power Plants Meeting Tomorrow's Recharging Needs

Many critics ask how this country could possibly support millions of EVs on today's existing power grid. The Electric Power Resource Institute (EPRI) estimates that this country has the ability to support 50 million EVs without building any more power plants.

Another study puts this number closer to 20 million. Even so, 20 million EVs is only 10 percent of today's fleet of 187 million cars. Thousands more could be added if they are charged at night during off-peak hours. Twenty million EVs, each with 100,000 miles on the odometer, would reduce CO_2 emissions in this country by 500 million tons without building more power plants.

Southern California Edison (SCE) estimates that it has enough off-peak capacity to refuel up to 2 million cars, 25 percent of the area's automobiles. SCE estimates that it will only need to add 200 MW of capacity to accommodate EVs.

Summary

In conclusion, EVs will have a considerable impact on reducing air pollution, improving fuel efficiency, and reducing our overall dependency on foreign oil. As power plants improve efficiency and turn to cleaner fuels, such as natural gas and zero-emission sources, EVs will continue to be the best solution toward attaining clean air.

CHAPTER 3

Electric Vehicle History

Precedent said, it cannot be done. Experience said, it is done.
—Darwin Gross, Universal Key

While modern technology has made electric vehicles (EVs) better, there is very little new in EV technology. Today's EV components would be instantly recognizable in those that roamed our streets a century ago. As a potential EV builder, buyer, or converter, you should be happy to know that EVs have a long and distinguished heritage.

Let's delve briefly into some basic historical facts about EVs:

- The electric motor came before the internal combustion engine.

- EVs have been around since the mid-1800s, were manufactured in volume in the late 1800s and early 1900s, and declined only with the emergence and ready availability of cheap gasoline.

- Even so, EV offshoots—tracked buses, trolleys, subways, and trains—have continued to serve in mass-transit capacities until the present day because of their greater reliability and efficiency.

One of the greater electric transportation accomplishments I have seen (and became more aware of when I worked for the New York Power Authority) is that most of the Metropolitan Transportation Authority (MTA) subways and trains are electric-powered. A minimal number of trains in the MTA fleet are diesel-powered. In addition, each year I worked for the State of New York, the MTA shifted more and more from compressed natural gas to hybrid electric transit buses. While such vehicles are not 100 percent electric, they prove the point that an electric drive is cleaner and more fuel efficient than just an alternative fuel.

1900s

The battery and electric motor combination borrowed from the EVs of the "novelty era" up until 1915 and applied as a starter for internal combustion engines was responsible for the great upsurge in the internal combustion engine vehicle's popularity. The starter motor systems employed in all of today's internal combustion engine vehicles are virtually unchanged from the original early 1920s concept.

Battery EVs also have been extremely popular in some limited-range applications. Forklifts have been battery electric vehicles (BEVs) since the early 1900s, and electric forklifts are still being produced.

The 1947 invention of the point-contact transistor marked the beginning of a new era for BEV technology. Within a decade, Henney Coachworks had joined forces with National Union Electric Company, the makers of Exide batteries, to produce the first modern electric car based on transistor technology, the Henney Kilowatt, produced in 36- and 72-V configurations. The 72-V models had a top speed approaching 96 km/h (60 mi/h) and could travel nearly an hour on a single charge. Despite the improved practicality of the Henney Kilowatt over previous electric cars, it was too expensive, and production was terminated in 1961. Even though the Henney Kilowatt never reached mass-production volume, its transistor-based electric technology paved the way for modern EVs, plus its name is a unit measure of electricity. Not bad!

Timeline of Vehicle History

Studying vehicle history is similar to looking at any economic phenomenon. iPods are a good example. The first iPod was a novelty; the one hundredth created a strong desire to own one. By the ten thousandth, you own one; by the one millionth, the novelty has worn off; and after the hundred millionth, they're considered ubiquitous. The same with vehicles—past events shift the background climate and affect current consumer wants and needs. The innovative Model T of the 1910s was an outdated clunker in the 1920s. The great finned wonders of the 1950s and muscle cars of the 1960s were anachronisms by the 1970s. A vehicle that was once in great demand is now only junkyard material because consumer wants and needs change.

The nearly 100 "golden years" of the internal combustion engine vehicle, which swamped the early steam and electric offerings in a wave of cheap gasoline prices and offered the ability to travel where there were no tracks, are declining. It should be noted, however, that this trend has played out in fits and starts repeatedly over the last several decades. Most prominently in the 1990s and now in the 2000s, EVs were made in response to California's zero emission vehicle mandate but were mostly reclaimed and destroyed by automakers a few years later and then resurrected under President Obama. While we are on the cusp of seeing EVs and plug-in hybrids built, it remains to be seen how these technologies will play out politically and in the marketplace.

A brief look at EV history is helpful in understanding why these vehicles came, went away, and are back again.

The Timeline of Electric Cars

Steam engines came first, followed by electric motors and, finally, by internal combustion engines. The close proximity of coal and iron deposits in the northern latitudes of what came to be known as the industrialized nations—the United States, Europe/England, and Asia—made the steam engine practical. The thriving post–industrial revolution economy provided by the steam engine created the climate for electrical invention. Electric devices made the internal combustion engine possible. Vehicles powered by them followed the same development sequence.

Up to 1915

This period marked the transition from the "novelty era" (most bystanders were amazed that these devices actually worked) to the "practical era," where early buyers just wanted a vehicle to go from one point to another with minimum hassle, and finally, to the "production era." Steam and electric offerings were overwhelmed by the dominance of cheap oil and gasoline and virtually disappeared as competitors to internal combustion engine vehicles after 1915.

Huff and Puff—Steam First, Then Electric, Then Oil, and Then Electric Again

While the Cugnot steam tractor is a far cry from the Stanley Steamer automobiles of the early 1900s (a streamlined version of the latter set the land speed record at 122 mi/h in 1906) and still further removed from the high-performance Lear steam cars of a few decades ago, the problem with steam vehicles remains the steam. Water needs a lot of heat to become steam, and it freezes at cold temperatures. To get around these and the basic "time to startup" problems, technical complexity was introduced in the form of exotic liquids to withstand repeated evaporation and condensation and exotic metals for more sophisticated boilers, valves, piping, and reheaters.

Thus steam-powered vehicles declined in favor of the other two vehicle types after the early 1900s. In the early 1900s, steam vehicles unquestionably offered smoothness, silence, and acceleration. But stops for water were typically more frequent than stops for kerosene, and steamer designs required additional complexity and a lengthy startup sequence.

While 40 percent of the vehicles sold in 1900 were steam (38 percent were electric), EVs offered simplicity, reliability, and ease of operation, whereas gasoline-powered vehicles offered greater range and fuel efficiency. Thus steamers declined, and only a handful operate today.

Simultaneously, when internal combustion engine vehicles were still decades away from dominance, discovery of the Los Angeles oil field in the 1890s, the Spindletop field near Beaumont, Texas, in January 1901, and the Oklahoma fields of the early 1900s saw boom and bust times that priced a 42-gallon barrel of crude oil (typically from 15 to 20 percent recoverable as gasoline) as low as 3 cents a barrel. While direct current (DC) and alternating current (AC) electrical distribution systems guaranteed that electric lighting would replace the kerosene lamp, cheap domestic oil, which kept gasoline prices between 2 and 10 cents a gallon between 1890 and 1910, guaranteed the success of internal combustion engine vehicles.

Like Rockefeller with oil, Henry Ford was the individual who was in the right place (Detroit) at the right time (October 1908) with the right idea (Model T Ford) at the right price ($850 FOB Detroit). Ford had attended neither the *Chicago Times-Herald* race nor the earlier World's Columbian Exposition of Chicago that opened on May 1, 1893, but written information derived from these events doubtless inspired his first creation—the 1896 Quadricycle 7.

General Motors (GM) innovations such as color; streamlining; smoother, more-powerful six-cylinder engines; and annual model styling changes made Ford's Model T obsolete despite its $290 price in the 1920s. Internal combustion engine vehicles were now on their way. In 1900, half of the 80 million people in the United States lived in a few large (mostly eastern) cities with paved roads, and the other half lived in towns linked by dirt roads or in countryside with no roads at all. Less than 10 percent of the 2 million miles of roads were paved. More than 25 million horses and mules provided mobility for the masses. Electric lighting in the larger cities was dwarfed by the use of kerosene lamps, popularized by the discovery of plentiful amounts of oil, in the countryside.

The Golden Age of Internal Combustion Engines

What did the allies learn from World War II? They relearned the lesson from World War I: *Whoever controls the supply of oil wins the war.*

Allies nearly lost the war to the Germans in the North Atlantic—the success of submarine wolf packs made it nearly impossible for allied oil tankers to resupply England, Europe, and Africa. Even in 1945, countless lives were wasted and the Russians moved toward Berlin while Patton's tanks sat without fuel in France, giving the Germans time to regroup and resupply.

While the United States provided six of every seven barrels of Allied oil during World War II, it was recognized by many in government that the United States would soon become a net importer of oil. More oil had been discovered in Bahrain in 1932 and in Kuwait (Burgan field) and Saudi Arabia (Damman field) in 1938. In 1943, as all eyes turned toward the Middle East, with its reserves variously estimated at around 600 billion barrels, the U.S. government proposed the "solidification plan" to assist the oil companies (i.e., share the financial risk) in Saudi Arabian oil development. While this plan and subsequent revisions to it were rejected with much indignation by the oil companies, decades later, with 20/20 hindsight, they would all gladly change their votes.

A World Awash in Oil After World War II

After World War II, pent-up consumer demand released another type of blitzkrieg, the internal combustion engine automobile. U.S. automobile registrations almost quadrupled, and the rest of the world's auto population grew more than 20-fold—from 13 million in 1950 to nearly 300 million in 1990 and 500 million in the 2000s.

Oil exploration in this period was in high gear. A nearly inexhaustible supply had apparently been found in the Middle East; gasoline prices bounced between 20 and 30 cents per gallon until the early 1970s. Aided by the convenience of the internal combustion engine automobile, America moved to the suburbs, where distances were measured in commuting minutes, not miles. Fuel-efficient automobiles were the last item on anyone's mind during this period. Gasoline was plentiful and cheap (reflecting underlying oil prices), and regular local retail price wars made it even cheaper.

Environment was an infrequently used word with an unclear meaning. Highway construction proceeded at an unprecedented pace, culminating with the Interstate Highway Bill signed by President Eisenhower in 1956, authorizing a 42,500-mile superhighway system. Public transportation and the railroads—the big losers in Japan and Europe owing to World War II damage—also became the big losers in the United States as the government formally finished the job that major industrial corporations, acting in conspiratorial secrecy and convicted of violating the Sherman Antitrust Act, had started in the 1930s and 1940s: ripping up the tracks, dismantling the infrastructure, and scrapping intercity and intracity light rail and trolley systems that could have saved consumers, cities, and the environment the expenditure of billions of dollars today.[1] Internal combustion engine vehicle growth in the United States exploded with World War I. After World War II, world internal combustion engine automotive growth was even more dramatic. What made all this possible was the unprecedented availability of oil and the relative price stability that allowed U.S. gasoline prices to move from roughly 10 to 30 cents per gallon during this 50-year period. Yet in terms of inflation-adjusted real dollars, the cost of gasoline actually went down. Is it any wonder that no one cared how large the cars were in the 1950s or how much gas they guzzled in the 1960s? Gasoline was cheaper than water.

But there also were some problems: no major discoveries since the Alaskan and North Sea oil fields of two decades ago, the increasing concern with oil supplies beginning during World War II, the introduction of nuclear and natural gas energy alternatives in the 1950s, and the hardening of public opinion with the increasing frequency of smog and air-quality problems, oil and nuclear environmental accidents, and foreign oil shocks. Already forced to comply with more stringent emission standards by the Clean Air Act of 1968, the first oil shock of 1973 caught the "big three" U.S. automobile manufacturers with their pants down. Japanese and European auto manufacturers had smaller, more fuel-efficient internal combustion engine vehicle solutions as a result of years of higher gasoline prices (owing to higher taxes earmarked for infrastructure rebuilding). As we have also seen recently in the Gulf of Mexico oil spill by British Petroleum, spills can be pretty ugly, costly, and environmentally destructive.

The market share lost by the big three to foreign automakers has never been regained. By the early 1990s, the wild oil party of the preceding 75 years was over.

Environmental problems, the need for energy conservation, and the instability of the foreign oil supply all signal that the sun is setting on the internal combustion engine vehicle. It will not happen overnight. In the near term, the industrialized nations of the world and emerging third world nations will consume ever greater amounts of foreign oil. But it's inevitable that a replacement of the fossil fuel–burning internal combustion engine vehicle will be found.

Twilight of the Oil Gods

After the first oil shock—the Arab embargo that followed the October 1973 Yom Kippur War—everyone knew there was a problem. By cutting production 5 percent per month from September levels and cutting an additional 5 percent each succeeding month until their price objectives were met, the Organization for Petroleum Exporting Countries (OPEC) effectively panicked, strangled, and subverted the industrialized nations of the world to their will. The panic was exacerbated by nations and oil companies scrambling for supplies on the world market and overbuying at any price to make sure that they had enough and consumers doing the same by waiting in lines to "top off their tanks" when weeks before they would have thought nothing of driving around with their gas gauges on empty. When the dust settled, U.S. consumers, who had paid about 30 cents a gallon for all the gas they could get a few months earlier, now paid $1 a gallon or more at the height of the crisis and waited in lines to get a rationed amount. Figure 3-1 shows the drastic change.

How could it happen? Easy. The oil crises in 1951 (Iran's shutdown of Abadan), 1956 (Egypt's shutdown of the Suez Canal), and 1967 (Arab embargo following the Six-Day War) were effectively managed by joint government and oil industry redirection of surplus U.S. capacity. By the early 1970s, though, there was no longer any surplus

Figure 3-1   The oil embargo—not fun at all! I remember it!

capacity to redirect—the U.S. production peak of 11.3 million barrels of oil per day occurred in 1970. Until the 1970s, the oil industry focused on restraining production to support prices. Collateral to this action, relatively low oil prices forced low investment and discovery rates, and import quotas kept a lid on supplies. But rising demand erased the need for production-restraint tactics and the surplus capacity along with it.

This shock occurred after the Iranian oil strikes began in November 1978. Iran was the second largest oil producer, exporting 4.5 million barrels per day. The strikes reduced this to 1 million in mid-November, and exports ceased entirely by the end of December. I personally remember this time while I was up in Beacon, New York, watching the news with my dad about the oil shocks. We use to watch Roger Grimsbey on Channel ABC7.

While loss of this supply was partially offset by other suppliers within OPEC, burgeoning TV network coverage, simultaneously broadcast throughout the world, of the Shah's mid–January 1979 departure and the Ayatollah Khomeini's arrival (along with other internal Iranian events) convinced the world that Iran would never return to its pro-Western ways and initiated a 1973-style hoarding and panic buying spree. What started as a 2 million barrel per day shortfall became 5 million barrels per day as governments, oil companies, and consumers scrambled for supplies. Hoarding at all levels exacerbated the problem, gas lines appeared again, and oil prices went from $13 to $34 a barrel.

Although Iranian exports returned to the market by March 1979, the ill-timed Three Mile Island nuclear accident of March 28, 1979, further intensified the panic surrounding energy awareness, in addition to forever altering public opinion on nuclear power.

Several other factors contributed to making the gasoline crisis that occurred during 1979 in most industrialized nations of the free world more severe than any previous crisis. Many refineries set up to process light Iranian crude could not deliver as much gasoline from the alternate heavier crude oil they were forced to accommodate. Uncooperative (and in some cases conflicting) policies by federal, state, and local governments and oil companies disrupted the orderly distribution of the gasoline supplies that were available. The appearance of oil commodity traders, who could make huge profits on the play between the long-term contract and spot prices for oil, artificially bid up the price of spot oil in response to the prevailing buy-all-you-can-get-at-any-price mentality. Lengthy gasoline lines and rationing interfered with all levels of business and personal life. By the time the Iran hostage crisis began, a state of anarchy existed in the world oil market that President Carter's subsequent embargo of Iranian oil and freeze on Iranian assets did little to ease. Today, during the Obama administration, things have really not changed with Iran, have they?

Then, at the OPEC meeting of June 1980, the official price averaged $32 per barrel, but OPEC inventories were high, and approaching economic recession caused price and demand to fall in consuming countries, further swelling their inventory surplus. When Iraq attacked Iran, one of the first steps in its war plan was to bomb the Abadan refinery (September 22, 1980). The net result after reprisals was that Iran oil exports were reduced during the war, but Iraq exports almost ceased. After Saudi Arabia raised prices from $32 to $34 a barrel in October 1981, an unprecedented boom was created, and drillers came out of the woodwork.

Electric Vehicles

EVs enjoyed rapid growth and popularity until about 1910 and then a slow decline until their brief resurgence in the 1990s. Internal combustion engine vehicles passed steam and electric vehicles early in the 1900s. More than any other factor, cheap and nearly unlimited amounts of domestic (and later foreign) oil, which kept gasoline prices between 10 and 20 cents a gallon from 1900 through 1920, suppressed interest in alternatives to internal combustion engine vehicles until more than 50 years later (the 1970s).

Forget Oil! Electric Cars Have the Need for Speed!

In Chapter 1 you read about the speed myth and learned that you can make EVs go as fast as you want. Each step involved a bigger motor, more batteries, a better power-to-weight ratio, and a more streamlined design.

The very first land speed record of any kind was set by an EV in 1899 at 66 mi/h. In the same year, the first speeding ticket awarded to any sort of vehicle went to a Manhattan electric cab zipping along at 12 mi/h. Just after the turn of the century, a Baker Electric automobile went 104 mi/h.

After interest in EVs resumed, Autolite reached 138 mi/h in 1968, Eagle Pitcher bumped it to 152 mi/h in 1972, and Roger Hedlund's "Battery Box" pushed it to its present 175 mi/h in 1974. In 1992, Satoru Sugiyama, in the Kenwood-sponsored "Clean Liner," was going to try for 250 mi/h at the Bonneville Salt Flats using a 650-hp Fuji Electric motor from a Japanese bullet train and 113 Panasonic lead-acid batteries. Unfortunately, a simple component failure prevented the first day's run, and wind or weather wiped out the next six days of his Speed Week window.

When you consider that GM's Impact dusted off a Mazda Miata and Nissan 300ZX in 0- to 60-mi/h standing-start races, there should be no question in your mind that today's EVs or dragsters from the National Electric Drag Racing Association (NEDRA) can go fast and get there quickly. If you have difficulty with the concept, remember that France's TGV electric train routinely goes 186 mi/h en route (it can go 223 mi/h), and the Swedish-built X2000 electric train rode the Washington-to-New York corridor at 125 mi/h (it can go 155 mi/h) starting in early 1993. An electric train is nothing more than an EV on tracks (with a long cord!). Meanwhile, no one is saying that EVs are wimpy in the speed department anymore.

National Electric Drag Racing Association

NEDRA was formed in 1997 as an organization to promote safe EV drag racing and education as well as to keep records for electric drag racing. In 1999, then NEDRA President John Wayland had Vice President Roderick Wilde and Tech Director Bill Dube work with the National Hot Rod Association (NHRA) to include EVs in NHRA drag racing. EVs were accepted and included in the 1999 NHRA rule book and were later adopted into the Federation of International Automobiles (FIA) drag racing rule book. This was a milestone event for NEDRA because the NHRA rules stated since 1953 that you must have an internal combustion engine in order to race on its tracks. Since its beginning, over 15 years ago, NEDRA has grown worldwide to include representatives in five other countries. NEDRA looks forward to a further constructive relationship with the NHRA.

In March 1996, after the Electric Vehicle Technology Competitions (EVTC), a small group of "ampheads" met at a local pizza eatery. They talked of forming an electric drag racing association over a few beers and pizza. In the spring of 1997, ampheads from around the country gathered in the Wilde Evolutions' offices in Jerome Arizona for two days of intense meetings to iron out the bylaws and class divisions. Present were John "Plasma Boy" Wayland from Portland Oregon, who became NEDRA's first president; Roderick "Wildman" Wilde, who became vice president; Lou Tauber from Portland Oregon, who became the secretary/treasurer; Bill Dube, an engineer from Denver, Colorado, who became the national tech director and who wrote all the safety rules; and Dean Grannes and Stephanie Matsumora from Fremont, California, who took on the duties of webmaster and membership secretary. Dennis Berube showed up to give his input on one of the days. This was the beginning of NEDRA, which is still growing today and putting on EV drag racing events around the country. In 1999, Bill Dube and Roderick Wilde lobbied the NHRA to include electric cars and electric motorcycles in NHRA racing. The NHRA had a rule in its book since its formation in 1953 that you must have an internal combustion engine. The new rule allowing electrics was approved and was first published in the 1999 NHRA rulebook. In 2008, NEDRA became a specialized chapter of the Electric Auto Association (EAA).

Then, on January 30, 2012, the National Electric Drag Racing Association (NEDRA), a nonprofit corporation and specialized chapter of the EAA, was proud to announce that it had become an NHRA Alternative Sanctioning Organization (ASO). NEDRA joined the ranks of other prestigious NHRA ASOs, including the All Harley Drag Racing Association (AHDRA) and the American Drag Racing League (ADRL). Of note is that only three of the six NHRA ASOs are automobile drag racing associations; the other five are for motorcycles. To become an ASO, NEDRA's class and safety rules were reviewed and approved by NHRA. According to Josh Peterson, NHRA vice president, the NHRA ASO program is designed to enable independent sanctioning organizations to conduct drag racing programs not offered by the NHRA at NHRA member tracks. See the NHRA's website for inclusion of NEDRA as an ASO.[2]

NHRA was founded in 1951 by Wally Parks and is the largest motorsports sanctioning body in the world, with more than 40,000 licensed NHRA drag racers who compete at 130 NHRA member tracks around the United States and Canada.

The Need for Distance

You also heard the range myth debunked in Chapter 1; it's possible to build EVs with ranges comparable with those of many traditional internal combustion engine vehicles.

- In 1900, a French BGS Electric went 180 miles on a single battery charge.
- A Nissan EV-4H truck went 308 miles on lead-acid batteries in the 1970s.
- Horlacher Sport pushed that to 340 miles using Asea Brown Boveri sodium-sulfur batteries in 1992, and BAT International bumped it to 450 miles later in the year using proprietary batteries/electrolyte in a converted Geo Metro. The Geo averaged nearly 75 mi/h during its 10-hour test period, an impressive speed.

There also should be no question in your mind that EVs can go far. However, still today, in the early 2000s, we are still debating the same issues regarding range for all-electric vehicles. As in Led Zepplin, "The Song Remains the Same"!

Casey Mynott Builds for Speed

Casey is a member of NEDRA and has the need for speed. He used a WARP motor (which will be discussed in Chapter 6) with a Zilla controller (discussed in greater detail in Chapter 8) that will show him some kick in a drag race. Since this vehicle was meant for speed, the underbelly or chassis (discussed in Chapter 5) needed additional reinforcement to push itself that fast without breaking the car. Steel rods and additional metal plating were needed to keep this truck moving fast. The double motor with extrastrong motor plate shows two WARP motors back to back to bring that extra torque needed on the speedway. This truck will haul for Casey. Just to give you an idea of the strength of this vehicle, this adapter motor plate was built by Ford Motor Company for trucks such as F-150s but that need extra stability, such as a Ford truck racing the Baha Desert!

Given the facts that most Americans average fewer than nine miles per trip and that electricity is ubiquitous in most American urban areas, an EV is no more likely to run out of juice than an internal combustion engine vehicle would run out of gasoline, certainly less likely than running out of diesel or any of the new alternate fuels that aren't carried in every filling station! Heck, the gas station down the hill from my house would love to have a charging station if it were profitable. When the number of EVs increases to a point of no return, Jack and his family will be enticed to install EV chargers. Until then, I just remind people that Mrs. Henry Ford drove a Detroit Electric car as now on display at the Henry Ford Museum in Detroit, Michigan (as seen in Figure 3-2).

The Need for an Association

Worldwide

The World Electric Vehicle Association (WEVA), led by Chairman Hisashi Ishitani, was formed by

- Electric Drive Transportation Association (EDTA)
- Electric Vehicle Association of Asia Pacific (EVAAP)

FIGURE 3-2 Mrs. Henry Ford's Detroit Electric now on display in the Henry Ford Museum.

- European Association for Battery, Hybrid and Fuel Cell Electric Vehicles (AVERE)
- Electric vehicles research (idtechex.com)
- Multilateral Cooperation to Advance Electric Vehicles
- The Implementing Agreement for Cooperation on Hybrid and Electric Vehicle Technologies and Programs (IA-HEV)

IA-HEV was formed in 1993 to produce and disseminate balanced, objective information about advanced electric, hybrid, and fuel cell vehicles. IA-HEV is an international membership group collaborating under the International Energy Agency (IEA) framework.

North America

- National Electric Drag Racing Association (NEDRA)
- Electric Auto Association (EAA, North America) and its chapter Plug In America
- East Coast Electric Drag Racing Association
- Magazine About Hybrid Cars and Electric Vehicles

Europe

- Avere-France
- eCars-Now!
- EV Cup
- Electrical Vehicle Charge Systems[3]

Lithium Ion Just Starts the EV Market

It was the advent of lithium-ion battery technology in 2004 that allowed the electric drive to become truly viable. The new batteries, already proven in laptop and consumer battery applications, solved the age-old issues of cycle stability and load resistance. Bundled together in series of 100, they were able to provide the currents that electric drive trains require (up to 400 A, about 25 times the available capacity from a standard household outlet). BMW Group seized the opportunity the new technology presented by initiating Project i—a small, dynamic think tank whose task was to develop sustainable mobility solutions for the future needs of the world's drivers. One such initiative was the Mini-E, which has been gathering customer experience in test fleets since mid-2009.

The Need for Events, Cars, Books, and Movies

The 1990s Until Today

Environmental and conservation concerns put real teeth back into EV efforts, and even General Motors (GM) got the message. Indeed, GM did a complete about-face and led the parade to EVs. Resumption of interest in EVs during this wave was led by unprecedented legislative, cooperative, and technological developments.

EVs of the 1990s also benefited from improvements in electronics technology because the 1980s mileage and emission requirements increasingly forced automotive manufacturers to seek solutions via electronics. Although EV interest suffered a lull during the 1980s, that same decade saw a hundred-fold improvement in the capabilities of solid-state electronics devices. Tiny integrated circuits allowed a computer that took up a whole room to be replaced with a computer on your desktop at the beginning of the 1980s and by one that could be held in the palm of your hand by the beginning of the 1990s. Development at the other end of the spectrum—high-power devices—was just as dramatic. Anything mechanical that could be replaced by electronics was, in order to save weight and power (energy). Solid-state devices grew ever more muscular in response to this ongoing need. Batteries became the focused targets of well-funded government-industry partnerships around the globe; lead-acid, sodium-sulfur, nickel-iron, and nickel-cadmium batteries advanced to new levels of performance. Everyone began quietly looking at the mouthwatering possibilities of lithium-polymer batteries.

The 1990s EV offered substantial technology improvements under the hood, were more energy-efficient, and were closer to being manufacturable products than engineering test platforms. Six trends highlight EV development:

- High levels of activity at GM, Ford, and Chrysler

- Increased vigor created by new independent or high-end manufacturers
- New and improved prototypes
- High levels of activity overseas
- High levels of hybrid activity
- A boom in individual internal combustion vehicle conversions

Regulation in California

Where It All Started—Low-Speed Vehicles Got EVs
More into Society in the 1990s and 2000s

It is ironic that BEV golf carts have been available for years. Golf carts have led to the emergence of neighborhood electric vehicles (NEVs) and low-speed vehicles (LSVs), which are speed-limited at 25 mph but are legal for use on public roads. NEVs were offered primarily by car companies during the end of the California zero-emission vehicle (ZEV) mandate.

As of July 2006, there were between 60,000 and 76,000 low-speed, battery-powered vehicles in use in the United States, up from about 56,000 in 2004, according to Electric Drive Transportation Association estimates. In fact, at the end of my tenure at the New York Power Authority, I managed the low-speed vehicle (LSV) donation programs from Ford and Chrysler of over 250 vehicles. I believe several thousand vehicles were donated by the car companies to receive ZEV credits for the amount of EVs placed on the road in 2003.

This period initially marked a brief resurgence of internal combustion engine vehicles, particularly larger cars, trucks, and sport utility vehicles (SUVs). The ZEV mandate in California created a backlash against EVs by automakers, who didn't want to be required by law to build anything, let alone something other than their core internal combustion engine products. After the automakers banded together with the federal government and started legal proceedings against the state of California, the California Air Resources Board gutted the ZEV mandate (now known as the ZEV program), effectively releasing automakers from having to build EVs at all. Once the automakers were no longer required to market EVs, EV programs were quickly terminated.

During this phase, other vehicle technologies were embraced to various degrees by the automakers. The Partnership for a New Generation of Vehicles program, under President Clinton's administration, stimulated the development of gasoline hybrid vehicles by domestic automakers. While Detroit never actually deployed hybrid cars during this phase, competitive spirit compelled Japanese automakers to do so. This led to popular vehicles such as the Honda Civic hybrid and the Toyota Prius, with most major automakers eventually offering at least one hybrid model. Among domestic automakers, hydrogen became the alternative fuel of choice for new-concept cars, which were accompanied by promises to mass-market these vehicles by 2010. As the end of the decade approached, approximately 175 hydrogen fuel cell vehicles had been deployed in test fleets, but none appeared in showrooms.

As public awareness of the issues surrounding petroleum dependence—climate change, political instability, and public health issues owing to poor air quality, to name a few—has increased, the tide seems to be turning back toward plug-in vehicles. This has been stimulated on a mass level by such pop-culture devices as the films *An Inconvenient Truth* and *Who Killed the Electric Car?* and *Revenge of the Electric Car* by Chris

FIGURE 3-3   Chris Paine, director of *Who Killed the Electric Car?* and *Revenge of the Electric Car.*

Paine (the first two were the number one and number three documentaries of 2006, respectively) and on a very personal level by rising gasoline prices (see Figure 3-3). Increasingly, plug-in vehicles, which were once seen as a crunchy, environmental choice, are gathering bipartisan support as those concerned with energy security are beginning to embrace the alternative of using cheap, clean, domestic electricity to power vehicles instead of foreign, expensive, comparatively dirty petroleum. With this broad coalition of support and declining auto sales, the automakers have little choice but to get on board with newer alternatives to internal combustion vehicles.

New technology also has stimulated enthusiasm; in addition to EV, the automakers have started working on low-speed electric vehicles, hybrid electric vehicles, plug-in hybrid electric vehicles (which combine a certain number of electric miles with the "safety net" of a hybrid propulsion system), and people building their own electric cars. Depending on the vehicle's configuration, drivers might drive only in electric mode for their weekly commutes and use gasoline only when driving long distance. This "best of both worlds" concept has renewed enthusiasm for EVs as well, and both types of plug-in vehicles are benefiting from newer lithium-ion battery technology, which stores more energy than previous lead-acid and nickel–metal hydride types, providing longer ranges.

Legislation provided both the carrot and the stick to jump-start EV development. California started it all by mandating that 2 percent of each automaker's new-car fleet be comprised of ZEVs (and only EV technology can meet this rule) beginning in 1998, rising to 10 percent by the year 2003. This would have meant 40,000 EVs in California by 1998 and more than 500,000 by 2003.

California was quickly joined in its action by nearly all the northeastern states (ultimately, states representing more than half the market for vehicles in the United States had California-style mandates in place)—quite a stick! In addition, for corporate average fuel economy (café) purposes, every EV sold counted as a 200 to 400 mi/gal car under the 1988 Alternative Fuels Act. But legislation also provided the carrot. California provided various financial incentives, totaling up to $9,000 toward the purchase of an EV, as well as nonfinancial incentives, such as the High Occupancy Vehicle HOV lane access with only one person. The National Energy Policy Act of 1992 allowed a 10 percent federal tax credit up to $4,000 on the purchase price of an EV. Other countries followed suit. Japan set a target of 200,000 domestic EVs in use by 2000, and both France and Holland enacted similar tax incentives to encourage EV purchase.

The California program was designed by the California Air Resources Board (CARB) to reduce air pollution and not specifically to promote EVs. The regulation initially required simple ZEVs but didn't specify a required technology. At the time, EVs and hydrogen fuel cell vehicles were the two known types of vehicles that would have complied; because fuel cells were (and remain) fraught with technological and economic challenges, EVs emerged as the technology of choice to meet the law. Eventually, under pressure from various manufacturers and the federal government, CARB replaced the zero-emissions requirement with a combined requirement of a very small number of ZEVs to promote research and development and a much larger number of partial ZEVs, an administrative designation for *super ultralow-emissions vehicles* (SULEVs), which emit about 10 percent of the pollution of ordinary low-emissions vehicles and are also certified for zero evaporative emissions. While effective in reaching the air pollution goals projected for the zero-emissions requirement, the market effect was to permit the major manufacturers to quickly terminate their public BEV programs.

Since the electric car programs were destroyed, the market has developed an expansive appetite for hybrid electric cars and cleaner gasoline cars. GM's EV1 and EV2, Chrysler's Epic minivan, and Ford Ranger, as well as Honda's EVPlus, Nissan's Hypermini (which now sells the Nissan Leaf shown in Figure 3-4 in droves), and Toyota's RAV4 and electric cars (shown in Figure 3-5), were recalled and destroyed. Roughly 1,000 of these vehicles remain in private hands owing to public pressure and campaigns waged by grassroots organizations such as dontcrush.com (now known as Plug In America), the Rainforest Action Network, and Greenpeace.

Contradicting automaker claims of anemic demand for EVs, these vehicles now often sell for more on the secondary market than they did when they were new. The whole episode became known as such a debacle that it spawned a feature-length documentary directed by former EV1 driver and activist Chris Paine, entitled *Who Killed the Electric Car?*, which premiered at the 2006 Sundance Film Festival and was released in theaters by Sony Pictures Classics.[4]

9/11, Oil, and Our New Understanding of EVs

Understandably, the attacks on the World Trade Center and the Pentagon and the plane that crashed in Pennsylvania on September 11, 2001 clearly show that our reliance on imported oils is damaging our national and financial security. Oil reaching over $140 a barrel or gas at well over $4.00 per gallon recently created the resurgence, acceptance, and understanding that we need electric cars. September 11 also ended the period of low and stable oil prices. I remember watching the towers in flames and saying to a colleague that things will never be the same again. Now, we are faced with a serious discussion about energy security, reducing our reliance on imported oil, and dealing with its environmental impact on our globe.

Figure 3-4   Nissan Leaf at the New York Auto Show in 2011.

Figure 3-5   Toyota RAV4 cannot be crushed! (*Toyota Motor Corporation.*)

GM's Awakening—The Volt

Figure 3-6 also shows how what comes around goes around. The electric car has come back to GM because the electric car has been part of our American automotive history since the beginning of the car. The ironic twist of events in terms of oil, national security, and climate change shows an interesting side to GM and its understanding of the need for electric cars to help stabilize our economy. While the Volt is planned to become a plug-in hybrid, it is starting out as GM's own EV.

In January 2006, Bob Lutz, vice chairman of GM, originally one of the main proponents of crushing the EV1 program (see Figure 3-7), became a pragmatist and

FIGURE 3-6   GM's EV1 next to a Detroit Electric Car from 1915. (*Russ Lemons.*)

FIGURE 3-7   GM's EV1 being crushed, as seen in *Who Killed the Electric Car?* (*http://ev1-club .power.net/archive/031219/jpg/after2.htm.*)

stated the he believes that electrification of the car is the only way to preserve American car culture. "We are agonizing over what to do to counter the tidal wave of positive PR for Toyota," Lutz said. That month, GM came up with the Volt.

Even Rick Wagoner, president of GM, was not a proponent of electric cars. But then Hurricane Katrina sent oil prices soaring. For Wagoner, this was a sign of how volatile the oil markets were becoming and a harbinger of bad times to come. Even the much-maligned energy policy of the Bush administration was changing: In his State of the Union address, President Bush urged Congress to impose tougher fuel economy rules. By January 2008, the Volt had become the centerpiece of GM's green strategy.

March 19, 2009

President Barack Obama today announced the launch of two major programs that will drive the development of the next generation of electric vehicles in the United States and support the growth of domestic jobs. As part of the American Recovery and Reinvestment Act, the U.S. Department of Energy announced the release of two competitive solicitations for up to $2 billion in federal funding for competitively awarded cost-shared agreements for manufacturing of advanced batteries and related drive components as well as up to $400 million for transportation electrification demonstration and deployment projects.

By contributing to the reduction of petroleum use and greenhouse gas emissions, these projects will advance the United States' economic recovery, national energy security, and environmental sustainability. Today's announcement will also help meet the president's goal of putting 1 million plug-in hybrid vehicles on the road by 2015.[5]

This funding has been divided between two new funding opportunity announcements:

- Recovery Act: Transportation Electrification (DE-FOA-0000026)
- Recovery Act: Electric Drive Vehicle Battery and Component Manufacturing Initiative (DE-FOA-0000028)

Does GM's CEO regret not moving faster? You bet he does. Wagoner wishes that he hadn't killed the EV1. And he acknowledges underestimating the emergence of consumer societies in China and India that would help to put the $100 floor under oil prices.

Today, all of this is beside the point. The looming question is whether Wagoner can keep his promises. "It's the biggest challenge we've seen since the start of the industry," he says. "It affects everything we think about."[6]

Wagoner added in a recent interview on PBS, "We had about 100 years of an auto industry in which 98 percent of the energy to power the vehicle has come from oil. We're really going to change that over the next time period, things like battery development and applying batteries to cars, as we're planning on doing with the Volt, are important steps, kind of, in the next 100 years of the auto industry."[7]

Besides GM seeing the light after 9/11, conversion car companies such as Steve Clunn's Grass Roots EV saw an increase in electric car conversion requests. More and more people have been buying electric car conversions, electric cars, low-speed EVs, hybrid electric cars, and plug-in hybrid electric cars. Car companies, the public, the media, and the free market are all starting to fully accept the fact that electric cars *need* to be part of our automotive future.

EVs for the Twenty-First Century

The "build your own" phenomenon now applies to twenty-first-century conversions, and car companies are converting their cars to all-electric. What goes around comes around, as the saying goes. People have been building commuter-based electric cars that are getting a lot of attention and are now increasing in sales. The Tango by Commuters Cars Corporation (seen in Figure 3-8), the Corbin Sparrow, and Phoenix Motorcars are great examples. While they look different from most cars, they are electric, go over 70 mi/h, and have a significant range (100+ miles).

As mentioned earlier, people have been successfully converting internal combustion engine vehicles to EVs for at least the past 35 years. This entire period has been marked by an almost total absence of comments about this activity from the naysayers. It is ironic that while electric cars were around before the internal combustion engine and also will be around after them, the common thinking of today is that the development of electric cars has followed a path from gas, to gas hybrid, to plug-in hybrid, to all-electric Tesla.

Tesla Motors is a Silicon Valley automobile startup company that unveiled the 185-kW (248-hp) Tesla Roadster on July 20, 2006. As of March 2008, Tesla had begun regular production of the Roadster. The Roadster has an amazing AC drive of new design, a new controller, a new motor, and a new battery subsystem.

The Roadster, shown in Figure 3-9, delivers full availability of performance every moment you are in the car, even while at a stoplight. Its peak torque begins at 0 rpm and

FIGURE 3-8   The Tango. (*Commuter Cars Corporation.*)

Figure 3-9 The Tesla Roadster. (*Tesla Motors.*)

stays powerful at 13,000 rpm. This makes the Tesla Roadster six times as efficient as the best sports cars while producing one-tenth the pollution with a range of 220 miles. The estimated cost is $60,000, with a $30,000 model planned later on.

The Tesla is another great "build your own" sports car for the masses to look at and be excited about the future of electric cars. The best part is that the company wants to build more vehicles that cost $30,000.

Toyota RAV4 and Scion

The RAV4 is the purest and greatest example of building your own EV using an existing vehicle platform! In addition, it was one of the EVs that was crushed in the 1990s and 2000s during the fights over the California mandates, as seen in the movie *Who Killed the Electric Car?* Take an existing EV, such as the Scion, trick it out with great batteries, a great motor, and a great controller, and watch it work!

Toyota RAV4 Electric Vehicle

Recent reports have incorrectly stated that the 2012 RAV4 EV will only be marketed to fleet and car-sharing programs.

We'd like to set the record straight. The 2012 RAV4 EV will definitely be sold to the general public. We anticipate robust public interest in the RAV4 EV and are keen to inform consumers that their future vehicle options include a battery electric Toyota.

Toyota is the only manufacturer bringing two battery electric vehicles to the market in 2012—the RAV4 EV and the Scion iQ EV. While the RAV4 EV will be available to the public, the Scion iQ EV will be marketed to fleet and car-sharing programs only.

Nissan Leaf

Nissan first started promoting its electrification strategy in 2010 at Bear Mountain State Park in New York. I tested one of their electric prototypes in a Nissan Cube chassis (by building their own electric vehicle) which became the Nissan Leaf. Now this vehicle has been selling well since the beginning of 2012. According to Masahiko Tabe, Nissan's manager of advanced vehicle engineering, the vehicles will be tested regionally, with California marked as a target market. Indeed, the state has been earmarked as a proving ground by a number of manufacturers looking to launch alternative-fueled vehicles owing to its stringent emissions standards.

Nissan sold 1,000 Nissan Leafs in Norway in just six months, taking almost 2 percent of the total car market in February 2010, demonstrating the impact of comprehensive incentives and developed charging infrastructure. The government support and charging infrastructure have helped the Nissan Leaf become the second best selling Nissan in Norway and the ninth best selling passenger car overall in February. Norway has the highest level of support in Europe for EV purchases with a zero value-added tax (VAT), no new-car tax, free parking, exemptions from some tolls, and the use of bus lanes in Oslo. The existing on-street charging infrastructure in Oslo currently has approximately 3,500 public charging points, many of them free to use.

Olivier Paturet, general manager of zero emission strategy at Nissan Europe, was delighted to attend the handover, commenting,

> We are very happy to see that the ambition of the Norwegian government has matched our own with strong support for the widespread introduction of electric vehicles. The Norwegian package of incentives is unsurpassed, and the recharging infrastructure is established and accessible. We can see that Norway is leading the way with its proactive approach to encouraging its citizens to drive electric vehicles. We hope it will continue with the further development and upgrading of the charging infrastructure.

The one-thousandth car handover took place on April 25, 2010 at 12.30 p.m. in front of the Ministry of Environment, where Minister Bård Vegar Solhjell handed over the keys to Nils Haugerud. Also attending was Birger N. Haug, the dealer responsible for selling the car and 550 of the 1,000 sold so far, making him one of the largest-volume Leaf dealers worldwide.

All the sales so far have come from a select group of dealers, and Nissan will be expanding that number, as Pål Simonsen, country manager for Norway explained,

> Leaf has been a success for us and for the nine dealers that started selling the car last year, so it is a natural step for us to open up for sales from other Nissan dealers as well. We are extremely satisfied with our Leaf sales performance in Norway. In a short period of time, the Leaf became our second best selling model, proving that EVs can be a valid and competitive alternative to combustion engine cars. We are absolutely committed to the Leaf, and we are working with our dealers on commercial strategy to go even further.

To learn more about Nissan's EVs, please visit www.nissan-zeroemission.com/, and build yourself your own Nissan Leaf!

Ford Focus Electric

Ford Motor Company selected Compact Power, Inc. (CPI), part of the of LG Chemical family of companies, as its supplier of lithium-ion battery packs for the 2011 Ford Focus Electric for the vehicles sold in the United States. CPI is located in Troy, Michigan. The company began battery pack assembly for the Focus Electric in 2011. The company also wants to "finalize a production site in the United States."[8] However, for starters, these lithium-ion cells will be built in Korea by CPI parent company LG Chem. LG Chem will then give the batteries to CPI to localize their cell production at its new site in Holland, Michigan!

Ford's EV strategy includes

- The Transit Connect Electric Vehicle
- The Ford Focus Electric Vehicle

"The Ford engineering team was able to capture and gain insight on how the batteries were performing from charging to depletion, the range the Focus Electric was getting per charge, as well as data on how the car was doing overall," explained Ford's Lightner.[9] Figure 3-10 shows the charging door you push in and rotate to the left on the Ford Focus Electric Car. Sweet!

It's a Real Car

When I wrote *Build Your Own Electric Vehicle*, I thought it was always best to keep it simple. Some people thought The Ford Focus had some plasticity strengthening tools (as referenced on Mother News Network). However, the EV has a top speed of only 84 mi/h (136 km/h); if you are going any faster, please let the police deal with you! It's a Ford Focus, just electric. This is what is going to make people love electric cars. I talked about this recently in a BBC story about electric cars. Once you build things people like, they'll buy them, clear and simple.

FIGURE 3-10   The charging door for the Ford Focus. Ford builds their own electric Focus! (*Go Further Conference, Detroit, Michigan.*)

Much of the Focus Electric's steering, handling, and braking feel is shared with the agile, sporty, gas-powered Focus models on which it's based, making the Focus Electric a dynamic driver's car. "More than any other electric vehicle on the market, Focus Electric loses none of the dynamics and quality of driving a traditional car," said Sherif Marakby, director of Ford's electrification programs and engineering. "It shares many of the same premium components and features as its gasoline-powered counterpart while delivering distinct efficiencies and a uniquely exciting driving experience."[10]

The car includes a standard 120-V outlet cord and will take from about 6 p.m. to 7 a.m. to charge because of its lithium-ion battery. In addition, you can buy an electric car charging station for your house that is really simple. It will likely recharge the car's advanced Ford-engineered lithium-ion battery pack at home on a daily basis using the recommended 240-volt wall-mounted charge station that will be sold separately or the 120-V convenience cord that comes with the vehicle. These batteries are serious business.

"We're very excited about the potential of Focus Electric in the marketplace. With so many of us accustomed to recharging mobile electronics on a daily basis, we're confident our customers will take to the vehicle recharging process just as easily because that's exactly what it is—easy," said Nancy Gioia, Ford director of global electrification. "Not only have we made the practice of plugging in simple and straightforward, we're working with leading technology companies and the utility industry to make the EV experience empowering and engaging."

Ford Transit EV

All-electric commercial vans built on the Ford Transit Connect vehicle body, equipped with Azure Dynamics' patented Force Drive battery electric power train, and assembled by AM General at its facility in Livonia, Michigan, are reaching the market 13 months after the collaboration to develop the ZEV was first announced. To date, all initial units have designated customers. Azure Dynamics' LEAD customer program includes seven companies that took delivery of their first units in 2010, and completely delivered in 2011. Customers that have been previously announced include AT&T, Southern California Edison, Xcel Energy, Johnson Controls, Inc., New York Power Authority, Canada Post, and Toronto Atmospheric Fund EV300. Additional LEAD customers will be identified by the end of 2012.

Ford first announced the collaboration in October 2009, with an agreement for Azure Dynamics to upfit the Transit Connect van with Azure's Force Drive battery electric drive train technology, including Johnson Controls and Saft advanced lithium-ion battery and a commitment to deliver the initial vehicles by the end of 2010 to the North American market. Initial production began in the fourth quarter of 2010, with full production of the Transit Connect Electric slated to ramp up in April 2011.

"Supplier collaboration is important on all Ford product programs, but it was especially key in this effort, which went from contract signing to vehicle production in 13 months," said Sherif Marakby, Ford director of electrification programs and engineering. "A strong teamwork environment established by Azure and Ford was critical to delivering this vehicle."[11]

"The Transit Connect Electric program is a great representation of our product development approach," said Scott Harrison, CEO of Azure Dynamics. "We have been able to focus on integration of our Force Drive power train solution and benefit from Ford's expertise in building ground-up global vehicle platforms. Meanwhile, capital costs are kept in check by contracting AM General for labor and assembly services—an area where the company excels."[12]

The Michigan Economic Development Corporation provided incentive funding for Azure Dynamics to encourage selection of a Michigan-based partner for final assembly. Azure chose AM General to produce the Transit Connect Electric in its facility in Livonia, Michigan. AM General, a long-established contract vehicle assembler and services provider, is responsible for final upfit of the Transit Connect Electric. Transit Connect Electric and the Ford Focus Electric have been out since 2011, and the plug-in hybrid electric and two next-generation lithium-ion battery-powered hybrid vehicles are the CMAX plug-in hybrid. Sales are doing much better than Ford expected.

Cadillac

The innovative Cadillac Converj Concept, a dramatic luxury coupe with extended-range EV technology, is moving forward as a production car that will be called the Cadillac ELR. Development of the ELR has just begun, so details on performance, price, and timing will be announced later.

The Cadillac ELR will feature an electric propulsion system made up of a T-shaped lithium-ion battery, an electric drive unit, and a four-cylinder engine/generator. It uses electricity as its primary source to drive the car without using gasoline or producing tailpipe emissions. When the battery's energy is low, the ELR seamlessly switches to extended-range mode to enable driving for hundreds of additional miles.

"The concept generated instant enthusiasm," said Don Butler, vice president Cadillac marketing. "Like other milestone Cadillac models of the past, the ELR will offer something not otherwise present—the combination of electric propulsion with striking design and the fun of luxury coupe driving."[13]

Cadillac selected the name ELR to indicate the car's electric propulsion technology, in keeping with the brand's three-letter international model-naming convention. "The Converj Concept sparked the idea of combining the desirability of a grand touring coupe with electrification," said Ed Welburn, GM vice president of Global Design. "There's no mistaking it for anything but a Cadillac, an aggressive, forward-leaning profile and proportion showcase a uniquely shaped, modern vision of a personal luxury 2 × 2," Welburn said. Cadillac also recently announced that it will add two new vehicles to its product lineup in 2012, the XTS large luxury sedan and an all-new luxury compact sedan codenamed ATS.

Volvo Only in Europe

Volvo Car Corporation's work on electrification technology includes a systematic approach to safety issues related to battery power. All Volvo cars' existing safety systems also will be available in the company's electric cars.

Volvo Cars and the battery manufacturers have far-reaching product responsibility as regards both production and recycling. This ensures proper handling of the battery when it comes to the end of its life in the car.

Smart Fortwo and Mercedes Going E

In the development of the new generation, the Smart Fortwo benefited from extensive experience and customer feedback gathered worldwide over the past years. Driven by a battery supplied by Deutsche Accumotive for the first time, the 55-kW electric motor passes the 120 km/h mark. With acceleration of 0 to 100 km/h in less than 13 seconds, even driving on urban motorways becomes enjoyable. The 17.6-kWh battery enables the lively city car to travel around 140 km in city traffic without producing any local emissions. (Electric energy consumption, combined cycle: 15.1 kWh/100 km; CO_2 emission, combined: 0 g/km; energy efficiency class: A+.)

When completely discharged, it takes a maximum of eight hours to fully charge the battery from a household socket or charging station using the electricity systems of most countries (i.e., overnight). For the greater freedom of quick charging, it is possible to optionally equip the vehicle with a quick-charging function. The 22-kW onboard charger enables a completely empty battery to be fully charged in less than an hour. A power cable is needed to connect the vehicle to a public quick-charging station or a wall box at home or work.

Since its components come from joint ventures with the German companies Bosch (EM-motive) and Evonik (Deutsche Accumotive), and since the vehicle is built in a plant in Hambach (France), the new Smart Fortwo electric drive is the first truly European electric car. The new "sale and care" model makes opting into electric mobility particularly attractive: it offers customers an opportunity to buy, finance or lease the vehicle at an attractive price and to rent the battery for a monthly fee. Purchase, financing or leasing of the vehicle including the battery is also possible.[14]

Honda Fit

As I wrote on my website, the Honda Fit EV uses the same five-passenger layout found in the popular Fit hatchback. When the Fit EV production model is introduced, it will be powered by a lithium-ion battery and coaxial electric motor with a top speed of 90 mi/h and a range of 100 miles (70 miles when applying the Environmental Protection Agency's adjustment factor). The system allows the driver to select between "Econ," "Normal," and "Sport" to instantly and seamlessly change the driving experience to maximize efficiency or improve acceleration. While in Econ mode, practical driving range can be increased by as much as 17 percent compared with driving in Normal mode and up to 25 percent compared with Sport mode. The Fit EV is designed to be easy and convenient to charge. Battery recharging can be accomplished in less than 12 hours using a conventional 120-V outlet and less than six hours using a 240-V outlet.

Chris Woodyard from *USA Today* recently wrote a story about how the Honda Fit EV is being leased at $399 per month because the purchase price is $36,325 before all those handy-dandy destination charges are added. Honda is focusing on leasing its vehicles—at least initially. For most people, this is a very strategically effective way to get an electric car. Well, frankly, I believe that it's the early adopter phase of this vehicle. There will be people who will pay for it just like any other technology and/or appliance. Then, after a year or so, the prices will come down. However, there are probably people who will buy the vehicle outright given the tax incentives out there for EVs.

The 2013 Fit EV comes billed as being capable of an impressive 123 city-mile-per-charge range or 76-mile range in combined adjusted city/highway. But it's going to bump up against plug-in electrics from Toyota, with the Prius, and Ford, with the plug-in Focus. And that's not even counting upstarts such as Coda.[15]

Hybrid and Plug-In Hybrid EVs

Right now, EVs are available at almost every price point. If you cannot figure out how to fit one of them into your life, you can buy a new hybrid EV, such as a Toyota Prius, for a bit more than $20,000 or a used one starting at about $10,000. Gasoline-powered hybrid EVs are not the ultimate answer to our energy problems, but they do provide an excellent platform for developing EV components, such as electric motors, batteries, and transmissions. They also use much less gas than their internal combustion engine–only brethren.

But there's more good news. Many people are not waiting until CAFE standards reach 35 mi/gal by 2020. They are driving vehicles now that get better than 35 mi/gal. Some people are starting to drive plug-in hybrids that achieve over 100 mi/gal. In the United States, 40,000 people drive EVs that use no gasoline and produce zero emissions. From the full performance to the low-speed electric vehicles—the total number of EVs has increased significantly within the past five to seven years.

Phoenix Motorcars (www.phoenixmotorcars.com), based in Ontario, California, plans to build both a midsized SUV and a sports utility truck (SUT) with 130-mile range for $45,000 using NanoSafe batteries from Altairnano. A version with a more than 250-mile range is also in development. The company's lithium-ion battery is enhanced by nanotechnology that permits recharging in under 10 minutes, the car can go a hundred miles on a charge, and it has a top speed of 95 mi/h.

Tesla Motors, a Silicon Valley automobile startup company, unveiled the 185-kW (248-hp) Tesla Roadster on July 20, 2006. As of March 2008, Tesla began regular production of the roadster. In 2012, the company sold and delivered its first Model S Electric Car. The estimated cost is $60,000, and a $30,000 model is planned for later on.

REVA Electric Car

REVA is an Indian-built city car. The REVA is sold in the United Kingdom as the G-Wiz as well as in several European countries. In the United States, it is classified as a neighborhood electric vehicle (NEV) or low-speed vehicle (LSV) and is limited for use in neighborhoods. The CityEl (www. cityel.de) is a three-wheeled EV produced in Germany. Then there is Modec (www.modec.co.uk/), which is based in the United Kingdom, which builds electric delivery vans. There are many independent companies that are selling electric cars. The free market even shows how internationally, electric cars are increasing in size. There are other companies out there, so research the best one for you!

Audi Electric Cars Going e-tron

The future of electric mobility at Audi is highly dynamic. The R8 e-tron has set a world record for a production vehicle with an electric drive system on the Nürburgring Nordschleife—the toughest test track in the world. Racing driver Markus Winkelhock piloted the high-performance all-electric-drive sports car around the demanding 20.8-km (12.92-mile) track in 8:09.099 minutes. With this vehicle, Audi has achieved yet another milestone in its history, following overall wins at the 24-hour races at the Nürburgring and Le Mans.

"The R8 e-tron has given a magnificent demonstration of its potential on the toughest race track in the world," said Michael Dick, Audi AG board member for technical development. Dick, who completed a fast lap himself in the R8 e-tron, added, "The record-setting drive confirmed that we are on the right track. To us, electric mobility has never been about sacrifice, but rather is about emotion, sportiness and driving pleasure."

A comparison with the current record lap driven by a combustion-engined production car shows just how impressive the 8:09.099-minute time really is: The record time of 7:11.57 minutes was achieved with a Gumpert Apollo Sport, which is powered by a 515-kW (700-hp) Audi V8 gasoline engine.

The drive system of the Audi R8 e-tron that Markus Winkelhock drove to the world record corresponds in every detail with that of the production model that will come on the market at the end of the year. Both the car's electric motors generate an output of 280 kW and 820 Nm of torque; more than 4,900 Nm (3,614.05 lb-ft) is distributed to the rear wheels nearly from a standing start. The Audi R8 e-tron accelerates from 0 to 100 km/h (62.14 mi/h) in just 4.6 seconds. Its top speed is normally limited to 200 km/h (124.27 mi/h); 250 km/h (155.34 mi/h) was approved for the record-setting lap.

The R8 e-tron's rechargeable lithium-ion battery stores 49 kWh of energy—enough for a distance of about 215 km (133.59 miles). Its T shape allows it to be installed in the center tunnel and in the area between the passenger compartment and the rear axle. It is charged by energy recovery during coasting and braking. The ultralight car body of the Audi R8 e-tron is made primarily of aluminum; this is a main reason why the high-performance sports car weighs just 1,780 kg (3,924.23 lb), despite the large battery.

In order to further underscore the production relevance of the R8 e-tron and the capability of its drive technology, Audi has set another record on the Nordschleife in addition to the single-lap record time. Immediately afterwards, Markus Winkelhock drove two fast laps in one go in a second R8 e-tron that was limited to 200 km/h (124.27 mi/h). At 8:30.873 and 8:26.096 minutes, both laps were well under the important 9-minute threshold.

"The record drives were a fantastic experience for me," said Markus Winkelhock. The 32-year-old, who lives near Stuttgart, has a high standard for comparison—the Audi R8 LMS ultra, in which he, along with Marc Basseng, Christopher Haase and Frank Stippler, won the 24-hour race at the Nürburgring a few weeks ago. Of course, the R8 e-tron is a production car, not a racing car with the assistance of aerodynamics," Winkelhock emphasized. "But with its low center of gravity and rear biased weight distribution, it brings with it a lot of sporty qualities. The torque with which the electric motors propel the car uphill beats everything that I know—even if they make hardly any noise in the process, which at the start was really a completely new experience for me. In places where I really need traction, the torque vectoring—the displacement of the torque between the powered wheels—really helps me."

Michael Dick proudly summed up events after the record-setting drive at the Nürburgring. "Within just a few weeks we've taken on some big challenges and in the process we've shown that we are at the forefront with all of our drive concepts," he said, adding: "We won the 24-hour race at the Nürburgring in May in the Audi R8 LMS ultra with a ten-cylinder engine. In mid-June we triumphed at the 24 Hours of Le Mans with the Audi R18 e-tron Quattro—the first overall victory for a hybrid electric vehicle in the toughest race in the world. And now we've set another record with the all-electric-drive R8 e-tron on the most demanding track there is."

Near-Future Trends for Electric Drives

There are many options today for electric cars. A Tesla Roadster costs about $100,000 and the company is expecting to launch a vehicle for around $60,000. You can order an eBox from AC Propulsion for $73,000 fully loaded. Also, Vectrix is now selling a freeway-legal electric motorcycle for $11,000. Or you can buy one of the many electric bikes online for less than $1,000 to about $2,000 for really high-performance electric bikes.

TH!NK is again starting to sell cars in Europe and has now started TH!NK North America. TH!NK has even developed a full-performance electric car, which can only improve its chances for success. Toyota, Ford, Volvo, and Saab have all-electric vehicles either for sale or in production. Other fleets are doing their own custom integration of plug-in hybrids from sedans to heavy vehicles.

In addition, there are big conversion companies building cars for car companies. This will be talked about throughout this book. AMP Electric Vehicles out of Ohio is converting GM and Chevy cars to all electric. On the international front, Israel has committed to EV infrastructure across the country. Israel is working directly with Renault-Nissan, who will manufacture an electric car. Israel is also working with a California startup, founded by former SAP executive Shai Agassi, that will build the electric vehicle infrastructure.

Israel is hoping to have over 500,000 charging stations across the country and up to 200 battery-exchange locations. While leaders plan to start with a pilot program of a few vehicles, they would like to mass-market them by 2011. In addition, Israel cut the taxes on EVs by 10 percent to incentivize consumers to buy them. We should applaud the efforts by Israel and Renault-Nissan to work together to create this EV market.

Since we are now headed directly toward gas prices of $5 per gallon, as shown on the bumper sticker on an eBox in Figure 3-11, the electric car is far from being killed. In fact, it has emerged in many different forms that will only increase over time.

Liberty Electric Cars Builds Europe's Electric Cars in the DELIVER Project

Liberty Electric Cars is an electric conversion car company in Europe that has been crowned the car companies by participating together with Volkswagen, Fiat, Michelin, IKA, POLIS, SP and HPL Proto in the prestigious EU program DELIVER to design a fully electric light commercial vehicle with the goal to reduce the environmental impact in urban areas by 40 percent. Liberty Electric Cars was selected to become one of the major partners of DELIVER, thanks to its extensive experience in electric commercial vehicle engineering and design. Its team of highly skilled experts played a crucial role in the development of the Modec truck, a range of 5.5-ton commercial vehicles that have been sold to a wide variety of customers across Europe. Operators of the Modec

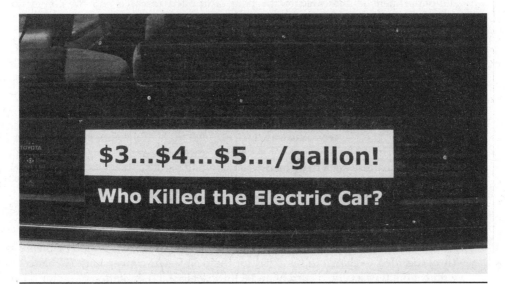

FIGURE 3-11 *Who Killed the Electric Car?* bumper sticker on an eBox. How appropriate!

truck include global companies such as FedEx, UPS, Tesco's, and Veolia. The unique team of engineers has created EVs that have been driven over 2.5 million miles—the greatest distance and thus one of the best experiences of any EV engineering team in the world. The company counts some 200 person-years of automotive engineering expertise and in particular 70 person-years of experience in EV technology.

"This exciting opportunity helps support our goal to optimize our commercial vehicle technology and to play a major role in the development of the new generation of electric commercial vehicles that will help to reduce CO_2 emissions in urban areas across Europe," said Colin Smith, head of vehicle engineering of the company. The DELIVER project is cofunded by the European Community's 7th Framework Programme for Research and Technological Development, which is the EU's main instrument for funding research in Europe.[16]

Liberty Electric Cars has demonstrated its technical superiority in EV platform technology with the launch of a concept demonstrator, the Liberty E-Range. This is the world's first pure electric luxury 4×4 that has a range of 200 miles on a single charge—this is the upper end of the range that any current EV is able to achieve. For any further information on DELIVER, please go to www.deliver-project.org.

Electric Mobility Still Is an Academic and International Move

The Department of Electric Mobility was founded in 2010 with the aim of transferring the concept of material flow management to consistently meet the challenges of future mobility. Because of high CO_2 and toxic emissions, enormous energy consumption, and dependence on the oil market, efficiency improvements in the mobility sector are a priority. New powerful batteries can here with electric mobility new perspectives emerge. The "Sustainable Mobility" deals with all stages of the electric mobility. These include studies on the ecological balance, the business model development, the regional value and the local energy supply integration in approximate structures. In addition, the department is dedicated to the field of applied research and the conversion of conventional vehicles to electric drive.[17]

- www.stoffstrom.org/en/institute/area/e-mobility/ is the website for the e-mobility department
- http://efa-mobile.de/ is the electric car club that Rudiger is part of in Germany

Here you can find more information about the origins of Rudiger's activities in the EV world: http://efa-mobile.de/eic.pdf.

However, the research that was done enabled many car companies to convert cars to electric. The students at the E-Mobility research institute in Germany built an inexpensive battery pack that TurnE is now selling because they worked on it together. The following subsection provides a pretty good description of what they did. The economics are talked about in Chapter 8.

TurnE

After one year and 7,000 km of experience with the first E-Smart prototype on the road, TurnE decided to change some parts for the next Smart conversions: (1) battery cells with continuous specified current and a lower height, (2) limitation of motor power to 25 kW (40-kW short peak periods are not necessary for this small car, and reducing peak current is better for the battery and motor), (3) a new BMS with active balancing function, and (4) improved motor cooling, and the company designed a nice battery case for Smart Fortwo (the company has a close collaboration with E-Car-Tech CEO Christian von Hösslin (www.turn-e.de/uber-turn-e/kontakt/); together they've been converting Smart Fortwo since 2010 (www.turn-e.de/die-zukunft-fahrt-elektrisch/city-253075/).

EVE (Italy)

This EV conversion company (www.electro-vehicles.eu/) has been converting Smart Fortwo, motorcycles, boats, etc. since 2009 (www.electro-vehicles.eu). All the Smart Fortwo conversion kits are based on the first prototype by Roberto Vezzi from Bergamo, Italy, in 2009, who's also the Italian representative of eCars-Now! (http://ecars-now.wikidot.com/ecars:electric-smart-fortwo).

Tesla Is Building Model X Electric Cars and They Are Selling Like Hot Cakes

One day after the introduction of the Model X shown in Figure 3-12, without any advertising, advance sales exceeded $40 million. The compelling nature of the product created massive media attention and resulted in "Model X" being the third most searched term on Google. The night of the introduction, traffic to teslamotors.com increased 2,800 percent. Two-thirds of all visitors were new to the website.

In proof that the Model S and Model X are complementary vehicles, new Model S reservations were up 30 percent. Model X is equipped with unique Falcon Wing doors

FIGURE 3-12 Tesla Model X Reveal: $40 million in orders. (*Tesla Motors Events.*)

that open up and out of the way, allowing one to stand up in the second row and step directly to the third row; as well as a great looking grill, as shown in Figure 3-12. Built on the same platform as the Model S, but with a longer wheelbase, the Model X has approximately the same external dimensions of an Audi Q7. It will offer dual-motor all-wheel drive for superior all-weather driving and the option of a 60- or 85-kWh battery. The Model X Performance version will accelerate from 0 to 60 mi/h in 4.4 seconds. This would make the Model X faster than many sports cars, including the Porsche 911 Carrera. Matched with superior handling resulting from its low center of gravity, the Model X will offer a remarkable combination of functionality, style, and performance.

The Model X will be priced competitively with other premium SUVs. It will be built at the Tesla Factory in Fremont, California. Production will begin at the end of 2013, and deliveries will begin early in 2014, with volumes targeted at 10,000 to 15,000 units per year.[18]

Summary

We must realize that electric drive technology is already and will continue to be part of everyday automotive technology. Everyone seems to be building their own EV. Whether it is a hybrid, plug-in hybrid, or pure EV, people are switching to electric every day. In addition, more and more people are converting their cars to electric internationally than ever before—from Germany to Switzerland to Australia and back to the United States, where TH!NK is being built with batteries from a U.S.-based company.

Every conversion company, such as the DeLorean car company, that makes all-electric cars has a backlog, and more business just keeps coming as seen in Figure 3-13. However, it is important to have plug-in hybrids and hybrid cars. They transition the market toward a fully electric drive vehicle. Even if people are not interested in ending our dependence on foreign oil or saving the environment, electric drive technology makes cars with better acceleration and torque that cost less to run than a regular gas-powered car today.

On that point, I saw a Tesla Roadster (shown in Figure 3-14) sitting charging at the Hilton Orlando on the Disney Park while at an event for Ford Motor Company showing off its electric vehicles. It was a very cool and reassuring sight to see.

FIGURE 3-13   DeLorean electric cars. (*Back to the Future.*)

FIGURE 3-14   Tesla Roadster charging outside a Hilton in Orlando, Florida. (*Seth Leitman.*)

CHAPTER **4**

The Best Electric Vehicle for You

In this chapter you'll learn about electric vehicle (EV) purchase tradeoffs, conversion tradeoffs, and conversion costs. You'll find out how to pick the best EV for yourself today—whether buying, converting, or building.

EV Purchase Decisions

When you go out to buy an EV today, you have three choices: buying a ready-to-run EV from (1) a major automaker, or (2) an EV conversion shop, or (3) buying a used EV conversion from an individual. Your two most basic considerations are how much money you can spend and how much time you have. Another dimension of your purchase tradeoff is where (and from whom) you can obtain your EV. This chapter will present your options and highlight why conversion is your best choice at this time.

Conversions Can Save You Money and Time

Saving money is an important objective for most people. It's also logical to assume that you want to arrive at a working EV in a reasonably short time. When you combine these two considerations, conversion emerges as the best alternative. Figure 4-1 shows you why at a glance.

A conversion:

- Costs less money than either buying a ready-to-run EV or building one from scratch.

- Takes less time than building an EV from scratch and only a little more time than buying a ready-to-run EV.

As you can see in Figure 4-1, buying a used EV clearly lowers the cost for that category.

A Tesla Model X can cost you about $100,000 but a converted car with a similar build can be built for $20,000. You probably can be 99 percent sure that this will happen. When Tesla Motors or the electric car industry can reach a reasonably priced $20,000 car people will listen. The Nissan Leaf is great but people want 300 miles and a drive system like none other. I love the Leaf and when Tesla gets to sell a $20,000 electric car, battery prices will come down dramatically, and it still will be cheaper to convert than to buy from a car company. However, these car companies are building their own vehicles and making

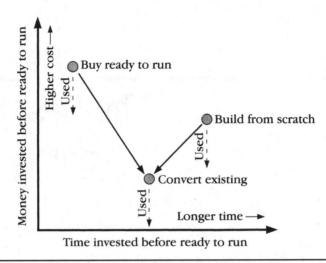

FIGURE 4-1 For time and money invested, converting is the best alternative over building and buying.

them all electric. Like Ford Motor Company with the Ford Focus electric car. Just a different drive system, but the same chassis. This is the amazing new way at looking at building your own electric vehicle with the help of a car company. They are building a car for you.

Currently, though, a new Nissan Leaf EV typically could cost from $20,000 to $30,000 after tax rebates, and maybe some local EV dealer can give you a type of EV (i.e., it looks like one but drives like a low-speed vehicle with a top speed of maybe 50 mi/h), but a car repair shop probably would do the electric vehicle conversion for $5,000 to $7,500 in today's market. Then, in a month or two, you could be driving around in your own EV conversion for under $15,000.

However, the best part about this book is the idea that car companies are building EVs for you. Just go and buy the car you want. If that car is in your price range, *please* go buy a Tesla, Nissan Leaf, or Corbin Sparrow. If you want a cheaper car, go for an EV conversion. This book shows both. Yet the Mazda Miata profiled in this book shows that you can do a conversion for about $15,000.

Let's look at the buy-build-convert alternatives more closely and then look at the who, where, and when tradeoff.

Buying or Leasing a Ready-to-Run EV Saves You a Lot of Time

Ready-to-run EVs are available today from most major automakers. Not even 10 years ago the major automakers had EVs for lease and sale, then they killed them, and now they have brought them back. In fact, to ensure that car companies would be in the business of EVs, back in 2011, seven auto manufacturers decided to work on a single charging port for every car they built. Bottom line:

- BMW, Daimler, Myers Motors, Ford Focus, General Motors (GM), Porsche, and Volkswagen agreed to support a harmonized single-port fast-charging approach for use on EVs in Europe and the United States.

- The system is a combined charging approach that integrates all charging scenarios into one vehicle inlet/charging connector and uses identical methods for the vehicle to communicate with the charging station.

- The seven auto manufacturers also agreed to use HomePlug as the communication protocol. This approach facilitates integration of the EV into future smart grid applications.

- By agreeing on a single, harmonized direct-current (DC) fast-charging system, the automakers will facilitate infrastructure planning, reduce vehicle complexity, and improve the ownership experience for EV customers.

Figure 4-2 shows the standard charger that the companies use. I caught this photo in 2010 at the New York North American Auto Show.

EV Names: Why an Ampera or a Leaf?

by Finlo Rohrer, BBC News Magazine

The U.K. government has announced the electric cars that will be eligible for a subsidy, and similar programs operate in the U.S. But what's behind the exotic names these futuristic vehicles have?

Would you drive a Nissan Leaf? Would a Vauxhall Ampera tempt you to part with your hard-earned cash?

FIGURE 4-2 Ford Focus electric car in New York City. (*Beth Fitemi.*)

There is a perception on both sides of the Atlantic, and in Asia, that electric cars are at a watershed—that one big push is all it would take to build up the momentum to persuade the world's motorists to start switching away from petrol in significant numbers.

The argument behind electric cars is simple. Countries want to control carbon emissions, and if they increase renewable energy generation, electric cars will become an increasingly ecological proposition.

Politicians also have an interest in making electric cars work as they try and make sure their areas manufacture both the vehicles and the lithium-ion batteries that are typically their heart.

Electric car enthusiasts have often simplified dark forces at work in the failure of previous models, but it seems now the mood is somehow different.

And you can tell a lot from the names of cars.

A slew of the new electric cars have names that sound, well, electric. There's the Chevrolet Volt and its sister vehicle the Vauxhall Ampera.

The Nissan Leaf makes a direct tilt at making the driver feel good about buying into something environmentally friendly. The Citroen C-Zero goes the same way, taking zero carbon and putting it right into the model name.

And why do the names matter? Because the task for the motor manufacturers is not just selling their own car, but tacitly selling the whole concept to a so-far ambivalent public.

At the moment, many of us are skeptical, notes electric car evangelist Michael Boxwell, author of the *Electric Car Guide 2011*, as well as a book on the Mitsubishi i-MiEV.

"People always ask 'how far does it go' and 'what happens if it runs out of charge.'"

The Government Is Subsidizing Several Vehicles

But when he puts someone in an electric car, they soon change their mind.

"When they go into the car and ride in it, they are always impressed," says Boxwell. "You have got a car with a huge amount of torque—it is quicker than the petrol equivalent. It is very easy to drive, you have got no gears to worry about and you don't have the noise and vibration of a petrol car."

One of the jobs of the car's name is to be part of a branding that will encourage the ambivalent to go to a showroom and sit in a car.

Normally when car manufacturers want to choose a name, they will use an array of focus group tests and other market research to make sure they're picking something safe. But what is clear is that no car maker wants to start a marketing drive with an iffy name.

They can be abstract, they can be innocuous real words and they can carry subtle meanings. The Nissan Qashqai refers to a once-nomadic Turkic tribe now settled in Iran, while the Murano is named after the island off Venice that historically housed glassmakers.

The Renault Megane's translation into Japanese as "spectacles/glasses" seems accidental, while the Toyota RAV4 has the feeling of an acronym, supposedly standing for "recreational active vehicle four-wheel drive."

The Renault Clio Is Named After One of the Greek Muses

Sometimes firms will go to branding experts like Chris Davenport, head of verbal identity at Interbrand. They use a network of offices around the world to check that the word isn't bad in any major market. Then they try and gauge the name's potential effectiveness.

But many of the electric car names seem something of a departure from the norm.

Despite being an expensive sports car, the Tesla has proved significant.

"Some of these names are pretty abstract," says Davenport. "Probably there will be a story behind it. It is very difficult to sell a name without a story behind it.

FIGURE 4-3   Interior of the Nissan Leaf EV at the New York Auto Show 2011.

"[Maker of the Leaf] Nissan (shown in Figure 4-3) has got other brands like the Cube—they have used real English words before and that has been reasonably successful."

The Leaf and the C-Zero Are Ostentatiously Green

"Leaf is going to make you think of the environment," says Fowler.

Then you have the Volt and the Ampera that hammer home the "electric-ness" of the cars.

"It's a route one approach—they could have been slightly more imaginative," says Fowler.

In fact, think about it in the context of how cars are usually branded, and having something electric-sounding as the main model name seems more than a little odd.

Can you imagine purchasing a Honda Petrolia or a Mercedes Dieselium?

"Do you see any other car out there or truck that has a gas [petrol] driven name?" says Seth Leitman, author of *Build Your Own Electric Vehicle*.

He doesn't believe most car manufacturers have really got to grips with how to sell an electric car. Waiting lists for new models are soon oversubscribed, but they are in the thousands, and makers must think in the millions.

"They are in a situation of not knowing how to market a car," says Leitman. "They should create cars that people want. They need to be calling them sexy things and doing the things they do with a regular car."

The Smart Fortwo electric car and the Tata Indica Vista EV are two examples where the "electric-ness" is a mere extension of a wider brand. It represents the likely path for the manufacturers, thinks Davenport.

"You will get to the point where all cars are electric and you don't have to talk about it anymore. Some of the names might have to be retired."

For Boxwell, the manufacturers need to look at the purchasing patterns for current electric cars, like the G-Wiz, and understand who they are selling to.

"They are all assuming people who are going to buy these things are greenies—the reality is its just people who are interested in technology."

Smart Forvision: A Look at the Future of Electric Mobility

Together with BASF, the largest automotive supplier in the chemical industry, Smart is demonstrating the concept vehicle Smart Forvision shown in Figure 4-4.

"With the Forvision, Smart is doing justice to its role as Daimler's think tank for urban mobility. We are presenting numerous world premieres that make uncompromising electric mobility possible. With the clear objective of greatly increasing the zero-emission range, we

Figure 4-4 Smart electric car getting smarter with BASF.

concerned ourselves with all factors that influence this on the vehicle. This resulted in completely new concepts and materials in the areas of insulation, reflection, lightweight design and energy management. In addition to transparent organic solar cells, transparent and energy-saving light-emitting diodes and infrared-reflective films and coatings, high-performance foams are used for insulation against cold and heat," says Dr. Annette Winkler, head of Smart.

"Smart is also setting new standards of lightweight design with the use of the first all-plastic wheels. In conjunction with Smart's revolutionary DNA, we have designed a vehicle that is so unique that we can't wait to take it out of the research laboratory and onto the roads!"

Daimler AG and BASF SE have developed a new concept vehicle that combines both companies' ideas for holistic electric mobility. The two companies have combined their technological competencies for the first time, developing a futuristic vehicle concept that offers decisive solutions to the challenges of the future. The new vehicle brings design, lifestyle and technology together to form a new functional whole.

"Cars of the future need materials and technologies which reduce energy consumption whilst also increasing the range and level of comfort. Our innovations make a decisive contribution to this," says Dr. Christian Fischer, head of BASF Polymer Research. "We are proud to have developed a holistic concept for sustainable urban mobility in cooperation with Smart. Together we are presenting a pioneering vehicle which is without parallel."

Thanks to the combination of Smart's automotive expertise and BASF's material and system competence, a vehicle has been created which showcases technologies for sustainable and holistic electric mobility of the future. The researchers and designers intentionally realized a mixture of visionary materials and technologies in the concept vehicle—some of these are still at a laboratory stage, while others have a realistic chance of entering series production.

Energy Efficiency: Light and Energy from Above

The hexagonal transparent areas on the roof of the Smart Forvision are an eye-catching feature—as the first light-transmitting roof that also generates energy. Transparent solar cells covering the entire roof surface are the technology used here. They are based on organic dyes embedded in a sandwich roof. The transparent dyes of the solar cells are light-activated. Even in diffused light and poor light conditions they generate enough energy to power the multimedia components and the three fans that assist with climate management in the vehicle interior. If the vehicle is standing in the sun, the ventilation is permanently operated with the help of these solar cells, keeping the car cool. This new photovoltaic technology opens up further efficiency potential, and the energy generated can be used for further applications in the car.

There is an additional new feature under the solar cells: Transparent OLEDs (organic light-emitting diodes) illuminate the vehicle interior when the door is opened or a button is pressed. When switched off, they allow for a clear view outside. This results in a glass roof effect during the daytime, whilst the areas are pleasantly illuminated without any dazzle at night. Thanks to a free choice of colors, the new OLEDs do not only offer more design freedom, they also consume less than half as much energy as conventional energy-saving lamps.[1]

EVs Have Some Big-Name Backers

The Peugeot iOn, with the gadgetry feel of the name, is along the right lines, thinks Boxwell.

Once all-electric cars have names with no indication of "electric-ness" in them, the evangelists will claim victory.

Other Cool Electric Cars

"Some people like pickups, some like SUVs. Once they start to become more integrated into the brands, once Ford and Toyota make a hybrid version of every car, more and more society will transition away from oil and toward electric," says Leitman.

In addition, most automakers are making hybrid electric cars, plug-in hybrid electric cars, and all electric vehicles. This includes the Ford Focus electric car, shown in Figure 4-2.

There are a few *new* automakers that are *developing* new EVs. This includes neighborhood electric vehicles or NEVs. NEVs (also called low-speed vehicles) are a class of vehicles defined in Vehicle Motor Safety Standards Rule No. 500, which stipulates that the vehicle can be driven on roads only posted at 35 mi/h or less.

Another category of electric car is the city electric vehicle (such as the TH!NK). The top speed of this vehicle is about 55 mi/h. It can be driven on roads posted from 40 to 65 mi/h where a NEV cannot go. This is a popular category overseas, where gasoline is very expensive. We also utilized this class of vehicle in the NYPA/TH!NK Clean Commute Program since the Ford TH!NK City was perfect for around-town commuting, and all of the major roads in these areas did not exceed 55 mi/h.

Tesla Motors

Then there is the Tesla. Tesla Motors and Athlon Car Lease are bringing the Model S, Tesla's premium electric sedan, into corporate fleets across Europe. We are talking

- Germany
- France
- Italy
- Spain
- Belgium
- Netherlands
- Sweden
- Portugal
- Poland

Athlon Car Lease also has reserved 150 Model S sedans to ensure early availability of the Model S for its customers.

The two companies first established a leasing program for the Tesla Roadster and Roadster Sport across Europe in July 2011. Both companies share a commitment to making EVs more easily available and affordable.

"We are excited to kick off Model S leasing with a company that shares our passion for a future of energy-efficient vehicles that are incredibly fun to drive," said George Blankenship, Tesla's vice president of sales and ownership experience. "Athlon Car Lease is a valuable partner and is instrumental in our efforts to show customers across Europe the benefits of driving electric."[2]

The Tesla Roadster shown in Figure 4-5 is the new Model S for sale.

FIGURE 4-5   Tesla Model S. (*Tesla Motors.*)

Toyota RAV4 EV

Recent reports have incorrectly stated that the 2012 RAV4 EV will be marketed only to fleet and car sharing programs. Toyota wanted to set the record straight on that one. In fact, the company reported: "The 2012 RAV4 EV [shown in Figure 4-6] will definitely be sold to the general public. We anticipate robust public interest in the RAV4 EV and are keen to inform consumers that their future vehicle options include a battery electric Toyota."[3]

Toyota is the only manufacturer bringing two battery electric vehicles to the market in 2012—the RAV4 EV and the Scion iQ EV. While the RAV4 EV will be available to the public, the Scion iQ EV will be marketed to fleet and car-sharing programs only.

FIGURE 4-6   Toyota RAV4 EV. It's like a build-your-own EV or a converted RAV4!

Nissan Leaf

At the New York International Auto Show, the 100 percent electric, zero-emission Nissan Leaf, shown in Figure 4-7, was named "2011 World Car of the Year," edging out the BMW 5-Series and the Audi A8 for the top spot. This award is the latest in a string of accolades for the world's first affordable mass-market, all-electric vehicle for the global market, which also was named "European Car of the Year."

"It is a great joy that the world's first mass-marketed electric vehicle, the Nissan Leaf, has won the prestigious award of 2011 World Car of the Year," said Nissan Chairman and CEO Carlos Ghosn. "This accolade recognizes Nissan Leaf, a pioneer in zero-emission mobility, as comparable in its driving performance, quietness, and superb handling to gas-powered cars. And it validates Nissan's clear vision and the values of sustainable mobility that we want to offer to customers around the world."[4]

The World Car Awards jurors observed:

> The Leaf is the gateway to a brave new electric world from Nissan. This 5-seater, 5-door hatchback is the world's first, purpose-built, mass-produced electric car. It has a range of over 100 miles on a full charge, claims Nissan, takes around 8 hours to recharge using a 220- to 240-V power supply and produces zero tailpipe emissions. Its low center of gravity produces sharp turn-in with almost no body roll and no understeer. The good news is that this electric car feels just like a normal car, only quieter.[5]

FIGURE 4-7 Nissan Leaf "2011 World Car of the Year" at the 2012 New York Auto Show. (*Seth Leitman.*)

Order books for the 100 percent electric Nissan Leaf opened in Spain, putting Spanish customers next in line in Europe to receive the award-winning car. At 29,950 euros after government incentives, the Nissan Leaf will cost about the same as a comparably equipped diesel or hybrid vehicle. The price includes the battery, which benefits from a 5-year/100,000-km warranty.

In order to make electric motoring even more accessible to individual customers in Spain, the compact family hatchback also will be available through a private finance scheme. For example, with a deposit of 4,500 euros, the Nissan Leaf will be available for less than 500 euros per month for three years. After that period, the customer has the option of keeping the car by making a final payment of approximately 12,600 euros, trading it in for a new one, or handing it back to Nissan.

"The 100 percent electric Nissan Leaf [shown in Figure 4-3] is the world's first mass-produced electric vehicle by a major manufacturer and is the first electric C-segment car to be made available in Spain. The Nissan Leaf opens up a new era in mobility in our country and demonstrates Nissan's commitment to the environment and new technologies," said Manuel de la Guardia, managing director of Nissan Iberia, speaking at a news conference on the first press day of the Barcelona International Motor Show.

Nissan Leaf in Japan

Nissan is even providing telematics and helping to provide hot spots while driving the Nissan Leaf EV for Japanese baseball spring training camp in 2012, "Camploo okinawa! Campaign 2012," organized by JTB Okinawa, with an information and communication technology (ICT) system designed for the Leaf EV to be connected 24/7. The Nissan Carwings Data Center (the Leaf's Hal) information control center and the onboard telematics communication unit (TCU) will be giving drivers information via a dedicated tablet PC about where to shop and go around town.

In addition, with a special tablet PC, rental car customers can use remote-control functions of the Nissan Leaf, such as remote climate control and battery status functionalities that currently are provided only to owners.

The Netherlands Is Building Its Own EV

The Renault-Nissan Alliance has signed a memorandum of understanding with the E-laad.nl Foundation, a consortium of Dutch grid operators, to jointly promote the spread of zero-emission vehicles in the Netherlands. Holy electric car news!

As part of the agreement, the alliance will provide a full range of 100 percent electric vehicles to private customers starting in June 2012 with the Nissan Leaf, this year's European Car of the Year, followed shortly by a full zero-emission range from Renault. At the same time, E-laad.nl Foundation will build a nationwide, affordable public charging infrastructure compatible with such cars. Together the efforts by the alliance and the E-laad.nl Foundation are expected to support the Dutch government's aim to have EVs account for 10 percent of the Dutch car market by 2020.

E-laad.nl Foundation has already begun rolling out the 10,000 charging stations it plans to install across the country. The first 2,000 points will be placed in towns with more than 10,000 inhabitants on request from the local city councils. The remaining 8,000 stations will be installed within close distance of the homes or workplaces of EV customers. Buyers of a Renault or Nissan EV can make the request for an E-laad.nl Foundation charging point when placing their order at the respective alliance dealer.

Membership access to the public charging infrastructure, including the electricity, will cost about 100 euros a year.

The launch of the five-seater Nissan Leaf in the Netherlands in the summer will be followed by Renault's Kangoo ZE van and Fluence ZE luxury sedan in the autumn. Both cars can be reserved at www.renault-ze.nl. The two-seater Twizy (already available for "prereservations") and the compact hatchback ZOE will be launched in 2012.

Thanks to the advanced navigation system built into their cars, Renault and Nissan EV customers will easily be able to find the location of the E-laad.nl Foundation charging points, as well as plot their journey according to them.

Mitsubishi i-MiEV

Mitsubishi Motors's 2012 i-MiEV that I saw a few years ago at the New York International Auto Show (Figure 4-8) earned the top spot on the "Most Fuel-Efficient 2012 Cars" list published by TheDailyGreen.com. According to TheDailyGreen.com: "The 2012 Mitsubishi i leads the pack both for fuel-efficiency and cost, and ranks as the most fuel-efficient car of the year." In addition, the Mitsubishi i scored first-place honors on the EPA's list of "Fuel Economy Leaders: 2012 Model Year" in the governmental agency's annual Fuel Economy Guide thanks to the electric vehicle's lofty 112 combined/126 city/99 highway miles per gallon equivalent. To be frank, with prices starting at $21,625 for the well-equipped entry-level ES model and the premium-grade SE version beginning at $23,625, all bets are off regarding electric cars in the United States.

Figure 4-8   Mitsubishi i-MIEV at the New York Auto Show. (*Seth Leitman.*)

REVA Is Loved in the United Kingdom and India. It's Selling!

The Indian-made rival REVA—sold as the G-Wiz in the United Kingdom—still seems to be going strong. Having upgraded its range and speed, REVA/G-Wiz sales saw another boost in performance with the launch of the REVA L-ion. The company claims that the new technology will greatly increase the appeal of the REVA/G-Wiz, making it a viable transport option for the suburbs as well as the city—read on for the impressive details.

The new version of the REVA will have a range of 120 km (75 miles) per charge and a maximum speed of 80 km/h (50 mi/h). The company is also launching a fast-charging station, offering a 90 percent charge in one hour. While these kinds of numbers are far from making the REVA a drop-in replacement for the petrol car, REVA points out that the vast majority of journey's are well within the vehicle's range.

Research conducted by Professor Julia King for 2008's *King Review* of low-carbon cars concludes that 93 percent of car journeys are less than 40 km (25 miles), and 97 percent are less than 80 km (50 miles). REVA's own data, based on 55 million kilometers of driving by customers in 20 cities worldwide, also reflect the nature of everyday city driving and the number of short trips. REVA's decision to offer this combination of increased range and speed extends the L-ion's usage to the suburbs and means that up to 95 percent of all car journeys can be completed without the requirement to recharge.[6]

EV Conversion Shops

Buying a Ready-to-Run EV from an Independent Manufacturer

There are a number of conversion shops around the country. Conversions shops will convert the vehicle of your choice. In many respects, this is an advantage because you get the vehicle you want converted by a professional. Electric Vehicles of America, Inc. (EVA), has a network of conversion shops across the country.

Ready-to-run EVs are available from independent manufacturers and their dealers today. EVs that use an internal combustion engine chassis, such as those from Tesla, cost $40,000 to $100,000. However, EVs are coming down in price as the number manufactured increases—their nonrecurring costs then can be amortized over a greater number of units. EVs are starting to be far lower in cost than gas-powered cars when economies of scale from increased production kick in.

EVA Signs with Flux Power

In July 2001, EVA signed a contract with Flux Power of Excondido, California. Flux Power and its lithium batteries are unique in a number of ways:

- Flux Power provides quality lithium batteries to original equipment manufacturers (e.g., Wheego, Epic) and will provide the same batteries to hobbyists. EVA will handle that.

- Flux Power can provide cells or 12-V lithium batteries, BMSs, and battery chargers. This means that there is one manufacturer responsible for the entire battery system.

- Flux Power provides automotive-grade BMSs and accessories.

- Flux Power also provides the controller area network (CAN) bus communication and capability.

If you are considering lithium cells or batteries, EVA can provide a quote for you.

Triton Trike. Affordable!

EVA is proud to be the EV component supplier to Triton Trikes (www.TritonTrikes.com). Triton is ready to put its two-seater electric into production. This is a unique 72-V commuter vehicle starting at $12,000.

Converting a Vehicle

Many people have converted their own vehicle to electric. It requires the least amount of money but the most amount of your time. A typical conversion can take 100 hours.

There are a number of advantages to doing your own conversion.

1. You are running the build (the converted vehicle and the components selected).

2. You have the ability to diagnose, troubleshoot, and repair any problems quickly.

3. If your desire is to custom-build an EV from the ground up, there is certainly nothing to stop you. Yet, given the time and money to do this, it is better to buy a new electric car or a conversion with an alternating-current (AC) propulsion drive system. Bottom line, you are spending a lot more time and money than in buying new.

If you have a donor car, here is what you can expect for costs to budget for the components alone:[7]

War P 9-inch motor (nonspeeding highway driver)	$1,395.00
Z1K-LV controller	$1,775.00
SW-200B contactor	$128.00
PB-6 pot box accelerator	$76.50
PC-30 DC-to-DC converter	$175.00
A13X400-4 safety fuse	$21.50
e-Meter with complete setup	$286.20
PCF- 20 battery charger	$1,499.00
Total	**$5,356.20**

Converting Existing Vehicles

Conversion is the best alternative because it costs less than either buying ready-made or building from scratch, takes only a little more time than buying ready-made, and is technically within everyone's reach (certainly with the help of a local mechanic and absolutely with the help of an EV conversion shop). Conversion is also easiest from a labor standpoint. You buy the existing gas-powered vehicle chassis you like (certain chassis types are easier and better to convert than others), put an electric motor in your chassis, and save a bundle. It's really quite simple; Chapter 10 covers the steps in detail.

To do a smart EV conversion, the first step is to buy a clean, straight, used gas-powered vehicle chassis. A vehicle from a salvage yard or a vehicle with a bad engine is one of the best choices because you do not know if the transmission, brakes, or other components and systems are satisfactory. Once you select the vehicle, then you add well-priced electrical parts or a whole kit from a vendor you trust and do as much of the simple labor as possible while farming out the tough jobs (i.e., machining, bracket-making, etc.).

Whether you do the work yourself and just subcontract a few jobs or elect to have someone handle the entire conversion for you, you can convert to an EV for a very attractive price compared with buying a new EV. As Bob Batson reported recently, a fully loaded EV conversion is still cheaper than any EV on the market today. As car company prices go down, so do the prices of conversions.

Figure 4-9 is from Joe Porcelli and Dave Kimmins of Operation Z. It shows their Nissan 280Z rear battery compartment. While we get into the conversion in Chapter 10,

FIGURE 4-9 Joe Porcelli and Dave Kimmins of Operation Z's Nissan 280Z rear battery compartment. (*Operation Z.*)

it's important to note that the trunk space was already built and a perfect match for the battery compartment. Another example why taking an existing chassis always works.

Even though this is step three in the total conversion process, I thought it would be good to go over it because it is important. Also, there have been critics of this book who didn't fully read the book to know that their questions were answered. More important, the book, *Build Your Own Electric Vehicle*, is not just about you converting a car; it is for the person who wants to just go out there and buy an EV. Such buyers need to be smart about the car. There is no better place to start than right here with *Build Your Own Electric Vehicle*.

So you go and buy a shell to convert to an all-electric car and buy it at a bottom-rock price. If you are not a mechanic or skilled in such matters, there is a list of conversion companies on the website (buildyourownelectricvehicle.wordpress.com) for this book. However, for conversions, future mechanics and automakers of the world, it is also about knowing your suppliers for the electric propulsion system and other components for your car! This is a universal issue!

So here we go from my friend Lynne Mason:

> The electric car conversion is the world's best kept secret! You can buy them already done, or you can do it yourself. (This is my favorite part! You can find someone who's done a conversion you like, and then just buy it.)
>
> I don't know about you, but I'm not going to convert an internal combustion engine (ICE) machine into an electric-powered rocket in my spare time. I think I mentioned before that my mechanical aptitude is measured in negative numbers.
>
> However, this is what I've found: There are people out there who actually *like* doing this sort of thing . . . and are willing to sell them to you when they're done! I'm thinking, let's leave the sweating and grease to them and just pony up the *cash* instead!

Even if you go out and buy an EV ready-made, you still want to know what kind of job the manufacturer has done so that you can decide if you're getting the best model for you. In all other cases, you'll be doing the optimizing, either by the choices you've made up front in chassis selection or by other decisions you make later on doing the conversion. In this section you'll be looking at key points that contribute to buying smart:

- Review why conversion is best—the pro side.
- Why conversion might not be for you—the con side.
- How to get the best deal.
- Avoid off-brand, too old, abused, dirty, or rusty chassis.
- Keep your needs list handy.
- Buy or borrow the chassis manual.
- Sell unused engine parts.

Converting Existing SUVs and Van-Type Cars

While the Ford Flex would be a great car to convert to all electric, it resembles a large van to me. I recently was in a car going up a mountain right outside Portland. That car got 21 mi/gal, which is efficient, but the all-electric version would handle much better because the batteries could be placed in the back and underneath the middle of the vehicle.

For these reasons and others, you will never find an 8,000-lb SUV recommended in this book as a potential conversion candidate. On the other hand, minivans—particularly the newer, lighter models—offer intriguing prospects for conversion, and you can look further into them as your needs require. But vans in general, even minivans, are usually more expensive, heavier, or take longer to convert than other chassis styles, so investigate before you invest.

High-End Conversions—We Convert, the Car Companies Come Building!

There are a bunch of vehicles that are immediately available for conversion, and it would triple the pages in this book to cover them all. We'll just look (for eye candy) at an electric Rolls-Royce, but the principle is simple: *You can convert an existing car to something new!* The Rolls-Royce Centurion uses a 2,000-A controller that moves this ship effortlessly around town or on the open road. The conversion even keeps all of the amenities, such as power brakes, power windows, video screen, automatic transmission, and air conditioning. These Centurions are heavy duty and manage a lot of low-slung weight while offering that unsurpassed ride of a Rolls-Royce (Figure 4-10).

Typical Centurion models include 26 100-Ah batteries. By using the custom Royal Centurion body (from the years 1987 to 2001) with 24-inch wheels, allowances are gained for a complete front-to-rear battery tray. This creates a very low center of gravity and excellent front-to-rear weight distribution.

Early feedback on the new Rolls-Royce electric concept car shows that drivers like the power and quiet performance but consider the driving range to be an issue. Known as the 102EX Phantom Experimental Electric, the new prototype incorporates a massive high-energy battery and offers a range of more than 120 miles. But it still may not be enough for the ultraluxury segment addressed by the company.

Design News reported that:

> UQM was recently awarded a $3 million grant from the U.S. Department of Energy that leverages the company's extensive experience in electric propulsion technology

FIGURE 4-10 Paul Liddle's custom-built electric Rolls-Royce with a standard transmission.

to explore the development of non-rare-earth-magnet motor technology. The impact of such technology could lead to a broadening of UQM's product portfolio with products that have higher performance at lower cost.

In addition to powering the CODA all-electric sedan, UQM PowerPhase electric propulsion systems have been selected to power the Saab 9-3 ePower, Audi A-1 e-Tron, and Rolls-Royce 102EX Electric Phantom preproduction test fleet vehicles. UQM is also powering Proterra's electric composite transit buses, as well as Electric Vehicles International's all-electric medium-duty truck and walk-in van. The company has a new facility with 40,000 units of annual production capacity.

According to BBC motoring writer Jorn Madslien, who was one of the first people outside Rolls-Royce to have driven the company's newly built electric car:

> The silence is, if not deafening, then at least spine-tingling as the 2.7 ton Rolls-Royce Phantom effortlessly takes off down the drive of the carmaker's Goodwood factory. The only sound is a slight tire noise, and even that is barely audible inside the luxury car's insulated cabin. Seated deep in the soft, hand-stitched leather seat, accelerating along narrow Sussex country lanes, the car feels marginally less stable than the conventional Phantom, especially as it conquers the bendy road beyond East Lavington toward Duncton.

Paul Liddle's conversion of a Porsche 959 (shown in Figure 4-11) is today's state-of-the-art conversion. It offers breathtaking performance and range. Paul Liddle's typical Porsche includes 16 to 28 batteries depending on range, weight, and speed desired. These electric cars have respectable performance, going from 0 to 60 mi/h in around five seconds. With a 2,000-A controller with 269 V and an 11-inch warp motor, in fourth gear at a 0.98 ratio, the car can achieve 1,200 ft-lb of torque.

FIGURE 4-11   Paul Liddle's Porsche 959 conversion electric vehicle.

Interview with Paul Liddle

as Lynne Mason Reported on Electric-Cars-Are-for-Girls.com
(So Don't Take My Word For It!)

A girl can hardly say "electric sports car" without the Tesla Roadster popping up in her thought balloon. But the best sports car in the world has already been made, sorry, by Porsche—the 911. So, since you can't improve on perfection, why not electrify it?

Paul Liddle from World Class Exotics, based in West Palm Beach, Florida, has been doing exactly that. He converts sports cars like the Porsche 911 (he's done several of these) and the Lamborghini Diablo into high-end, performance-oriented electric sports cars, and he found you can build your dream EV for a lot less than the 100K price tag of the Tesla.

Me: You convert a variety of cars to electric, don't you? I see a bunch of high-end performance vehicles on the CoolGreenCar website, and then a couple of serious exotics, but also bus conversions and EMIS hybrids, too. What was your first conversion, and when?

Paul: Eight years ago, a 911 with my 959 conversion body. I have been dealing with Porsches and driving daily since 1986.

[For those who don't speak Porsche, the 959 was a supercar that Porsche built on the framework of the 911. It is sort of a limited-edition car; Jay Leno may have one, but they're quite rare. For those of us who are not lucky enough to own one of these amazing Porsche supercars, there's a body kit you can buy that will turn a 911—there are plenty of those around—into a 959.]

Me: What was the most exotic car you've ever converted to electric?

Paul: The Diablo.

Me: On your EV Photo Album pages, a couple of your cars are attributed to World Class Exotics. Is that you?

Paul: That's me.

Me: I found your electric sports car business on the EV Photo Album and wanted to ask you a few things about them.

Paul: Okay.

Me: First, it seems like those electric sports car conversions, the Porsches, could be termed "performance cars" based on your choice of components. Did you initially build your electric sports cars to race?

Paul: I've raced at several tracks. The cars I convert can be made as race cars or as daily drivers.

[Batteries: Do you want lead or lithium with that?]

Me: Those 12-V batteries, I can't help thinking this is a drag racers choice, not a daily driver looking for range. How do you get 100 miles out of an electric sports car? Admirable restraint?

Paul: The lead acids go more like 50 miles in a Porsche.

Me: So why does it say "100-mile range" on your website?

Paul: One-hundred miles is with a 32-kWh pack of lithium–iron phosphate.

Me: I thought you were using lithium-polymer batteries.

Paul: We are using LiFePO4 now. We built the first lithium Porsche with the lithium-polymer, but since then changed over.

Me: I see that you're offering two-motor conversions. Interesting! A lot of people have been asking me about that. How are they controlled?

Paul: Either with a Zilla or two controllers with two pots for two motors.

Me: The 911 you converted that's called the Blue One? That's got a Kostov 11-inch motor. Compared with the Warfield/NetGain motors you've installed more recently, how do you like it? Is it still performing well?

Paul: That car is sold, but generally there are not any problems with any of the motors; they just go and don't need servicing for a very long time.

Me: You've kept the Porsche drive train in most of those conversions. How is it holding up? I suppose the Porsche drive train might be more tolerant of the combination of DC series-wound motor, plenty of amps, and lead feet than some cars' drive trains.

Paul: The Porsche transaxle is strong!

Me: It had better be, huh?

Paul: I haven't had any failures as of yet.

Me: Do you do any direct-drive conversions, or is it better to keep the existing transmission for an electric sports car, or what? Another question I get a lot is about in-wheel motors.

Paul: We mostly use Porsche transaxles but have done direct drive several times. In-wheel motors are awesome, just expensive; $20k each.

Me: Ouch. Do you do any AC conversions for electric sports cars?

Paul: We are going to be trying three manufacturers.

Me: Did you go with direct drive on this one? There isn't any drive train info on the EV Photo Album page [www.evalbum.com/819].

Paul: No. That's because I lost my hard drive after the Diablo. I've got other pictures of the car, but none of the motor mounted. We customized the interior to have a custom third seat in the middle when we got rid of the gas tank from the middle and then used that area to strengthen the chassis.[8]

Electric Vehicles of America (EVA) was one of the first to convert small pickup trucks such as the Chevy S-10 (Figure 4-12). The design puts the batteries under the bed for a low center of gravity and leaves the bed completely free for trips to the dump or local lumberyard. With a greater payload than most cars, trucks can have an even greater range. Pickup trucks are a great EV conversion choice—one you'll get a chance to examine in detail in Chapter 10. Table 4-1 covers the other dimensions of the buy, convert, or build tradeoff; what vehicles or parts are available; who you can get them from; where they are located; and when you will get to assemble them. There are many car dealerships offering electric cars but they are probably more expensive than a conversion.

Table 4-1 covers the other dimensions of the buy, convert, or build tradeoff: what vehicles or parts are available, who you can get them from, where they are located, and

FIGURE 4-12 Ken Watkin's S-10 conversion. (*Bob Batson at EV America.*)

when you will get/assemble them. There are not many places to buy new EVs today. If you don't reside in a place with easy access to an EV manufacturer or dealer, adding the weeks it might take to get your new EV shipped to you puts the conversion option—based on an internal combustion engine vehicle chassis obtained locally—into an even more attractive light.

Conversion assembly time is then measured in hours or days. Starting with a used chassis will take you longer for cleaning, preparation, and so on. The months it takes to build from scratch are consumed by your finding the vehicle kit, chassis, and parts you like and having them shipped to you. Build-from-scratch assembly times start where conversion times leave off and can range up into the thousands of hours depending on how exotic you get.

EV Conversion Decisions

When you do an EV conversion today, you have many chassis choices: small car, sports car, compact car, crossover, SUV (not recommended), and small truck. Small cars and sports cars may weigh less, but they also have less room for batteries and minimal payload. Most vehicle models gain 25 to 50 lb. each year as the manufacturers add more auxiliaries or sound-deadening materials. Cars and crossovers have an advantage of less aerodynamic drag, something that SUVs and trucks do not have.

	Who	Where	When
Buy New	Independent Manufacturers & Dealers	Regional Locations	Buy – Now Shipping – Weeks
Buy Used	EAA Classifieds Manufacturers & Dealers	Anywhere	Buy – Now Shipping – Weeks
Convert New	New Auto Dealers Electric Vehicle Parts Dealers	Local Auto Dealer Mail Order EV Parts	Buy – Weeks Assembly – 80–100 Hours
Convert Used	Used Auto Dealers Electric Vehicle Parts Dealers	Local Auto Dealer Main Order EV Parts	Buy – Weeks Assembly – 100–200 Hours
Build From Scratch	Manufacturers & Dealers Electric Vehicle Parts Dealers	Local Raw Material Mail Order EV Parts	Assembly – Months Assembly – 200–??? Hours

TABLE 4-1 Electric Vehicle Purchase Decisions Compared

Whatever vehicle you choose, select one you like. Why spend $10,000 and approximately 100 hours on a vehicle that you do not like? Your EV is a vehicle you will want to proudly show.

You have additional choices of an AC or DC drive system. Typically, AC systems operate at higher voltages and give faster performance, but they may cost more than a simpler DC system. Each system requires a motor, controller, batteries, and charger (don't forget the switches and fans, but we'll get there). This section will look at these choices and prepare you for the guidance given by the rest of the book.

To help you make the best decisions, it is necessary to first identify your requirements. Are you looking for a small commuter vehicle for yourself, or do you need an in-town vehicle for a small family? The number of people, performance, and range are all basic considerations. The cost of the drive system is directly related to your requirements.

If you are on a limited budget, it is critical to distinguish between requirements and desires. Many people want pickup of 0 to 60 mi/h in four seconds and the 150-mile range available from the Tesla EV, but not many are willing to pay the $100,000 required. Yet a $35,000 to $40,000 car right now is a great conversion from the best mechanics and the best batteries out there on the market today. You know, A123 Systems of Massachusetts is selling the best lithium-ion and nanophosphate batteries for the car companies. While they just filed bankruptcy and were just purchased by Johnson Controls, the technology will not be lost and can only gain strength with utilizing a well respected company like Johnson Controls. We'll get into that in Chapter 8.

Also, as Lynne Mason reported:

I've been asking around about the best AC electric motors to put in your electric car conversions and have discovered something wonderful. There's a new breed of AC motors for cars, this time for lower voltage systems. In a curvy fiberglass Speedster replica, it really flies!

Once upon a time, a lot of us wanted AC electric motors because they were better for electric cars in many respects, but they were considered by the wise old wags to be too expensive and too complicated for the average Joe to use. This is best left to the experts and the high-dollar hobbyist, although we came up with some really good examples of people who did great AC conversions that are shown on my website: www.electric-cars-are-for-girls.com/ac-conversion.html.

However, Hi Performance Electric Vehicle Systems, formerly known as HPGC, has come up with three-phase AC electric motors that are suitable for a full-size electric car, but for a fraction of the price we're used to (less than 5,000 USD for the electric power train instead of more than 15,000 USD). They're calling it the HPEVS AC-50. And yes; it *is* as good as you've heard.

Jack Rickard from EVTV Motor Verks (www.evtv.me) in Missouri has built a couple of cute, fast little fiberglass replica electric sports cars using HPEVS AC electric motors and lithium batteries. I got a chance to ask him a few questions about his conversions the other day—it went something like this:

Lynne: Your electric sports cars are beautiful, Jack. Can you tell me about them?

Jack: I like them too. I never get over it. Both the Spyder 550 and the Ivory 356 Speedster use the HPEVS system, the HPEVS AC-50 electric motor paired with a Curtis 1238 controller. This controller is very robust and quite configurable. We like it a lot. For example, we use a hydraulic pressure transducer . . .

Lynne: Like this one?

Jack: . . . Something like that one, yes, that puts out 0 to 5 V DC based on how much pressure you have in your brake line, in turn a function of how hard you press on the brake. The Curtis will use this input to modulate the regenerative braking.

[Translation: The result is probably the best "feel" in regenerative braking we've ever experienced on any electric car.]

Lynne: In your cars, you're running the HiPer AC electric motors at 120 V? How is the acceleration at that voltage?

Jack: These cars will do 130 mi/h. The acceleration is 0 to 60 in about 11.2 seconds. In the case of the Speedster, this matches the performance of the original 1957 model it replicates and exceeds it slightly at all speeds from 0 to 10 to 0 to 120 km/h.

Lynne: How heavy a car can use these type of AC electric motors?

Jack: They're ideal for light vehicles less than 2500 lb, I think. You could do a 3,000-lb vehicle, but it would not have stellar performance. At 2,000 lb, or in the case of the Spyder at 1,840 lb, it is very sprightly.

Lynne: Yes, the system is kind of bulletproof. I've seen people get the more high-end AC system at 300 or 320 V. Seems like a big difference in cost if these AC electric motors work well!

Jack: We're very happy with the performance, and it's relatively inexpensive at $4,600 for the controller-motor pair.

[I asked Jack about the horsepower of an AC electric motor compared with a DC motor and compared with internal combustion in terms I could relate to.]

Lynne: So, theoretically, this motor has 100 electric hp. Usually I guess I translate horsepower as "How fast will my car get up and go from 0 to 45?" but I don't know if that's really more true of the SW DC motor or if it's the same with the AC motors because I've hardly seen any of them. How would you say AC horsepower compares with DC horsepower, or am I thinking of this correctly at all?

Jack: AC horsepower does compare with DC horsepower, and you're in deep water with horsepower thinking at all. But I'll try. An internal combustion engine is usually rated in horsepower. They list the *peak* or *highest* horsepower the motor can achieve, and this is at a very specific rpm. Something like 3,600 rpm for a V8 style engine or 4,600 rpm for a smaller sports car engine. Just a few hundred rpm off peak, the engine rarely produces half that much horsepower. And at 1,000 rpm, it might be 15 horsepower—from a 400-hp engine. Electric motors too are often rated in horsepower. The horsepower rating of an electric motor almost always refers to the amount of power you can run the motor at *continously* for one hour without burning up the windings. If it is a 10-hp motor, you could put 7,460 W of power in it for an hour before it burned up. Beyond that, we think of an internal combustion engine as developing or *producing* horsepower. We think of electric motors as consuming power or converting power. So while you can put 10 horsepower *into* that electric motor, the motor can actually take an unknown amount of power for a shorter period of time. It is not precisely a linear function. But if it *were* a linear function, you could say that that same 10-hp motor could do 20 horsepower for 30 minutes, 40 horsepower for 15 minutes, and 80 horsepower for 7½ minutes or 160 horsepower for 3¾ minutes, and well, you get the idea.

In a car, the typical acceleration if you are in a *drag race* might be 14 seconds. In real traffic, you might accelerate for 7 or 8 seconds. And so as a practical matter, the motor will do whatever we put *into* it without limit for a brief burst.

And so generally, we think of electric motors as producing *torque* rather than horsepower. (Actually, the two are related by the equation hp = torque × rpm/5,252.) The torque output for a given power input will hold constant in an electric motor from 0 rpm usually out to somewhere around 4,000 to 4,500 rpm. At that point it begins to taper off for various reasons I won't go into here. This flat torque curve is what makes the feel of an electric car so much more exhilarating than an internal combustion engine vehicle.

The result is pretty much *continuous* acceleration up to about 4,000 rpm—a very different driving experience. Different indeed. This is why we love to drive our electric cars.

Lynne: Thank you, Jack![9]

Your Chassis Makes a Difference

If you're going ahead with the conversion alternative, your most important choice is the chassis you select. Evaluate the vehicle. What is its curb weight? What is its payload? Can it handle the weight and space required for the necessary batteries? For example, a Mazda Miata cannot handle the twenty 6-V batteries often used in a pickup truck. Use the Internet to find curb weight and payload capacity of the chassis you're considering. Once you've identified a few cars you like, you should work with the conversion specialist or make cardboard mock-ups of the batteries you might use. Then you'll be sure to find a vehicle that fits.

Minimizing weight is always the number one objective of any EV conversion. While a van weighs more, a car typically is the most time-consuming conversion (because there is less room to mount EV options, thus increasing the problem of getting parts to fit). An internal combustion engine pickup truck chassis is actually an outstanding EV conversion choice because:

- A single-cab pickup truck's curb weight, minus its internal combustion engine components, is typically not much more than that of a car, yet it has considerably more room to add the extra batteries that translate into better performance.

- A manual transmission, no-frills, four- or six-cylinder internal combustion engine pickup truck used to be one of the least expensive new or used conversion platforms you could buy.

- The additional battery weight presents no problem for the pickup's structure.

- Its sturdy frame is specifically designed to carry extra weight, and extra or heavier springs are readily available if you need them.

- The pickup isolates the batteries from the passenger compartment very easily. This is really important for safety purposes and is not found in car or van conversion platforms.

- The pickup is much roomier. The engine compartment and pickup box or bed space offer flexibility for your component design and layout, and a front-wheel-drive model gives you additional flexibility for battery mounting.

- A late-model, compact, or intermediate pickup or even an SUV offers a frontal area comparable with equivalent-sized cars, yet front grill and engine areas can be more easily blocked or covered to reduce wind resistance and engine compartment turbulence.

Your Batteries Make a Difference

As I reported on my website and in Chapter 2, lithium-ion batteries, which store three to four times more energy per unit mass than traditional batteries, are now used extensively in portable electronic devices (e.g., computers, cell phones, MP3 players, etc.). The positive electrode materials in these batteries are highly effective but too expensive to be used in the large batteries needed for EVs and second-generation hybrid vehicles. In the future, these applications may rely on lithium–iron phosphate: It is environmentally friendly and has exceptional properties combined with low cost and good thermal stability (important for safety reasons). All these qualities make it the best candidate to be used in lithium batteries for future EVs.

More batteries means higher voltage, which dramatically improves your performance (shown in Figure 4-13). A 120-V battery string consisting of twenty 6-V batteries (about 1,200 lb) typically delivers a top speed of 60 mi/h or more and a 60-mile range (at reduced speed) in a 3,000-lb curb-weight pickup truck with a four-speed transmission. You might get more or less depending on your design and components. Increasing voltage will increase maximum motor speed, and that equates to being able to drive faster. Increasing battery ampere-hours will extend range. Adding batteries will add voltage and weight. But it is possible to increase voltage without changing weight by changing to a different battery.

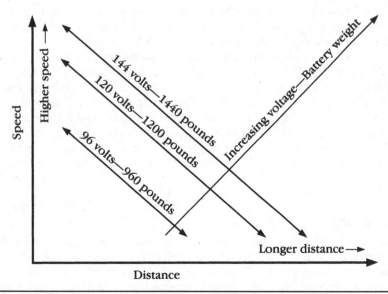

FIGURE 4-13 More batteries are always better than fewer—up to a point.

What	OK	Ideal	Best	Unlimited money
Chassis	Van	Pickup truck	Car	Custom-built Starship Electrocruiser
Motor	DC series 19 hp	DC series 22 hp	DC compound 30 hp	AC induction 50 hp
Controller	Curtis MOSFET PWM	Custom IGBT PWM	Custom IGBT PWH + Regen	Custom AC IGBT PWM + Regen
Battery Pack	96 volts – 16ea 12 volt Lead-Acid	120 volts – 20ea 6 volt Lead-Acid	144 volts – 6 or 12 volt Lead-Acid	120+ volts – NiCd, NaS, Lithium Polymer
Charger	120 volts, 20 amps	120 volts, 20 amps + Onboard	240 volts, 50 amps + Onboard	240 volts, 50 amps + Power boost trailer

TABLE 4-2   Electric Vehicle Conversion Decisions Compared

Table 4-2 presents all the dimensions of the conversion tradeoff: chassis, motor, controller, batteries, and charger versus okay, ideal, best, or unlimited money alternatives. If money is no object, you customize your chassis, add the latest powerful AC induction motor with custom controller, and tool around the countryside powered by nickel–metal hydride or lithium-polymer batteries towing your power-boost trailer.

The rest of us have to take it a bit slower; our movement toward the best category can proceed only as fast as our pocketbooks allow: a pickup, readily-available series DC motor, Curtis PWM controller, lead-acid batteries, and an onboard 120-V, 20-A charger. Actually, anything from the okay or the ideal column is acceptable. And while I am obviously prejudiced toward pickups, you don't have to be.

The point is this: Use what's available. AC propulsion, DC motors, Curtis controllers, and lithium-ion, nickel, and lead-acid batteries are with us in abundance today and at a reasonable prices. While not state of the art, they are proven to work and are available from many sources, and you can get lots of help if something goes wrong. By starting with known quantities, your chances of initial success and satisfaction are very high. You can get up and running quickly. When you're ready for your next conversion, you can experiment a bit, "push the outside of the envelope." Come on now, you can do this! You know what it's all about, what works, and where you'd like to make the changes.

The Procedure

Chapters 5 through 9 introduce you to chassis, motors, controllers, batteries, and chargers, and the new website for sources provides you with some sources to get you started. In Chapter 10 you'll look over my shoulder while I convert a Ford Ranger, following step-by-step instructions you can adapt to nearly any conversion you want. Chapter 11 shows you how to maximize the enjoyment of your EV once it's up and running. Use the website for sources—don't just take my word for it. Join the Electric Auto Association (EAA), and subscribe to its newsletters, twitter feeds, and Facebook pages. The website is buildyourownelectricvehicle.wordpress.com.

Electric Car Motors II: The AC Versus DC Debate

Jeff from Ohio wrote me the other day: "What, if any, are the advantages of AC electric car motors over DC motors?" Well, this is a debate worthy of a page of its own.

Metric Mind, a well-respected supplier of AC electric drive systems for electric cars, says that for general drivability, the AC system is far superior to the DC system. The company says that your first clue that this is true is that when the major automakers—General Motors (GM), Ford, and Toyota—made their first electric car, they *all* went with AC electric motors. This is probably not a coincidence.

These are the advantages of AC electric car motors that Metric Mind discusses on its website in the FAQ section, paraphrased liberally:

- **Regenerative braking without any extra stuff**—Because the AC system is set up nicely to "take back" energy from braking as readily as it "gives out" energy under acceleration, you can recover a *lot* of your battery power during the normal driving process. Metric Mind points out that a few DC systems can do this to some extent also, but they don't do it nearly as well, and it always makes them more complex and expensive.

- **Favorable torque**—In the capsule version, this means that AC electric car motors can be well matched to your gas-guzzler's torque curve so that you don't inadvertently murder your transmission with your normal driving habits. The gas-guzzler's engine is considerably different from a DC electric motor, and the drive system your donor was born with is not designed to withstand the low-end workout your series DC motor is capable of delivering. AC electric car motors are much more diplomatic, and your transmission will last longer. How does Metric Mind put it? "Broken gear teeth, stripped splines, twisted shafts, damaged CV joints . . . AC setups don't have these issues." Oh, the expensive carnage.

- **No motor brushes**—A lot of people, including my Uncle Chuck, say that brushes last a long time with no problems. Have them serviced every 50,000 miles or so, and you're good to go. Me, I say that they're just one more thing that can fail (Metric Mind seems to agree with me: "There are all these issues with brush advancing, seating, and commutator arcing and self-destroying at high rpm"), and no matter what Uncle Chuck says, they're no good in a regenerative braking environment. Ditto for electric reverse (Metric Mind again: "If electric reverse is used, the requirements for brush advance for forward and reverse rotation direction are contradictory"). This is not a problem for AC electric motors, which have no brushes.

- **Programmability**—Now a DC controller *could* be programmable but usually isn't. AC inverters are perfectly matched to the motor they're sold with, and you can set all the software parameters yourself to best fit your driving style and your batteries. Want to stretch out your range? Set the battery voltage to maximum for regen and minimum for driving. This is just a small sample of what the sophisticated AC inverter is capable of, all displayed on a laptop screen in real time as you drive.

- **Safety**—Want to hear something terrifying? If a DC controller's power stage fails, the entire pack voltage—all 120 or 140 V or whatever—is applied to the motor. In Metric Mind's colorful prose: ". . . you better have good circuit breakers, fuses, kill switches, and good reaction time if partial failure happens to be at the intersection while you are waiting your green." The worst that happens if an AC inverter fails is that you'll roll to a very boring stop and have to walk.

- **Electric reverse is as easy as adding a small toggle switch on the dash**—"All it takes for an inverter is to swap the sequence of two phases so that the rotor runs in the opposite direction," Metric Mind says. It's a bit more complicated with DC electric motors.

- **Ease of installation**—Contrary to popular opinion, AC systems are easier to install than DC systems, not harder. Yes, the AC inverter itself is a complicated instrument, but then so is the DC controller. They both use computers, but that's nothing unusual because so does your typical gas-gulper. As to the wiring, Metric Mind says: "To wire a Siemens AC system, you have to make six connections: three phases to the motor, two cables to the battery, and a plug encoder cable. The rest is low-voltage wiring: +12-V DC-DC output to 12-V wiring system in the car, three wires to the ignition switch, and three to the throttle pot, three to the direction switch, and two to the start inhibit switch. All this wiring harness is prefabricated and included." On the other hand, "A typical setup for a DC system with series-wound motor: two cables from the battery to the main contactors, three jumper cables for reversing contactors, two cables from the main contactors to the controller, two cables from the controller to the motor, one cable to jump the field and armature of the motor, and two wires to the DC-DC converter input. Low voltage side: three wires to the ignition switch, two wires to the throttle pot, two wires to the precharge contactor coil, two wires to the reversing contactors coils, two wires to the power resistor precharging capacitors in the controller, and two wires for start inhibit switch. One heavy wire to ground the DC-DC converter negative side and one to connect its output to the 12-V system in the car, and two wires from the motor temp switch to the light on the dash. Typically, no harness exists or is included; you come up with your own."

Where AC Electric Motors Aren't the Best Fit

There are two situations where a DC drive system is preferable to an AC system, though. The first is an EV drag racer who wants to blow the doors off an internal combustion engine machine sitting next to him or her and won't be going more than a quarter-mile at a time. This driver definitely will prefer a DC motor. And the other is the "quick and dirty" hobbyist converter who just needs to get something running on batteries as cheaply as possible. There's no shame in this; a lot of the EVs on the road today belong to this person. I'm just saying that there's something better out there, and the electric cars of the future will doubtless have three-phase AC electric motors. Eventually, electric car enthusiasts will have take a deep breath and conquer their fear of the unknown. We'll be better off in the long run. Come on in; the water's fine!

How Much Is This Going to Cost?

Now we come to the part where, as they say, the rubber meets the road—your wallet.

Let's look at an actual quote, add vehicle and battery costs, and then analyze the results. While you should not consider the costs presented here to be the last word, you can consider them typical for today's EV conversion efforts. In any event, they will give you a good idea of what to expect for a 144-V car system.

Notice also that the professionals, such as Electric Vehicles of America, Inc. (EVA), will tell you what performance you can expect, when you're going to get it, how much it's going to cost, and how long the quoted prices are valid. Be sure you get the same information, in writing, from any supplier you choose. It is important to select the components that will perform as a system. Don't expect to buy random components from the Internet and then get them to function properly as an integrated system. In addition, professionals should be available after the sale to assist you with any problems. Bob Batson of EVA has stated that the company's services remain with the converted vehicle. Thus, even the second, third, or fourth owner of its conversions is assured of the company's professional services.

Table 4-3 adds the pickup truck chassis and battery costs, shows what savings you might expect with a used and older chassis, and shows what extra costs to expect when using the latest new chassis and a few extra bells and whistles. Amounts you might obtain for selling off the internal combustion engine components were omitted from the comparisons; you can expect the vehicle costs to be lower if you do sell them.

Item	Economy $1500	Typical $3000	High $10,000	Customer Percent Allocated 28–31–57
Chasis Pickup truck	Earlier model used	Late model used	Last year new	8–8–4
Motor adapter plate Custom	$400 Local or do-it-yourself	$800 Professionally done	$800 Professionally done	15–16–10
Motor Advanced DC 22 hp	$800 Used	$1500 New	$1500 New	8–8–4
Controller Curtis PWM	$400 Used	$750 New	$750 New	12–19–14
Wiring & components Switchers, meters, wire	$600 Used	$1850 New	$2500 New	23–12–7
Battery pack 20ea 6 volt lead-acid	$1200 New	$1200 New	$1200 New	6–6–4
Charger 120V, WA, onboard	$300 Used	$550 New	$800 New	
Total	$5200	$9650	$17,550	

TABLE 4-3   Electric Vehicle Conversion Costs Compared

EV Economics

by Bob Batson

Introduction by Electric Vehicles of America

Spring 2011 is proving to be quite exciting for EV enthusiasts as gasoline prices increase to $4/gal. At the close of 2010, the Chevrolet Volt and the Nissan Leaf became available. Both have been highly publicized. There are two questions:

1. Does the hype live up to reality?
2. What is the true cost?

You will have to answer the first. The Volt has already been criticized for being a hybrid instead of a true electric. The Leaf has raised questions because buried in the fine print it may only do 100 miles at 22 mi/h. This is not much different from a conversion costing much less.

In our economic analysis, six vehicles were compared on a cost-per-mile basis. The vehicles were a 2011 Toyota Camry (gasoline), a Prius hybrid, a 144-V EV conversion with lead-acid batteries, an EV conversion with lithium batteries, the Chevrolet Volt, and the Nissan Leaf.[10]

Analysis

Table 4-3 presents the initial costs for the four vehicles. For the conversions, the additional cost of components and batteries were included. Then fuel costs, maintenance costs, insurance costs, and interest costs were estimated. These are estimates only because the actual costs may vary from state to state (e.g., insurance costs and fuel costs). Maintenance costs for the Volt and Leaf are unknown. Although it was boldly assumed that they would be equal to the Camry, they might be 50 percent higher, adding 2 to 3 cents per mile. The maintenance costs for the conversions were assumed to be equal to those of the other vehicles, even though most of the maintenance will be done by the owner.

The Toyota Camry life-cycle cost is 47 cents per mile at $4 per gallon of gasoline. Each 50 cent increase in the cost of gasoline adds 2 cents per mile. For the Chevy Volt, use of gasoline was not assumed. This could add up to 8 cents per mile if the car is totally run on gasoline, similar to the Toyota Camry.

For the Nissan Leaf, the additional costs associated with the charger and special installation procedures are not included in the evaluation. These costs may be $2,000 to $4,000. The Nissan Leaf has many claims regarding range, from 62 to 138 miles, based on various driving cycles.[11]

Do your own analysis; see if you come to the same conclusions? Please let me know.

Conclusion

The general conclusions from this analysis are as follows:

- Both conversions have a lower cost per mile than a new vehicle. This should not be a surprise.

- The lithium conversion costs less than the lead-acid conversion. This assumes a five-year life for the lithium battery. The warranty is typically only two years.

- The Chevrolet Volt has the highest cost per mile compared with the other vehicles, even when no gasoline is consumed. The cost is 62 percent higher than that of a simple conversion.

- The Nissan Leaf did better than the Chevrolet Volt only because of its lower initial costs.

Anyway, enjoy!

My thanks go to Bob Batson for his contributions to EV! He is really a great source when in doubt. However, I would not leave you with just one example of how great EV conversions are. You need the brass tacks. The costs. Well, check this out!

Eric Tischer's AC Conversion

by Lynne Mason

I just got a chance to talk to Eric Tischer, who's created the best homebuilt AC conversion I've ever seen, I don't mind saying. Sure, he's got a degree in "mechanotronic engineering" (mechanical engineering, computer-driven control—sound familiar?) and has a hobby of making major changes to cars. But he says that anybody could do a decent AC conversion with the right stuff, and it doesn't have to cost so much.

Lynne: Why did you decide to build an electric car?

Eric: I had done a few other engine swaps before my electric conversion. In 1998, I did my first engine conversion. I was 21 at the time. I bought a 1973 MG midget convertible and a 1985 RX-7 and decided to do a motor swap. It took me just over a month from the time I bought both cars until the day the MG was drivable with the RX-7 rotary engine and five-speed transmission. The car was dangerously fast, especially since I had the stock 155/80 R13 tires on it. The engine would wind up until the carbs could no longer supply enough fuel, about 8,000 rpm. The car sounded so sweet, but it was constantly snapping axles, and I decided to move on to something better. My second conversion was a Subaru turbo flat four in a Porsche 914. I designed the cooling system and heater core. I built the motor mounts and exhaust and fuel injection harness and even designed a cable-shift setup to replace the crude shift rods. Believe it or not, it only took me three days to convert the car from an air-cooled Porsche, to water-cooled Subaru. The car was not superfast, and after destroying several turbo motors, I decided to upgrade. The last turbo motor went with a bang. It had a leaky head gasket and was constantly backfiring. One backfire exploded the top of my intercooler off, and in the rear view mirror all I could see was a minifireball blowing though the rear deck lid! I eventually switched over to the 2.7 flat-six Subaru motor and ran that reliably for several years until I sold it. The car sounded just like a mean 911. After I sold these cars, I wanted to do another conversion, but something different, and something where I can put my "mechanotronic" skills to good use. At the time, gas was around $4.50 a gallon, so building an electric car was perfect.

Lynne: It just happened to coincide with your interests to build an AC conversion then, and the price of gas made it even more logical? No "save the whales" or "peak oil" motivation?

Eric: The price of gas certainly sparked my interest in making an electric conversion. It is a shame that there are no affordable electric cars on the market; I think they would sell quite well. For now, the only way to get one is to build it.

How Much Did It Cost?

Lynne: How much did the project cost you, if you don't mind me asking. The reason I ask is that the control portion of an AC system is the most intimidating part of building an AC conversion, and I'm dying to know if the DIY factor brought the costs down. Not to pry or anything. Not me.

Eric: I have spent around $20,000, including the car. Part of that $20,000 was used to build three motor controllers. The first prototype blew up after about a week of testing. The second one was limited to controlling only speed. The third one, which I am still currently using, can control speed and torque. If I were to do the conversion again from scratch, it could probably be done for about $15,000. Half that would be used for batteries, battery management, and a charger.

Lynne: This is a very good price for a good AC conversion, anyway.

All That Voltage!

Lynne: You've got a 300-V system in your AC conversion. Dude, that's a whopper with cheese compared with most of our modest little 120-V DC conversions. I've heard a lot of folks complain about their AC conversions having "bus-like acceleration," but they've gone with 120-V systems. Your AC conversion, on the other hand, doesn't resemble a bus in its acceleration *at all*, and I suspect that the 300-V system might have something to do with it.

Eric: The reason my car accelerates faster than some DC electric vehicles is because it has more usable power. Power = rpms(torque)/5,252. Most DC motors can only spin as fast as the battery voltage will allow. For example, 100 V might generate 2,000 motor rpms. If you want more speed, 140 V will get you to 2,800 rpms. An AC motor can spin at any speed regardless of voltage because the rpms are related to frequency. This is the frequency of the three-phase sine wave being generated by the motor controller (inverter). My AC motor speed is limited by the bearings, about 10,000 rpms.

Commuting

Lynne: You commute with this car?

Eric: This car gets me to work every day. I drive about 20 miles at 65 mi/h, and this requires me to charge the car for about 3.5 hours at 18 A and 115 V. The car will easily break 100 mi/h, but the fastest I've gone so far is 98.6 mi/h. The car weighs about 500 lb more than stock, with most of that weight in the trunk. I've driven the car about 450 miles in the last two weeks. I love how quiet it is. My heart was beating so hard the first time I drove it on the street that I was quite surprised it was actually working.

Lynne: I can imagine! Is this the first EV you've driven? They're addictive, look out!

Eric: This is the first. It was especially exciting because I had been working so hard to get the controller to work.

Lynne: Next, I want to talk for a minute about the motor you chose for your AC conversion.[12]

CHAPTER 5

Chassis and Design

A 20-hp electric motor will easily push
your 4,000-lb vehicle at 50 mi/h.

Even though this is step three in the total conversion process, I thought it would
be good to go over it because it is important.

In addition, there have been critics of the book who didn't read enough of the
book to know that their questions were answered. More important, *Build Your Own
Electric Vehicle* now is not just about you converting a car; it includes manufacturers, car
companies, and educational institutions to help solve the world's transportation
problems through the electrification of transportation.

Moreover, each individual wants to build his or her own electric vehicle (EV).
Whether you build it (which I *know* most people will not), a manufacturer builds it, or
a conversion shop builds it, we all want to get our own EVs. This book is intended to
provide a more global look at how people have caught the bug of *building their own EV*.
This is why you see electric cars in Australia, India, China, the United Kingdom, France,
and the United States. You can buy a converted EV from someone else, or if you want,
you can buy a Nissan Leaf, a Ford Focus, a BMW ActiveE, a Tesla Roadster, a Model S
or Model X, or a Toyota RAV4. There are many more people in this world who are not
mechanics and are unable to convert cars to all-electric, so if you are not a mechanic or
skilled in such matters, there is a list of conversion companies on our website which is
advertised in the back of the book.

The key is to know your suppliers for the electric propulsion system and other
components for your car. For example, the basic perspective is that a battery-powered
Volvo must be as safe as any other new Volvo when it comes to everyday driving as
well as in the event of an accident. "We apply the same high safety standards to all our
products, but the safety-related challenges may differ depending on the driveline and
fuel being used. To us, electrification technology is another exciting challenge in our
quest to build the safest cars on the market," says safety expert Jan Ivarsson, senior
manager of safety strategy and requirements at Volvo Cars. He adds: "It is
understandable that a lot of questions about electrification safety are related to what
will happen in an accident. But considering that less than 1 percent of our cars are
involved in an accident during their life on the roads, we must adopt a more holistic
approach, including all the aspects of day-to-day usage of the car."[1]

Everything from the way the cars are produced, used, and serviced to the way they are recycled is analyzed thoroughly and the information obtained is used to shape the development of the final production car.

Comprehensive Testing Under Way—and There's More on the Way

"We have made tests on [the] component level to see how the battery is affected by harsh braking and the subsequent collision. We have also carried out advanced full-scale crash tests to evaluate the technology used in electrically powered cars," reveals Jan Ivarsson. He adds: "The lithium-ion batteries are packaged in the center of the vehicle, removing them from the crumple zones."[2]

So here we go: Even if you go out and buy an EV ready-made, you still want to know what kind of job the manufacturer has done so that you can decide if you're getting the best model for you. In all other cases, you'll be doing the optimizing, either by the choices you make up front in chassis selection or by other decisions you make later on during the conversion. In this chapter you'll be looking at key points that contribute to buying smart:

- Why conversion is best—the pro side.
- Why conversion might not be for you—the con side.
- How to get the best deal.
- Avoid off-brand, too old, abused, dirty, or rusty chassis.
- Keep your needs list handy.
- Buy or borrow the chassis manual.
- Sell unused engine parts.

The chassis is the foundation of your EV conversion. While you might never build your own chassis from scratch, there are fundamental chassis principles that can help you with any EV conversion or purchase—things that never come up when using internal combustion engine vehicles, such as the influence of weight, aerodynamic drag, rolling resistance, and drive trains.

This chapter will step you through the process of optimizing, designing, and buying your own EV. You'll become familiar with the chassis tradeoffs involved in optimizing your EV conversion. Then you'll design your EV conversion to be sure that the components you've selected accomplish what you want to do. When you have figured out what's important to you and verified that your design will do what you want, you'll look at the process of buying your chassis.

Knowledge of all these steps will help you immediately (when reading about Chapter 10's conversions) and eventually (when picking the best EV chassis for yourself). The principles are universal, and you can apply them whether buying, building, or converting.

Choose the Best Chassis for Your EV

The chassis you pick is the foundation for your EV—choose it wisely. This is the message of this chapter in a nutshell. Since you're likely to be converting rather than building from scratch, there's not a lot you can do after you've made your chassis selection. The secret is to ask yourself the right questions and be clear about what you want to accomplish before you make your selection.

Provided that the converted EV does not greatly exceed the original vehicle's gross vehicle weight requirement (GVWR) weight-per-axle specifications, all systems will continue to deliver their previous performance, and stability and handling characteristics will be fine. In addition, the EV converter inherits another body bonus—the original bumpers, lights, safety-glazed windows, etc., which are already preapproved and tested to meet all safety requirements. There's still another benefit—you save more money. Automobile junkyards make money by buying whole cars (trucks, vans, etc.) and selling them off piece by piece for more than they paid.

When you build (rather than convert) an EV, you are on the other side of the fence. Unless you bought a complete kit, building from scratch means buying chassis tubing, angle braces, and sheet stock, as well as axles/suspension, brakes, steering mechanisms, gauges, instruments, dashboard/interior trim/upholstery, etc.—parts that are bound to cost you more à la carte than buying them already manufactured and installed in a completed vehicle such as the one shown in Figure 5-1 from TurnE conversions.

You want a chassis with an aerodynamic shape and thin wheels so that you can just give a shove and it runs almost forever. However, the frame also must be big enough and strong enough to carry you and your passengers, along with the motor, drive train/

Figure 5-1   Wheel, shock, and brake components from a soon to be all-electric Porsche Speedster. (*TurnE Conversions.*)

controller, and batteries. In addition, if you want to drive your EV on the highway, federal and state laws require it to be roadworthy and adhere to certain safety standards. For example, the BMW i3 Concept features a horizontal-split variant of the LifeDrive concept. Here, the drive module provides the solid foundation for the Life cell, which is simply mounted on top. The reason for this functional rendition of the LifeDrive architecture is the large battery. In order to ensure the greatest possible electric range, the battery in the BMW i3 Concept is correspondingly large. The most space-efficient place to store the battery cells was under the car's floor section, where they occupy the whole of the module's central section, giving the car optimal weight distribution and a low center of gravity. The battery is housed in an aluminum frame, which protects it from external impacts. Crash-active structures in front and rear provide the necessary energy absorption in the event of a front- or rear-end collision. The eDrive system is, as a whole, much more compact than a comparable combustion engine, cleverly accommodating the electric motor, gear assembly, and drive electronics—in space-saving fashion—within a small area over the rear axle.

The integration of all the drive components within the Drive module removes the need for a center tunnel bisecting the interior through which power previously would have been transferred to the rear wheels. The BMW i3 Concept therefore offers significantly more interior space than other vehicles with the same wheelbase and—through solutions such as a full-width seat bench—also allows the interior to be adapted extremely effectively to the needs of urban mobility. The BMW i3 Concept offers comfortable accommodation for four passengers and, with around 7 cubic feet 200 liters of trunk space, room for their luggage as well.

The chassis of the BMW i3 Concept is also ideal for city driving. Its enviably small turning circle and direct steering responses result in outstanding agility, notably at low speeds.[3]

Know Your Options!

The first step is to know your options. Your EV should be as light as possible if you are looking for not too many batteries. If you want a sport utility vehicle (SUV), expect more batteries. The motor–drive train–battery combination must match the body style you've selected. It also must be capable of accomplishing the task most important to you: a high-speed vehicle, a long-range vehicle, or a utility commuter vehicle midway between the two.

Thus step two is to design for the capability that you want. Your EV's weight, motor and battery placement, aerodynamics, rolling resistance, handling, gearing, and safety features also must meet your needs. You now have a plan.

Step three is to execute your plan—to buy the chassis that meets your needs. At its heart, this is a process no different from any other vehicle purchase you've ever made, except that the best solution to your needs might be a vehicle that the owner or dealer can't wait to get rid of—one with a gas-guzzling, diesel, or otherwise polluting engine—so the tables are completely turned from a normal buying situation.

A used chassis is usually the least expensive, but don't go for something too used. You want to feel confident about converting the vehicle you choose before you leave the lot. If it's too small or cramped to fit all the electrical parts, let alone the batteries, you know you have a problem. If it's particularly dirty, greasy, or rusty, you need to think twice. For example, with an output of 170 hp, a maximum torque of 184 lb-ft from a standstill, and acceleration of 0 to 60 mi/h in under nine seconds, the all-new

BMW ActiveE delivers an electrically powered adrenaline rush. It's the kind of exhilaration that both boldly embraces its heritage and paves a new road toward the future of mobility.

BMW engineers have developed everything from its impressive new motor to the enhanced aerodynamics and dedicated power electronics. The BMW ActiveE boasts brake energy regeneration, which converts kinetic energy into electrical energy for the battery; dynamic stability control for steadfast traction; rear-wheel drive; and near-perfect 50:50 weight distribution.

To further enhance its already impressive driving dynamics, BMW engineers have created a lowered center of gravity by optimizing the mounting of the batteries in the BMW ActiveE. The result is a vehicle that is as graceful as any BMW and one that clearly sets a new benchmark. In fact, the BMW ActiveE will redefine the way drivers measure efficiency and performance.[4]

The rest of this chapter covers the details. Let's get started. Figure 5-2 will give you an overall picture.

Optimize Your EV

Optimizing is always step number one. Even if you buy your EV ready-made, you still want to know what kind of a job has been done so that you can decide whether you're getting the best model for you. In all other cases, you'll be doing the optimizing—either by the choices you make up front in chassis selection or by your conscious optimizing decisions later on. In this section you'll be looking to minimize the following resistance factors:

- Weight and climbing and acceleration
- Aerodynamic drag and wind
- Rolling and cornering resistance
- Drive train system

You'll look at equations that define each of these factors, and construct a table of real values normalized for a 1,000-lb vehicle and nine specific vehicle speeds. These values should be handy regardless of what you do later—just multiply by your own EV's weight ratio and use them directly or interpolate between the speed values.

You'll immediately see a number of values reassembled in the design section of this chapter, when a real vehicle's torque requirements are calculated to see if the torque available from the electric motor and drive train selected is up to the task. This design process can be infinitely adapted and applied to whatever EV you have.

Conventions and Formulas

This book uses the U.S. automotive convention of miles, miles per hour, feet per second, pounds, pound-feet, and so on rather than the kilometers, newton-meters, and so on in common use overseas. Any formulas borrowed from the Bosch handbook have been converted to U.S. units.

Is there an EV conversion calculator somewhere that I can use to figure out how much motor output I will need for my desired performance? Yes, there's an excellent EV performance calculator at EV Convert (www.evconvert.com/tools/evcalc/). Down the right-hand side of the page you'll find your car's make and model from the

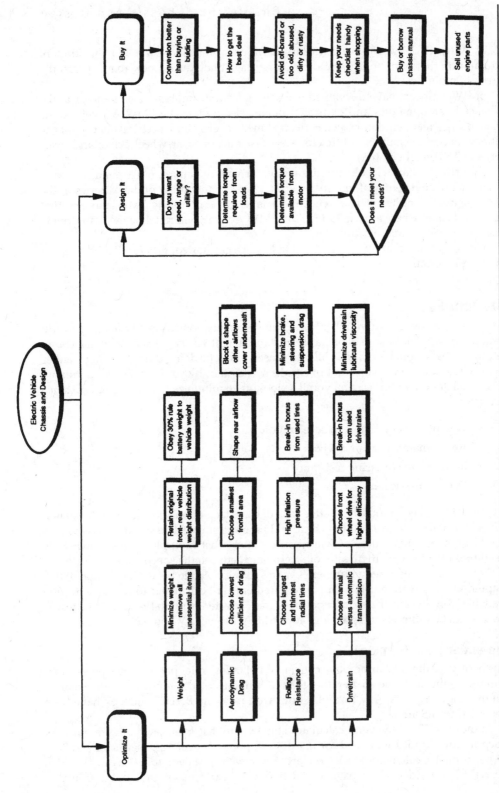

Figure 5-2 For a great EV chassis, optimize, design, and then buy it.

122

dropdown box, then your electric motor type in the second dropdown box, your batteries in the next, and so forth. [Note that the calculator was designed for direct-current (DC) motor systems, so if you're planning an alternating-current (AC) conversion, it won't be helpful.] Below that you put in the adjustments to the calculations, things the voltage of your system and the rolling resistance of your tires and how much weight you've removed from your donor vehicle—things like that. It's a very useful tool, and many thanks to Jerry Halstead for making it available on the Internet.

1. Power (ft-lb/sec) = Torque (ft-lb) × Speed (radians/sec) = Force in feet per second (FV)   (5.1)

   One horsepower (1 hp) equals 550 ft-lb/s.

Applying this to Equation (5.1) gives

2. Horsepower (hp) = FV/550   (5.2)

where V is speed expressed in feet per second and F is force expressed in pounds eighty-eight feet per second (88 ft/s) equals 60 miles per hour. Multiply feet per second by (60 × 60)/5,280 to get miles per hour.

3. Horsepower (hp) = FV/375   (5.3)

where V is speed expressed in miles per hour and F is force in pounds.

4. Horsepower (hp) = (torque × rpms)/5,252 = $2\pi/60$ × FV/550   (5.4)

5. Wheel rpm = (mi/h × rev/mi)/60   (5.5)

6. Power (kW) = 0.7457 × hp   (5.6)

The standard gravitational constant g equals 32.16 ft/s^2 or almost 22 mi/h per second. Thus

7. Weight W = mass M × g/32.16   (5.7)

For the rest of this book, we will refer to a vehicle's mass as its weight.

8. Torque = [F(5,280/2π)]/(rev/m) = 840.34 × F/(rev/mi)   (5.8)

Revolutions per mile refers to how many times a tire rotates per mile. Thus

9. Torque$_{wheel}$ = torque$_{motor}$ × (overall gear ratio × overall drive train efficiency)   (5.9)

10. Speed$_{vehicle}$ (in mi/h) = (rpm$_{motor}$ × 60)/[overall gear ratio × (rev/m)]   (5.10)

It Ain't Heavy, It's My EV

In real estate they say that the three most important things are location, location, and location. In EV conversion, the three most important things are weight, weight, and weight.

Remove All Unessential Weight

You don't want to carry around a lot of unnecessary weight. But unless you're starting with a build-from-scratch design, you're inheriting the end result of someone else's weight tradeoffs. This means that you need to carefully go over everything with regard to its weight versus its value at three different times.

Before You Purchase for a Conversion or Test Build

Think about the vehicle's weight-reduction potential before you buy it. Is it going to be easy (pickup) or difficult (van) to get the extra weight out? What about hidden-agenda items? Has a previous accident resulted in a filled fender on your prospective purchase? (Take along a magnet during your examination.) Does its construction lead to ease of weight removal or substitution of lighter-weight parts later? Think about these factors as you look.

During Conversion

As you remove the internal combustion engine parts, it's likely that you'll discover additional parts that you hadn't seen or thought of taking out before. Parts snuggled up against the firewall or mounted low on the fenders are sometimes nearly invisible in a crowded and/or dirty engine compartment. Get rid of all unnecessary weight, but do exercise logic and common sense in your weight-reduction quest. Substituting a lighter-weight cosmetic body part is a great idea; drilling holes in a load-bearing structural frame member is not, as shown in Figures 5-3A and 5-3B.

FIGURE 5-3A   Part of the subframe in need of replacement from an old chassis.

FIGURE 5-3B Now doesn't that look cleaner?

After Conversion

Break your nasty internal combustion engine vehicle habits. Toss out all extras that you might have continued to carry, including the spare tire and tools.

After all your work, give yourself a pat on the back. You've probably removed from 400 to 800 pounds or more from a freshly cleaned-up former internal combustion engine vehicle chassis that's soon to become a lean and mean EV machine. The reason for all your work is simple—weight affects every aspect of an EV's performance: acceleration, climbing, speed, and range.

Weight and Acceleration

Let's see exactly how weight affects acceleration. When Sir Issac Newton was bonked on the head with an apple, he was allegedly pondering one of the basic relationships of nature—his second law: $F = Ma$, or force F equals mass M times acceleration a.

For EV purposes, this formula can be rewritten as

$$F_a = C_i W a$$

where F_a is acceleration force in pounds, W is vehicle mass in pounds, a is acceleration in mi/h per second, and C_i is a unit conversion factor that also accounts for the added inertia of the vehicle's rotating parts. The force required to get the vehicle going varies directly with the vehicle's weight; twice the weight means that twice as much force is required.

a (in mph/sec)	a'= a/21.95	F_a (in pounds) C_i = 1.06	F_a (in pounds) C_i = 1.1	F_a (in pounds) C_i = 1.2
1	0.046	48.3	50.1	54.7
2	0.091	96.6	100.2	109.3
3	0.137	144.8	150.3	164.0
4	0.182	193.1	200.4	218.6
5	0.228	241.4	250.5	273.3
6	0.273	289.7	300.6	328.0
7	0.319	338.0	350.7	382.6
8	0.364	386.3	400.8	437.3
9	0.410	434.5	450.9	491.9
10	0.455	482.8	501.0	546.6

TABLE 5-1   Acceleration Force, F_a (in Pounds), for Different Values of C_i

The mass factor depends on the gear in which you are operating. For internal combustion engine vehicles, the mass factor is typically high gear = 1.1; third gear = 1.2; second gear = 1.5; and first gear = 2.4. For EVs, where a portion of the drive train and weight typically has been removed or lightened, the mass factor is typically 1.06 to 1.2.

Table 5-1 shows the acceleration force F_a for three different values of C_i for 10 different values of acceleration a and for a vehicle weight of 1,000 pounds. Notice that an acceleration of 10 mi/h per second, an amount that takes you from 0 to 60 mi/h in 6 seconds, nominally requires extra force of 500 pounds; 5 mi/h per second, moving from 0 to 50 mi/h in 10 seconds, requires 250 pounds.

Weight and Climbing

When you go hill climbing, you add another force:

$$F_h = W\sin \phi$$

where F_h is hill-climbing force, W is the vehicle weight in pounds, and ϕ is the angle of incline, as shown in Figure 5-4. The degree of the incline—the way hills or inclines are commonly referred to—is different from the angle of the incline, but Figure 5-2 should clear up any confusion for you. Again, weight is directly involved, acted on this time by the steepness of the hill.

Table 5-2 shows the hill-climbing force F_h for 15 different incline values for a vehicle weight of 1,000 pounds. Notice that the tractive force required for acceleration of 1 mi/h per second equals that required for hill climbing of a 5 percent incline, 2 mi/h per second for a 10 percent incline, and so on up through a 30 percent incline. This handy relationship will be used later in the design section.

To use Table 5-2 with your EV, multiply by the ratio of your vehicle weight. For example, a 3,800-pound Ford Ranger pickup truck going up a 10 percent incline would require 3.8 × 99.6 = 378.5 lb.

Weight Affects Speed

Although speed also involves other factors, it's definitely related to weight. Horsepower and torque are related to speed per Eq. (5.2):

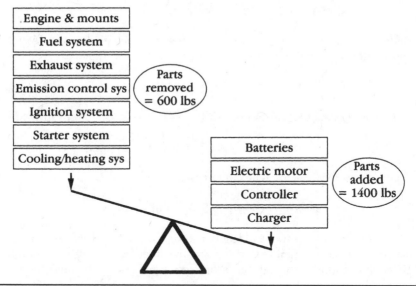

Figure 5-4   Front-to-rear weight distribution tradeoff.

Degree of incline	Incline angle O	sin O	F_h (in pounds)	a (in mph/sec)
1%	0° 34'	0.00989	9.9	
2%	1° 9'	0.02007	20.1	
3%	1° 43'	0.02996	29.6	
4%	2° 17'	0.04013	40.1	
5%	2° 52'	0.05001	50.0	1
6%	3° 26'	0.05989	59.9	
8%	4° 34'	0.07062	79.6	
10%	5° 43'	0.09961	99.6	2
15%	8° 32'	0.14838	148.4	3
20%	11° 19'	0.19623	196.2	4
25%	14° 2'	0.24249	242.5	5
30%	16° 42'	0.28736	287.4	6
35%	19° 17'	0.33024	330.2	
40%	21° 48'	0.37137	371.4	
45%	24° 14'	0.41045	410.5	

Table 5-2   Hill-Climbing Force F_h for 15 Different Values of Incline

$$Horsepower\ (hp) = FV/550$$

where F is force in pounds, and V is speed in feet per second. Armed with this information, Newton's second law equation can be rearranged as

$$a = (1/M) \times F$$

and because $M = W/g(10)$ and $F = (550 \times hp)/V$, they can be substituted to yield

$$a = 550(g/V)(hp/W)$$

Finally, a and V can be interchanged to give

$$V = 550(g/a)(hp/W)$$

where V is the vehicle speed in feet per second, W is the vehicle weight in pounds, g is the gravitational constant ($32.2\ ft/s^2$), and the other factors you've already met. For any given acceleration, as weight goes up, speed goes down because they are inversely proportional.

Weight Affects Range

Distance is simply speed multiplied by time:

$$D = Vt$$

Therefore,

$$D = 550(g/a)(hp/W)t$$

So weight again enters the picture. For any fixed amount of energy you are carrying onboard your vehicle, you will go farther if you take longer (drive at a slower speed) or carry less weight. You already encountered the practical results of this tradeoff in Chapter 4.

Besides the primary task of eliminating all unnecessary weight, there are two other important weight-related factors to keep in mind when doing EV conversions: front-to-rear weight distribution and the 30 percent rule.

Remove the Weight but Keep Your Balance

Always focus on keeping your vehicle's front-to-rear weight distribution intact and not exceeding its total chassis and front/rear-axle weight loading specifications. Figure 5-4 shows the magnitude of your problem for a Ford Ranger pickup truck similar to the one used in the Chapter 10 conversions. You have pulled out 600 pounds in engine, fuel, exhaust, emission, starter, and heating/cooling systems, but you're going to be putting 1,400 pounds back in, including 1,200 pounds of batteries (20 at about 60 pounds each). How do you handle it?

Table 5-3 provides the answers. Notice that the first row shows the 3,000-pound curb weight normally distributed 60 percent front (1,800 pounds) and 40 percent rear (1,200 pounds) with a 1,200-pound payload capacity. The second row shows that most

Item	Curb weight (lbs)	Front axle weight (lbs)	Rear axle weight (lbs)	Payload weight (lbs)
Ford Ranger pickup before conversion	3000	1800	1200	1200
Less IC engine and system parts	<600>	<500>	<100>	
Subtotal before conversion	2400	1300	1100	
Plus electric vehicle batteries, motor, etc.	1400	400	1000	
Ford Ranger pickup after conversion	3800	1700	2100	400
Battery weight 20ea 6-volt @ 60 lbs	1200			
Ratio battery weight to vehicle weight	32%			

TABLE 5-3 Electric Vehicle Conversion Weights Compared

of the weight you took out came from in and around the engine compartment—the 600 pounds you removed took 500 pounds off the front axle and 100 pounds off the rear axle. The secret is to put the weight back in a reasonably balanced fashion. This is accomplished by mounting four of the batteries—approximately 240 pounds—up front in the engine compartment along with the motor, controller, and charger. This puts about 400 pounds up front and about 1,000 pounds worth of batteries in the rear. The fifth row shows the results. You're up to a curb weight of 3,800 pounds, with a 1,700- to 2,100-pound front-to-rear weight distribution, but you're still inside the gross vehicle weight rating (also gross vehicle mass, GVWR, GVM), front/rear gross axle weight rating (GAWR), weight distribution, and payload specifications. Furthermore, when you drive the vehicle, it exhibits the same steering, braking, and handling capability that it had before the conversion.

As long as you keep within your vehicle's original internal combustion engine weight loading and distribution specifications, suspension and other support systems will never notice that you've changed what's under the hood. Overloading your EV chassis makes no more sense than overloading your internal combustion engine vehicle chassis. The best and safest solution is to get another larger or more heavy-duty chassis.

A postscript: Some owners actually prefer to adjust the shocks and springs of their conversion vehicle at this point to give a slightly firmer ride or (in the case of pickup truck owners) to return it to its previous firmer ride.

Remember the 30 Percent Rule

The 30 percent or greater rule of thumb (i.e., battery weight should be at least 30 percent of gross vehicle weight when using lead-acid batteries) is a very useful target to shoot for in an EV conversion. You'll want to do even better if you're optimizing for either high-speed or long-range performance goals. Table 5-3 shows that battery weight was 32 percent of gross vehicle weight for this conversion. Notice that if you opt for a 144-V battery system (adding four more batteries and 240 pounds of additional weight), the ratio goes up to 1,440/4,040, or 36 percent. Going the other way (taking out four batteries and 240 pounds), the ratio drops to 960/3,560, or 30 percent for a 96-V system. If you take out four more batteries and go to a 72-V system, the ratio drops to 720/3,320, or 22 percent. The rule of thumb proves to be correct in this case because you'd be unhappy with the performance of a 72-V system in this vehicle; even 96 V would be marginal.

Streamline Your EV Thinking

Until fairly recently, most of the automobile industry's wedge-front designs, while attractive, are actually 180 degrees away from aerodynamic streamlining. Look at nature's finest and most common example, the falling raindrop: Rounded and bulbous in front, it tapers to a point at its rear—the optimal aerodynamic shape. In addition, new bicycle helmets adhere perfectly to this principle as well.

While airplanes, submarines, and bullet trains have for decades incorporated the raindrop's example into their designs, automakers' design shops have eschewed this idea as unappealing to the public's taste. With plenty of internal combustion engine horsepower at their disposal, they didn't need aerodynamics—they needed style. Because batteries provide only 1 percent as much power per weight as gasoline, you and your EV do need aerodynamic awareness.

In this section you'll look at the aerodynamic drag and learn about the factors that come with the turf when you select your conversion vehicle and the items that you can change to help any EV conversion slip through the air more efficiently.

Aerodynamic Drag Force Defined

Mike Kimbell, an EV consultant that Bob Brant once knew, said it best: "Below 30 mi/h you could put an electric motor on a brick and never notice the difference." The reason is simple: Aerodynamic drag force varies with the square of the speed. If you're not moving, there's no drag at all. Once you get rolling, drag builds up rapidly and soon swamps all other factors. Let's see exactly how.

The aerodynamic drag force can be expressed as the aerodynamic drag force in pounds, is the coefficient of drag of your vehicle, which is the frontal area in square feet, and V is the vehicle's speed in miles per hour. To minimize drag for any given speed, you must minimize the coefficient of drag, and its frontal area.

Choose the Lowest Coefficient of Drag

The coefficient of drag has to do with streamlining and air turbulence flows around your vehicle, characteristics that are inherent in the shape and design of the conversion vehicle you choose. The coefficient of drag is not easily affected or changed later, so if you're optimizing for either high-speed or long-range performance goals, it's important that you keep this critical performance factor foremost in your mind when selecting your conversion vehicle. Figure 5-5 shows the value of the coefficient of drag for different shapes and types of vehicles.

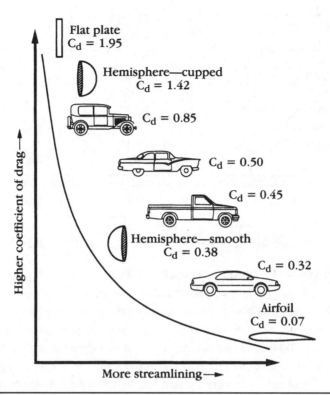

FIGURE 5-5 Coefficient of drag summary for different vehicle types and low ground clearance, no projections, and only two small openings to the engine compartment.

Notice that drag declined significantly with the passage of time—the 1920s Ford sedan had a coefficient drag of around 0.85, whereas today's Ford Taurus has a drag of 0.32. The values for typical late 1980s cars, trucks, and vans are as follows:

- Cars: 0.30 to 0.35
- Vans: 0.33 to 0.35
- Pickup trucks: 0.42 to 0.46

While car drag values typically have declined from 0.5 in the 1950s, to 0.42 in the 1970s, to 0.32 today, don't be misled. That snazzy contemporary open-cockpit roadster can still have a drag of 0.6. This is so because the drag has everything to do with air turbulence caused by open windows, cockpits, and pickup box areas, not with streamlining alone.

Table 5-4 shows how different areas—in this case taken for a 1970s vintage car—contribute to the drag of a vehicle. Contrary to what you might think, the body's rear area contributes more than 33 percent of the drag on the vehicle by itself, followed by the wheel wells at 21 percent, the underbody area at 14 percent, the front body area at 12 percent, projections (i.e., mirrors, drip rails, and window recesses) at 7 percent, and engine compartment and skin friction at 6 percent each.

Car Area	C_d Value	Percentage of total
Body–Rear	0.14	33.3
Wheel wells	0.09	21.4
Body–Under	0.06	14.3
Body–Front	0.05	11.9
Projections and indentations	0.03	7.1
Engine compartment	0.025	6.0
Body–Skin friction	0.025	6.0
Total	0.42	100.0

TABLE 5-4   Contribution of Different Car Areas to Overall Coefficient Drag for a Typical 1970s Vintage Car

Nissan to Produce 100 Percent Electric Van in Barcelona

Electric Variant of NV200 to Launch in 2013

- All-new 100 percent electric e-NV200 to enter production in 2013
- Nissan's second all-electric vehicle allocated to Barcelona plant
- Major project investment by Nissan of 100 million euros in Spain
- More than 700 jobs to be created at Nissan and its Spanish suppliers
- Multiuse vehicle providing families and businesses with unique driving experience
- Interior functionality and cargo capacity similar to existing NV200 with class-leading running costs

Nissan announced that a 100 percent electric compact van, the e-NV200, will go into production at its Barcelona plant in the 2013 fiscal year. The new model will be based on the existing award-winning NV200 currently produced at the plant and will be an important and innovative addition to Nissan's global range of light commercial vehicles (LCVs).

The e-NV200 will become Nissan's second all-electric vehicle following the launch of the multi–award-winning Nissan Leaf in 2010, underlining Nissan's long-term commitment to zero-emission mobility. The new model, which represents a 100 million euro investment by Nissan in Spain, is expected to create around 700 new jobs at the plant and throughout its local supplier base.

As the sole producer of the e-NV200, Barcelona will supply world markets, highlighting the growing competitiveness of Nissan's Spanish industrial operations. Last week a new medium-duty truck was allocated to Nissan's Avila plant in central Spain.

The e-NV200 will provide both businesses and families with functional and roomy interior space and will retain the host of innovative and practical features of the current NV200 lineup combined with the most advanced power train components of the Nissan Leaf. This will deliver a driving range similar to the Nissan Leaf on a single charge and similar performance, together with best-in-class running and maintenance costs.

Andy Palmer, Nissan executive vice president, said: "The e-NV200 represents a genuine breakthrough in commercial vehicles and further underlines Nissan's leadership within the electric vehicle segment. The new model will offer all the spaciousness, versatility and practicality of a traditionally powered compact van, but with zero CO_2 emissions at the point of use and also provide an outstanding driving experience that is unique to EVs. Crucially, it will also offer class-leading running and maintenance costs, which makes it an exceptionally attractive proposition to both businesses and families."

The e-NV200 also will make a significant contribution to Nissan's aim of becoming the world's largest LCV manufacturer by 2016 as a key pillar of the Nissan Power 88 midterm plan.

Last year Nissan achieved a historic milestone of selling more than 1 million LCVs world-wide in a single year.

Andy Palmer added: "The e-NV200 represents a bold and innovative addition to our commercial vehicle range, which is already one of the broadest of any manufacturer. I would like to thank the Spanish and Catalonian Government for their continued support of Nissan in Spain and congratulate the Barcelona workforce for earning the right to produce what will be an extremely important model for Nissan globally."

The current NV200 lineup has already sold around 100,000 units. Its popularity led it to be chosen as the next-generation New York City "Taxi of Tomorrow" following a rigorous selection process.

Currently, intensive evaluations of prototype NV200-based EVs are being carried out in Europe. Trials will continue in coming months in order to provide real-world feedback from the most demanding usage. Feedback will help Nissan to fulfill exact customer requirements ahead of the start of production expected in fiscal year 2013.[5]

Why do I show you this? Because if the car companies can figure it out, so can conversion companies by taking old gliders or chassis and updating, reusing, repurposing, and adding to the EV they are about to make. This is important.

Relative Wind Contributes to Aerodynamic Drag

Drag force is measured nominally at 60°F and a barometric pressure of 30 inHg in still air. Normally, those are adequate assumptions for most calculations. But very few locations have still air, so an additional drag component owing to relative wind velocity has to be added to your aerodynamic drag force calculation.

This is the additional wind drag pushing against the vehicle from the random local winds. The equation defining the relative wind factor is where w is the average wind speed of the area in mile per hour, V is the vehicle speed, and the relative wind coefficient that is approximately 1.4 for typical sedan shapes, 1.2 for more streamlined vehicles, and 1.6 for vehicles displaying more turbulence or sedans driven with their windows open.

Table 5-5 shows *the relative wind factor* calculated for seven different vehicle speeds—assuming the U.S. average value of 7.5 mi/h for the average wind speed—for the three different wind factor values.

C_{rw} at average	C_w factor	C_w factor	C_w factor	C_w factor	C_w factor	C_w factor	C_w factor
wind= 7.5 mph	at V= 5 mph	at V= 10 mph	at V= 20 mph	at V= 30 mph	at V= 45 mph	at V= 60 mph	at V= 75 mph
1.2	3.180	0.929	0.299	0.163	0.159	0.063	0.047
1.4 avg sedan	3.810	1.133	0.374	0.206	0.185	0.082	0.062
1.6	4.440	1.338	0.449	0.250	0.212	0.101	0.076

TABLE 5-5 Relative Wind Factor at Different Vehicle Speeds for Three Wind Speed Values

Aerodynamic Drag Force Data You Can Use

Table 5-6 puts the drag and *A* values for actual vehicles together and calculates their drag force for seven different vehicle speeds. Notice that drag force is lowest on a small car and greatest on the small pickup, but the small car might not have the room to mount the batteries to deliver the performance that you need. Notice also that an open-cockpit roadster, even though it has a small frontal area *A*, has drag force identical to the pickup truck.

To use Tables 5-5 and 5-6 with your EV, pick out your vehicle type in Table 5-6, and then multiply its drag force number by the relative wind factor at the identical vehicle speed using the appropriate relative wind factor for your vehicle type. For example, the 3,800-pound Ford Ranger pickup of Chapter 10 has a drag force of 24.86 pounds at 30 mi/h using Table 5-6. Multiplying by the relative wind factor of 0.250 from the bottom row = 1.6) of Table 5-5 gives you 6.22 pounds. Your total aerodynamic drag forced thus is 24.86 + 6.22 = 31.08 pounds.

Shape Rear Airflow

If you've seen a movie of a wind tunnel test with smoke added to make the air currents visible, you may have noticed a vortex or turbulence area at the rear of most vehicles tested. Those without access to wind tunnels notice the same effect when a trailer truck blows past them on the highway.

As with the falling raindrop shape, a boat-tail or rocket-ship-nose shape is ideal. While this is difficult to achieve, and no production chassis designs are available to help, you can benefit from rounding your vehicle's rear comers and eliminating all sharp edges. As you saw in Table 5-4, the body rear by itself makes up approximately 33 percent of the drag on a car, so any change you can make here should pay big dividends. In the practical domain of our pickup truck example, putting on a lightweight and streamlined (rounded edges) camper shell should be your first choice. A pickup truck with a streamlined camper shell is very close to a station wagon shape. The next choice is to put a cover over the pickup box. If even this is not feasible, leave the tailgate down and use a cargo net or wire screen instead, or simply remove the tailgate if this presents you with no functional problems.

Shape Wheel Well and Underbody Airflow

Next, we pay attention to the wheels and wheel well area. Table 5-4 showed that the tire and wheel well area by itself contributes approximately 21 percent of the drag of the car,

Vehicle	C_d	A	V= 5 mph	V= 10 mph	V= 20 mph	V= 30 mph	V= 45 mph	V= 60 mph	V= 75 mph
Small car	0.3	18	0.35	1.38	5.52	12.43	27.97	49.72	77.69
Larger car	0.32	22	0.45	1.80	7.20	16.20	36.46	64.82	101.28
Van	0.34	26	0.57	2.26	9.04	20.35	45.78	81.39	127.17
Small pickup	0.45	24	0.69	2.76	11.05	24.86	55.93	99.44	155.37
Roadster	0.6	18	0.69	2.76	11.05	24.86	55.93	99.44	155.37

TABLE 5-6   Aerodynamic Drag Force at Different Vehicle Speeds for a Typical Vehicle's Front Area of the Vehicle

so small streamlining changes here can have large benefits. Using smooth wheel covers, thinner tires, rear wheel well covers, no mud flaps, and even lowering the vehicle height all come to mind as immediate beneficial steps. Every little bit helps; anything you can do to reduce drag or turbulence is good.

The next obvious area is underneath the vehicle. Table 5-4 showed that this contributes approximately 14 percent of your coefficient drag. The immediate solution is simple: Close the bottom of the engine compartment.

There are no longer any bulky internal combustion engine components in the engine compartment, so cover the entire open area with a lightweight sheet of material. You probably still have a transmission—and you definitely have steering/suspension—components to deal with, so it might take several sheets of material. Whatever material you use, you don't want too thin a material for the body when the onrushing air strikes it, so choose a thickness that eliminates this possibility, and fasten the sheet (or sheets) securely to the chassis.

Make the area from behind the underside of your vehicle's nose to underneath the firewall region (or just beyond it) as smooth as possible using as lightweight a material as you can. A fully streamlined underside reduces drag and turbulence and can only help you.

Block and/or Shape Front Airflow

As you saw in Table 5-4, the body front and the engine compartment combined comprise approximately 18 percent of your drag. While you cannot make significant changes to the drag of a car or any motorized vehicle you inherit when you purchase your conversion chassis, you can replace or cover its sharp-cornered front air entry grill and block off the flow of air to the engine compartment. An EV doesn't need the massive air intake required in internal combustion engine vehicles to feed cooling air to the radiator and engine compartment. A 3-inch-diameter duct directing air to your electric motor is perfectly adequate. So anything you can do to round or streamline your vehicle's nose area (the smooth Ford Taurus or Thunderbird grillwork shapes of the 1980s and earlier) is fair game. What you want is the maximum streamlined effect with minimum weight, so use modern kit-car plastic and composite materials and techniques—autobody filler noses, if you please.

Air entering your EV conversion's now somewhat vacant engine compartment has the negative effect of creating under-the-hood turbulence, so blocking the incoming airflow with a sheet of lightweight material (such as aluminum) placed behind the grill works wonders. Whatever material you choose, just make sure that it's heavy enough and fastened securely enough not to buckle, rumple, or vibrate when the air strikes it. Also remember to leave a small opening for your electric motor's cooling air duct.

Roll with the Road

As they said in the movie *Days of Thunder*, "Tires is what wins the race." Today, tires are fat, have wide tread, and are without low-rolling-resistance characteristics; they've been optimized for good adhesion instead. As an EV owner, you need to go against the grain of current tire thinking and learn to roll with the road to win the performance race.

In this section you'll look at the "Rolling Resistance" row of Figure 5-1 and learn how to maximize the benefit from those four (or three) tire-road contact patches that are no bigger than your hand.

Rolling Resistance Defined

Again, vehicle weight is a factor, this time modulated by the vehicle's tire friction. The rolling resistance factor might at its most elementary level be estimated as a constant. For a typical under-5,000-pound EV, it is approximately

- 0.015 on a hard surface (concrete)
- 0.08 on a medium-hard surface
- 0.30 on a soft surface (sand)

Pay Attention to Your Tires

Tires are important to an EV owner. They support the vehicle and battery weight while cushioning against shocks, develop front-to-rear forces for acceleration and braking, and develop side-to-side forces for cornering. Tires are almost universally of radial-ply construction today. Typically, one or more steel-belted plies run around the circumference (hence radial) of tire. These deliver vastly superior performance to the bias-ply types (several plies woven crosswise around the tire carcass, hence, *bias* or *on an angle*) of earlier years that were replaced by radials as the standard in the 1960s. A tire is characterized by its rim width, the size wheel rim it fits on, section width (maximum width across the bulge of the tire), section height (distance from the bead to the outer edge of the tread), aspect ratio (ratio of height to width), overall diameter and load, and maximum tire pressure. In addition, the Tire and Rim Association defines the standard tire-naming conventions:

- **5.60 × 15 (typical VW bug) bias tire size**—5.60 denotes section width in inches, and 15 denotes the rim size in inches.
- **155R13 (typical Honda Civic) radial tire size**—155 denotes section width in millimeters and 13 denotes the rim size in inches.
- **P185/75R14 (typical Ford Ranger pickup) metric radial tire size**—*P* denotes passenger car, 185 denotes section width in millimeters, 75 denotes aspect ratio, *R* denotes radial, *B* denotes belted, *D* denotes bias, and 14 denotes the rim size in inches.

Doing a conversion today means using what's readily available. Any Goodyear dealer can give you a comparison of the published characteristics for the Goodyear Decathlon tire family. This is an economical class of all-weather radials (other tire makers have similar families) that should be more than adequate for most EV owners' needs. (*Note:* Tires with low rolling resistance are an excellent choice too.)

The calculated value is slightly lower than a measured value when new, and as tread wears down, you are looking at a difference of 0.4 to 0.8 inch less in the tire's diameter, which translates into even more revolutions per mile. The difference might be 30 revolutions out of 900—a difference of 3 percent—but if this figure is important to your calculations (all dimensions in inches), measure your tire's actual circumference in your driveway. A chalk mark on the sidewall and a tape measure and one full turn of your tire is all you need to tell if you're in the ballpark.

Ideally, you want soapbox derby or at least motorcycle tires on your EV: thin (little contact area with the road), hard (little friction), and large diameter (fewer revolutions per mile and thus higher mileage and longer wear). The point is that the rolling resistance force is affected by the material (harder is better for EV owners), the loading (less weight is better), the size (bigger is better), and the aspect ratio is better.

The variables in more conventional tire rolling resistance equations are usually tire inflation pressure (resistance decreases with increasing inflation pressure—harder is better), vehicle speed (increases with increasing speed), tire warm-up (warmer is better), and load (less weight is better).

Use Radial Tires

Radial tires are nearly universal today, so tire construction is no longer a factor. But you might buy an older chassis that doesn't have radials on it, so check to be sure because bias-ply or bias-belted tires deliver far inferior performance to radials in terms of rolling resistance versus speed, warm-up, and inflation.

Use High Tire Inflation Pressures If You Can

While you don't want to overinflate and balloon out your tires so that they pop off their rims, there is no reason not to inflate your EV's tires to their limit to suit your purpose. The upper limit is established by your discomfort level from the road vibration transmitted to your body. Rock-hard tires are fine; the only real caveat is not to overload your tires.

Brake Drag and Steering Scuff Add to Rolling Resistance

In addition to tires, rolling resistance comes from brake drag and steering and suspension alignment "scuff." Brake drag is another reason used vehicles are superior to new ones in the rolling resistance department—brake drag usually goes away as the vehicle is broken in. Alignment is another story. At worst, it's like dragging your other foot behind as you turned 90 degrees to the direction you're walking in. You want to check and make sure that front wheel alignment is at manufacturing spec levels and you haven't accidently bought a chassis whose rear wheels are tracking down the highway in sideways fashion. But neither of these contributes an earth-shattering amount to rolling resistance: The brake drag coefficient can be estimated as a constant 0.002 factor and steering/suspension drag as a constant 0.001 factor. Taken together, they add only 0.003—or an additional 3 pounds of force required by a vehicle weighing 1,000 pounds travelling on a level surface.

Rolling Resistance Force Data You Can Use

For most purposes, the nominal rolling resistance of concrete is 0.015 with nominal brake and steering drag of 0.003 added to it (total 0.018) is all you need. This generates 18.0 pounds of rolling resistance force for a 1,000-pound vehicle. The 3,800-pound Ford Ranger pickup truck of Chapter 10 would have a rolling resistance of 68.4 pounds. At 30 mi/h, the aerodynamic drag force on the Ford Ranger pickup truck of Chapter 10 is 31.08 pounds—less than half the contribution of its 68.4 pounds of rolling resistance drag.

Figure 5-6 shows the aerodynamic drag force and rolling resistance force on the Ford Ranger pickup truck of Chapter 10 plotted for several vehicle speeds. These two forces, along with the acceleration or hill-climbing forces, constitute the *propulsion* or *road load*.

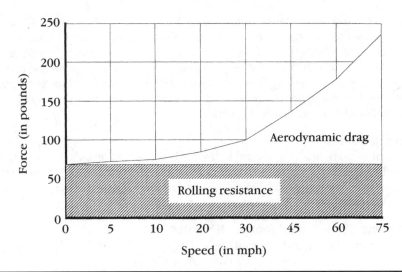

FIGURE 5-6   Rolling resistance and aerodynamic drag versus speed.

Less Is More with Drive Trains

In this section you'll look at the "Drive Train" row of Figure 5-1 and see how to get the most out of the internal combustion engine vehicle drive train components you adopt for your EV conversion. The drive train in any vehicle comprises components that transfer its motive power to the wheels and tires. The problem is that two separate vocabularies are used when talking about drive trains for electric motors as opposed to those for internal combustion engines. This section will discuss the basic components, cover differences in motor versus engine performance specifications, discuss transmission gear selection, and look at the tradeoffs of automatic versus manual transmission, new versus used, and heavy versus light fluids on drive train efficiencies.

Drive Trains

Let's start with what the drive train in a conventional internal combustion engine vehicle must accomplish. In practical terms, the power available from the engine must be equal to the job of overcoming the tractive resistances discussed earlier for any given speed. The obvious mission of the drive train is to apply the engine's power to driving the wheels and tires with the least loss (highest efficiency). But overall, the drive train must perform a number of tasks:

- Convert torque and speed of the engine to vehicle motion/traction
- Change directions, enabling forward and backward vehicle motion
- Permit different rotational speeds of the drive wheels when cornering
- Overcome hills and grades
- Maximize fuel economy

The drive train layout shown in simplified form in Figure 5-7 is used most widely to accomplish these objectives today. The function of each component is as follows:

- **Engine (or electric motor)**—Provides the raw power to propel the vehicle.

- **Clutch**—For internal combustion engines, separates or interrupts the power flow from the engine so that transmission gears can be shifted, and once engaged, the vehicle can be driven from standstill to top speed.

- **Manual transmission**—Provides a number of alternative gear ratios to the engine so that vehicle needs—maximum torque for hill climbing or minimum speed to economical cruising at maximum speed—can be accommodated, as seen in Figure 5-3A.

- **Driveshaft**—Connects the drive wheels to the transmission in rear-wheel-drive vehicles; not needed in front-wheel-drive vehicles.

- **Differential**—Accommodates the fact that outer wheels must cover a greater distance than inner wheels when a vehicle is cornering and translates drive force 90 degrees in rear-wheel-drive vehicles (might or might not in front-wheel-drive vehicles depending on how the engine is mounted). Most differentials also provide a speed reduction with a corresponding increase in torque.

- **Drive axles**—Transfer power from the differential to the drive wheels. Table 5-7 shows that you typically can expect 90 percent or greater efficiencies (slightly better for front-wheel-drive vehicles) from today's drive trains.

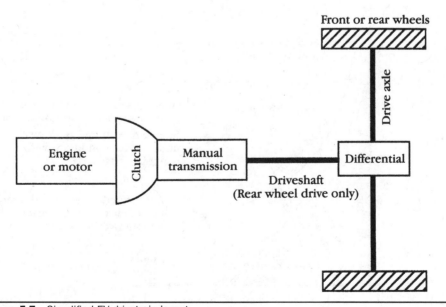

FIGURE 5-7   Simplified EV drivetrain layout.

Drivetrain type	Manual transmission	Driveshaft	Differential drive	Drive axle	Overall efficiency
Front wheel drive	0.96	not required	.097	.098	0.91
Rear wheel drive	0.96	0.99	.097	.098	0.90

TABLE 5-7 Comparison of Front- and Rear-Wheel Drive Train Efficiencies

Internal combustion engine vehicle drive trains provide everything necessary to allow an electric motor to be used in place of the removed gas-powered engine and its related components to propel the vehicle. But the drive train components are usually complete overkill for the EV owner. The reason has to do with the different characteristics of internal combustion engines versus electric motors and the way they are specified.

Difference in Motor Versus Engine Specifications

Comparing electric motors and internal combustion engines is not an apples-to-apples comparison. If someone offers you either an electric motor or an internal combustion engine with the same rated horsepower, take the electric motor—it's far more powerful. Also, a series-wound electric motor delivers peak torque on startup (0 rpms), whereas an internal combustion engine delivers nothing until you wind up its rpms.

An electric motor is so different from an internal combustion engine that a brief discussion of terms is necessary before going further. There is a substantial difference in the way an electric motor and an internal combustion engine are rated in horsepower. The purpose of Figure 5-8 is to show at a glance that an electric motor is more powerful

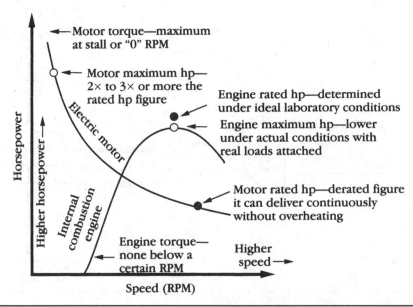

FIGURE 5-8 Comparison of electric motor versus internal combustion engine characteristics.

than an internal combustion engine of the same rated horsepower. All internal combustion engines are rated at specific rpm levels for maximum torque and maximum horsepower. Internal combustion engine maximum horsepower ratings are typically derived under idealized laboratory conditions (for the bare engine without accessories attached), which is why the rated horsepower point appears above the maximum peak of the internal combustion engine horsepower curve it can maintain without overheating. As you can see from Figure 5-9, the rated horsepower point for an electric motor is far down from its short-term output, which is typically two to three times higher than its continuous output in Table 5-7.

There is another substantial difference. While an electric motor can produce a high torque at zero speed, an internal combustion engine produces negative torque until

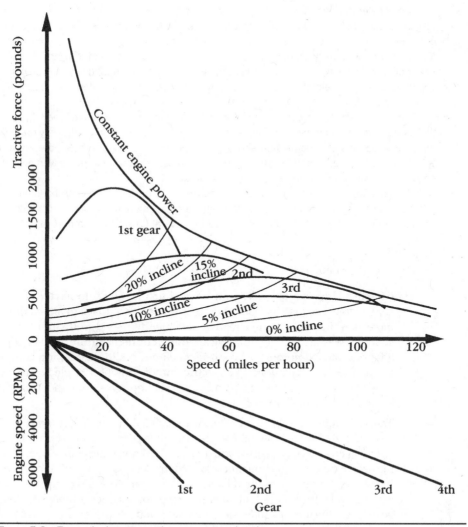

FIGURE 5-9 Transmission gear ratio versus speed and power summary.

some speed is reached. An electric motor therefore can be attached directly to the drive wheels and accelerate the vehicle from a standstill without the need for the clutch, transmission, or torque converter required by the internal combustion engine.

BMW ActiveE Stability Control Deemed Important to Its Build

Being in control is as important as being able to let go. With that in mind, the dynamic stability control system of the BMW ActiveE has been adapted to its electric drive train (Figure 5-10). Dubbed *Stability Management for Regeneration*, it works with the vehicle's motor management and brake energy regeneration to automate safe handling when decelerating in any situation.

The BMW ActiveE is designed to perform. But not all performance is related to speed. With its innovative brake energy regeneration system, you can go further—up to 20 percent further—than without this technology.

The accelerator pedal in the BMW ActiveE is more like a speed controller—with one foot, you can either speed the car up or slow the car down. As the driver eases off the accelerator, brake energy regeneration is activated to convert kinetic energy into electrical energy that is fed back into the battery. During this process, braking torque is created by the electric motor to effectively slow down the vehicle through its rear wheels. In most day-to-day driving, around 75 percent of all deceleration can be accomplished without using the brake pedal—increasing energy efficiency and significantly extending brake component life.[6]

Everything can be accomplished by controlling the drive current to the electric motor. While an internal combustion engine can only deliver peak torque in a relatively narrow speed range and requires a transmission and different gear ratios to deliver its power over a wide vehicle speed range, electric motors can be designed to deliver their power over a broad speed range with no need for a transmission at all.

All these factors mean that current EV conversions or car companies building their own EVs put a lighter load on their borrowed-from-an-internal-combustion-engine-vehicle drive trains, and future EV conversions eliminate the need for several drive train components altogether.

Let's briefly summarize:

- **Clutch**—Although basically unused, a clutch is handy to have in today's EV conversions because its front end gives you an easy place to attach the electric motor and its back end is already conveniently mated to the transmission. In short, it saves the work of building adapters, etc. In the future, when widespread adoption of AC motors and controllers eliminates the need for a complicated mechanical transmission, the electric motor can be coupled directly to a simplified, lightweight, one-direction, one- or two-gear-ratio transmission, eliminating the need for a clutch.

- **Transmission**—Another handy item in today's EV conversions, the transmission's gears not only match the vehicle you are converting to a variety of off-the-shelf electrical motors but also give you a mechanical reversing control that eliminates the need for a two-direction motor and controller—again simplifying your work. In the future, when widespread adoption of AC motors and controllers provides directional control and eliminates the need for a large number of mechanical gears to get the torques and speeds you need, today's transmission will be able to be replaced by a greatly simplified (and even more reliable) mechanical device.

BMW ActiveE Technical Data

European model shown

Body	
Doors/Seats	2/4
Length/Width/Height (in)	171.65/68.82/56.6
Wheelbase (in)	104.7
Track Front/Rear (in)	58.03/59.33
Turning Circle (ft)	35.1
Transmission Oil Incl. Axle Drive	Permanent Filling
Weight, Unladen (lbs)	4001
Max. Load (lbs)	739
Max. Permissible Weight (lbs)	4740
Luggage Compartment Capacity (l)	200

Running Gear	
Front Wheel Suspension	Two-joint Strut Suspension
Rear Wheel Suspension	Five-arm Axle Adapted to the Electrical Power Unit
Front Brakes	Single Piston Sliding Caliper /Ventilated Disc Brake
Diameter (in)	11.8
Rear Brakes	Sliding Caliper/Ventilated Disc Brake
Diameter (in)	11.8
Driving Stability Systems	ABS incl. Brake Assistance and Cornering Brake Control (CBC) Dynamic Stability Control (DSC) with Additional Functions
Steering	Rack-and-Pinion Steering with electric steering assistance
Overall Steering Ratio	14.4:1
Weight Distribution (%)	Front 49/Rear 51
Tires	205/55R16 with Emergency Running Properties/Rolling Resistance Optimized

Engine	
Type	Electric Motor
Motor Type	Synchronous Motor
Output (kW/bhp)	125/170
Torque From Idle (Nm)	250
Max RPM (min^{-1})	12000

High Voltage Electrical System	
Battery Capacity (kWh)	32
Battery System Weight (lbs)	992.1
Battery Charge Time at 240 V/32 A (7.7 kW)	4 to 5 hours
Battery Design	192 cells at 40Ah
Battery Cooling	Active Liquid Cooling
Battery Location	Tank, Tunnel, Front Wall
Peak Current (A)	400

Performance Ratings	
Power-to-Weight Ratio DIN (kg/kW)	9.76
Acceleration 0–62 (mph)	9 seconds
Top Speed[1] (mph)	90
Range FTP 72[2] (mls)	150
Operational Range[3] (mls)	100

Consumption (FTP 72)	
Total (kWh/mls)	0.19
Emitted CO_2 (g/mls)	0

[1]With electrical cut-off
[2]U.S. Federal Test Procedure, also referred to as UDDS (Urban Dynamometer Driving Schedule).
[3]Results may vary.

FIGURE 5-10 BMW ActiveE specifications.

- **Driveshaft, differential, drive axles**—These components are all used intact in today's EV conversions.

Going Through the Gears

The transmission gear ratios, combined with the ratio available from the differential (or rear end, as it's sometimes called in automotive jargon), adapt the internal combustion engine's power and torque characteristics to maximum torque needs for hill climbing or maximum economy needs for cruising. Please note that your selection applies to engine speed capabilities. Normally, the overall gear ratios are selected to fall in a geometric progression: first/second = second/third, etc. Then individual gears are optimized for starting (first), passing (second or third), and fuel economy (fourth or fifth). Table 5-8 shows how these ratios turn out in an actual production car—in this case a 1989 Ford Taurus SHO. Well, some things have changed as seen in Figure 5-11 since the new Taurus SHO, but most have not. I rode in the 2013 model in April.

Notice the first two gear pairs are in a 1.5 ratio, whereas the next two move to 1.35. Figure 5-8 also calculates the actual transmission, differential, and overall gear ratios (overall equals transmission times differential) for the 1987 Ford Ranger pickup truck that will be used later in the design section. Notice that the Ranger is optimized at both

FIGURE 5-11   A 2013 Ford Taurus SHO in Portland, Oregon. Not too many changes affect the calculations in this book. (*Seth Leitman.*)

Transmission Gear	1989 Taurus SHO Overall	Ratio to next gear	1987 Ranger Pickup transmission	1987 Ranger Pickup differential	1987 Ranger Pickup Overall
1st	12.01	1.5	3.96	3.45	13.66
2nd	7.82	1.5	2.08	3.45	7.18
3rd	5.16	1.35	1.37	3.45	4.73
4th	3.81	1.35	1.00	3.45	3.45
5th	2.79	N/A	0.84	3.45	2.90

TABLE 5-8   Comparison of Overall Transmission Gear Ratios for Ford 1989 Taurus SHO and 1987 Ranger Pickup

ends of the range but lower in the middle versus the Taurus, reflecting the difference in car versus truck design.

Automatic Versus Manual Transmission

The early transmission discussion purposely avoided automatic transmissions. The reason is simple: EV owners need efficiency, and automatic transmissions are tremendously inefficient. There's another good reason—even with off-the-shelf components, you are not going to shift gears as much with an EV. If you're driving around town, you might use only one or two gears. You just put your EV in the gear you need and go. There's far less need for the clutch, too. Remember, when you're sitting at an intersection, your electric motor is not even turning.

Two ordinary household fans can demonstrate the torque converter principle that automatic transmissions use. Set them up facing each other about two feet apart, and turn one on; the other starts to rotate with it. An automatic transmission uses transmission fluid rather than air as the coupling medium, but the principle is the same. An automatic transmission has a maximum efficiency of 80 percent, which drops dramatically at lower engine speed or vehicle torque levels. Automatic transmissions are also typically matched in characteristics to a given family of internal combustion engines with limited peak torque range—exactly opposite in behavior to electric motors. In short, choose a model with a clutch and a manual transmission for your internal combustion engine conversion vehicle.

Use a Used Transmission

There's a bonus here for the efficiency-seeking EV converter. Drive trains take thousands of miles to wear into their minimum-loss condition. When added to the fact that used tires have less rolling resistance, brake drag reduces as drums and linings wear smooth, and oil and grease seals have less drag after a period of break-in wear, you can expect about 20 percent less rolling and drive train resistance from a used vehicle than from a brand-new one (assuming the used vehicle is not badly worn or abused). And a used vehicle costs less, so it's a real bonus as seen in Figure 5-12 from TurnE Conversions using the old transmission from a gas Smart car!

Heavy Versus Light Drive Trains and Fluids

Efficiency-seeking EV converters should avoid not only automatic transmissions but also oversize axles, transmission, clutches, or anything that adds weight and reduces

FIGURE 5-12   Can I get a transmission please! (*TurnE Conversions.*)

efficiency. Even a manual transmission–based drive train will exhibit higher losses when operating at a low fraction of the gear's design maximum torque, which is the normal EV mode. The lower EV load will result in a lower efficiency from a heavy-duty part from a regular or economy part. Look for a vehicle with a lighter engine and manual transmission (e.g., four cylinders rather than six, etc.) in your purchasing search. Earlier vintage models in a series are preferable because manufacturers tend to introduce higher-performance options in successive model years.

The corollary of all this is the lubricant you choose. Using a lighter-viscosity fluid in your differential lets things turn a lot easier. You're not breaking any rules here. Instead of shoving 500 hp through your drive train, you're at the opposite extreme—you're putting in an electric motor that lets you cruise at 10 percent of the peak torque load used by the internal combustion engine you just replaced. You're shifting less, using a lower peak torque, and probably using it less often. As a result, your electric motor is putting only the lightest of loads on your internal combustion vehicle drive train, and you're probably using 50 percent (or less) of your drive train's designed capability. Thus less wear and tear on the gears means that you can use a lighter-viscosity lubricant and recover the additional benefit of further increased efficiency.

Design Your EV

This is step two. Look at your big picture first. Before you buy, build, or convert, decide what the main mission of your EV will be: a high-speed dragster to quietly blow away unsuspecting opponents at a stoplight, a long-range flyer to be a winning candidate at

Electric Auto Association meetings, or a utility commuter vehicle to take you to work or grocery shopping with capabilities midway between the former two. Your EV's weight is of primary importance to any design, but high acceleration off the line will dictate one type of design approach and gear ratios, whereas a long-range flyer will push you in a different direction. If it's a utility commuter EV you seek, then you'll want to preserve a little of both while optimizing your chassis flexibility toward either highway commuting or neighborhood hauling and pickup needs.

Even though it's 100 percent electric, the all-new BMW ActiveE carries itself in a body style based on the dynamic 1 Series Coupe. Based on pure BMW DNA, it offers its drivers athletic design reflective of its advanced agility. From the top of a new and distinctive power dome to its unique closed rear apron, the all-new BMW ActiveE boasts rear-wheel drive and a lowered center of gravity for the power-driven handling enjoyed in every BMW. Throughout the vehicle, features of a 1 Series Coupe are combined with details that highlight electromobility.[7]

In this section you'll learn how to match your motor–drive train combination to the body style you've selected by going through the following steps:

- Learn when to use horsepower, torque, or current units and why.
- Look at a calculation overview.
- Determine the required torque needs of your selected vehicle's chassis.
- Determine the available capabilities of your selected electric motor and drive train torque.

The design process described here can be infinitely adapted to any EV you want to buy, build, or convert.

Horsepower, Torque, and Current

Let's start with some basic concepts. Horsepower is a rate of doing work. It takes 1 hp to raise 550 pounds one foot in one second. But the second equation, which relates force and speed, brings horsepower to you in more familiar terms. It takes 1 hp to move 37.5 pounds at 10 mi/h. Great, but you also can move 50 pounds at 7.5 mi/h with 1 hp. The first instance might describe the force required to push a vehicle forward on a level slope; the second describes the force required to push this same vehicle up an incline. Horsepower is equal to force times speed, but you need to specify the force and speed you are talking about. For example, since we already know that 146.19 pounds is the total drag force on the 3,800-pound Ranger pickup at 50 mi/h, and equation 5 relates the actual power required at a vehicle's wheel as a function of its speed and the required tractive force:

$$\text{Horsepower (hp)} = (146.19 \times 50)/375 = 19.49 \text{ or } \textit{approximately 20 hp}$$

Thus only about 20 hp is necessary—at the wheels—to propel this pickup truck along at 50 mi/h on a level road without wind.

In fact, a rated 20-hp electric motor will easily propel a 4,000-pound vehicle at 50 mi/h—a fact that might amaze those who think in terms of the typical rated 90- or 120-hp internal combustion engine that might have just been removed from the pickup.

The point here is to condition yourself to think in terms of force values, which are relatively easy to determine, rather than in terms of a horsepower figure that is arrived at differently for gas-powered engines versus electric motors and that means little until tied to specific force and speed values anyway. Another point (covered in more detail in the discussion of electric motors in Chapter 6 and the discussion of the electrical system in Chapter 9) is to think in terms of current when working with electric motors. The current is directly related to motor torque. Through the torque-current relationship, you can directly link the mechanical and electrical worlds. (*Note:* The controller gives current multiplication. In other words, if the motor voltage is one-third the battery voltage, then the motor current is slightly less than three times the battery current. The motor and battery current would be the same only if you used a very inefficient resistive controller.)

Calculation Overview

Notice that the starting point in the calculations is the ending point of the force value required. Once you know the forces acting on your vehicle chassis at a given speed, the rest is easy. For your calculation approach, first determine these values, and then plug in your motor and drive train values for the vehicle's *design center* operating point, be it a 100 mi/h speedster, a 20 mi/h economy flyer, or a 50 mi/h utility vehicle. A 50 mi/h speed will be the design center for our pickup truck example.

In short, once you go through the equations, worksheets, and graphed results covered in this section and repeat them with your own values, you'll find the process quite simple. The entire process is designed to give you graphic results you can quickly use to see how the torque available from your selected motor and drive train meets your vehicle's torque requirements at different speeds. If you have a computer with a spreadsheet program, you can set it up once and afterwards graph the results of any changed input parameter in seconds.

For the mathematical courses for automotive technicians, in equation form, what I am saying is

Available engine power = tractive resistance demand

Power = (acceleration + climbing + rolling + drag + wind) resistance

You've determined every one of these earlier in the chapter. Under steady-speed conditions, acceleration is zero, so there is no acceleration force. This is the propulsion or road-load force you met at the end of the rolling resistance section and graphed for the Ford pickup in Figure 5-6. You need to determine this force for your vehicle at several candidate vehicle speeds and add back in the acceleration and hill-climbing forces. This is easy if you recall that the acceleration force equals the hill-climbing force over the range from 1 to 6 mi/h per second.

You can now calculate your electric motor's required horsepower for your EV's design center. Plugging the values for torque, speed, and revolutions per mile (based on your vehicle's tire diameter) into the equation will give you the required horsepower for your electric motor.

After you have chosen your candidate electric motor, the manufacturer usually will provide you with a graph or table showing its torque and current versus speed performance based on a constant voltage applied to the motor terminals. From these figures or curves you can derive the rpm at which your electric motor delivers closest to its rated horsepower.

With all the other motor torque and rpm values, you can then calculate wheel torque and vehicle speed using the following equations for the different overall gear ratios in your drive train.

You now have the family of torque-available curves versus vehicle speed for the different gear ratios in your drive train. All that remains is to graph the torque-required data and the torque-available data on the same grid. A quick look at the graph tells you whether you have what you need or you need to go back to the drawing board.

Torque-Required Worksheet

You've met all the values going into the level drag force before, but not in one worksheet. Now they are converted to torque values using equation 11 and new values of force and torque are calculated for incline values from 5 to 25 percent. Conveniently, these correspond rather closely to the acceleration values for 1 to 5 mi/h per second, respectively, and the two can be used interchangeably. If you were preparing a computer spreadsheet, all this type information would be grouped in one section so that you could see the effects of changing chassis weight on other parameters. You also might want to graph speed values at 5 mi/h intervals to present a more accurate picture.

Torque-Available Worksheet

There are a few preliminaries to go through before you can prepare the torque-available worksheet. First, you have to determine the horsepower of the electric motor.

Plugging in real values for the Ranger pickup with P185/75R14 tires and using the torque value from the torque-required worksheet at the 50 mi/h design center speed.

This corresponds quite nicely to the capabilities of the Advanced DC Motors Model FB1-4001, rated at 22 hp. From the manufacturer's torque-versus-speed curves for this motor driven at a constant 120 V and using this equation:

$$hp = (torque \times rpms)/5{,}252 + (25 \times 4{,}600)/5{,}252 = 21.89$$

This motor produces approximately 22 hp at 4,600 rpms at 25 ft-lb of torque and 170 A. Next, calculate wheel rpms using the following equation:

$$Wheel\ rpms = (mi/h \times rev/mi)/60 = (50 \times 808)/60 = 673.33$$

You then can calculate the best gear using the next equation:

$$Overall\ gear\ ratio = rpm_{motor}/rpm_{wheel} = 4{,}600/673.33 = 6.83$$

Now you are ready to prepare the torque-available worksheet shown in Table 5-9. This worksheet sets up motor values on the far left. Wheel torque and vehicle speed values for the first through fifth gears go from left to right across the worksheet and at higher values of torque and current from top to bottom of the worksheet within each gear. The net result is that you now have the wheel torque available at a given vehicle speed for each of the vehicle's gear ratios. A spreadsheet would make quick work of this; you could see at a glance the effects of varying motor voltage, tires size, etc.

Torque-Required and Torque-Available Graph

How do you read Table 5-10 and Figure 5-13? What does it tell you? The usable area of each gear is the area to the left and below it, bounded at the bottom by the torque required at the level condition curve. You want to work as far down the torque-available

Vehicle speed (mph)	10	20	30	40	50	60	70	80	90
Tire rolling resistance C_r	0.015	0.015	0.015	0.015	0.015	0.015	0.015	0.015	0.015
Brake and steering resistance C_r	0.003	0.003	0.003	0.003	0.003	0.003	0.003	0.003	0.003
Total rolling force (lbs)	68.4	68.4	68.4	68.4	68.4	68.4	68.4	68.4	68.4
Still air drag force (lbs)	2.76	11.05	24.60	44.19	69.05	99.43	135.35	176.78	223.73
Relative wind factor C_w	1.338	0.449	0.25	0.169	0.126	0.101	0.083	0.071	0.062
Relative wind drag force (lbs)	3.70	4.96	6.21	7.47	8.73	9.99	11.25	12.51	13.77
Total drag force, level (lbs)	74.86	84.40	99.47	120.07	146.19	177.83	215.00	257.69	305.91
Total drag torque, level (ft-lbs)	77.85	87.78	103.45	124.87	152.04	184.95	223.60	268.00	318.15
Sin ϕ, ϕ = Arc tan 5% incline	0.0500	0.0500	0.0500	0.0500	0.0500	0.0500	0.0500	0.0500	0.0500
Cos ϕ, ϕ = Arc tan 5% incline	0.9988	0.9988	0.9988	0.9988	0.9988	0.9988	0.9988	0.9988	0.9988
Incline force WSin ϕ (lbs)	190.04	190.04	190.04	190.04	190.04	190.04	190.04	190.04	190.04
Rolling drag force C_r WCos ϕ (lbs)	68.31	68.31	68.31	68.31	68.31	68.31	68.31	68.31	68.31
Total drag force, 5% (lbs)	264.81	274.36	289.43	310.02	336.14	367.78	404.95	447.64	495.86
Total drag torque, 5% (ft-lbs)	275.41	285.34	301.01	322.43	349.59	382.50	421.16	465.56	515.71
Sin (Arc tan 10% incline)	0.0996	0.0996	0.0996	0.0996	0.0996	0.0996	0.0996	0.0996	0.0996
Cos (Arc tan 10% incline)	0.9950	0.9950	0.9950	0.9950	0.9950	0.9950	0.9950	0.9950	0.9950
Incline force WSin ϕ (lbs)	378.52	378.52	378.52	378.52	378.52	378.52	378.52	378.52	378.52
Rolling drag force C_r WCos ϕ (lbs)	68.06	68.06	68.06	68.06	68.06	68.06	68.06	68.06	68.06
Total drag force, 10% (lbs)	453.04	462.58	477.65	498.25	524.37	556.01	593.18	635.87	684.08
Total drag torque, 10% (ft-lbs)	471.17	481.10	496.77	518.19	545.35	578.26	616.92	661.32	711.46

TABLE 5-9 Torque-Required Worksheet for 1987 Ford Ranger Pickup at Different Vehicle Speeds and Inclines*

Vehicle speed (mph)	10	20	30	40	50	60	70	80	90
Sin φ, φ = Arc tan 15% incline	0.1484	0.1484	0.1484	0.1484	0.1484	0.1484	0.1484	0.1484	0.1484
Cos φ, φ = Arc tan 15% incline	0.9889	0.9889	0.9889	0.9889	0.9889	0.9889	0.9889	0.9889	0.9889
Incline force WSin φ (lbs)	563.84	563.84	563.84	563.84	563.84	563.84	563.84	563.84	563.84
Rolling drag force C_rWCos φ (lbs)	67.64	67.64	67.64	67.64	67.64	67.64	67.64	67.64	67.64
Total drag force, 15% (lbs)	637.94	647.49	662.56	683.16	709.27	740.92	778.09	820.78	868.99
Total drag torque, 15% (ft-lbs)	663.48	673.41	689.08	710.50	737.66	770.57	809.23	853.63	903.77
Sin φ, φ = Arc tan 20% incline	0.1962	0.1962	0.1962	0.1962	0.1962	0.1962	0.1962	0.1962	0.1962
Cos φ, φ = Arc tan 20% incline	0.9806	0.9806	0.9806	0.9806	0.9806	0.9806	0.9806	0.9806	0.9806
Incline force WSin φ (lbs)	745.67	745.67	745.67	745.67	745.67	745.67	745.67	745.67	745.67
Rolling drag force C_rWCos φ (lbs)	67.07	67.07	67.07	67.07	67.07	67.07	67.07	67.07	67.07
Total drag force, 20% (lbs)	819.20	828.75	843.82	864.41	890.53	922.18	959.34	1002.0	1050.3
Total drag torque, 20% (ft-lbs)	851.99	861.92	877.59	899.01	926.18	959.08	997.74	1042.1	1092.3
Sin φ, φ = Arc tan 25% incline	0.2425	0.2425	0.2425	0.2425	0.2425	0.2425	0.2425	0.2425	0.2425
Cos φ, φ = Arc tan 25% incline	0.9702	0.9702	0.9702	0.9702	0.9702	0.9702	0.9702	0.9702	0.9702
Incline force WSin φ (lbs)	921.46	921.46	921.46	921.46	921.46	921.46	921.46	921.46	921.46
Rolling drag force C_rWCos φ (lbs)	66.36	66.36	66.36	66.36	66.36	66.36	66.36	66.36	66.36
Total drag force, 25% (lbs)	994.28	1003.8	1018.9	1039.5	1065.6	1097.3	1134.4	1177.1	1225.3
Total drag torque, 25% (ft-lbs)	1034.1	1044.0	1059.7	1081.1	1108.3	1141.2	1179.8	1224.2	1274.4

* Values computed for 1987 Ford Ranger pickup; weight = 3800 lbs; coefficient of drag, C_d = 0.45; frontal area, A = 24 square feet; relative wind factor, C_{rw} = 1.6; relative wind, w = 7.5 mph; tires = P185/75R14; revolutions/mile = 808; torque multiplier = 840.34/(revolutions/mile) = 1.04.

TABLE 5-9 Torque-Required Worksheet for 1987 Ford Ranger Pickup at Different Vehicle Speeds and Inclines* *(Continued)*

TABLE 5-10 Torque-Available Worksheet for 120-V DC Series Motor–Powered 1987 Ford Ranger Pickup at Different Motor Speeds and Gear Ratios

Gear parameters:

Vehicle gear	1st	2nd	3rd	4th	5th
Overall gear ratio	13.66	7.18	4.73	3.45	2.9
Overall gear ratio	13.66	7.18	4.73	3.45	2.9
Motor torque multiplier, equation (12)	12.294	6.462	4.257	3.105	2.61
RPM multiplier, equation (13)	165.56	87.02	57.33	41.81	35.15

Current in amps, torque in ft-lbs, vehicle speed in mph:

Motor Current	Motor Torque	Motor RPM	1st Wheel Torque	1st Vehicle Speed	2nd Wheel Torque	2nd Vehicle Speed	3rd Wheel Torque	3rd Vehicle Speed	4th Wheel Torque	4th Vehicle Speed	5th Wheel Torque	5th Vehicle Speed
100	10	7750	122.94	46.81	64.62	89.06	42.57	135.19	31.05	185.34	26.10	220.50
125	15	6400	184.41	38.66	96.93	73.54	63.86	111.64	46.58	153.06	39.15	182.09
150	20	5000	245.88	30.20	129.24	57.46	85.14	87.22	62.10	119.58	52.20	142.26
170	25	4600	307.35	27.78	161.55	52.86	106.43	80.24	77.63	110.01	65.25	130.88
190	30	4100	368.82	24.76	193.86	47.11	127.71	71.52	93.15	98.05	78.30	116.65
210	35	3900	430.29	23.56	226.17	44.82	149.00	68.03	108.68	93.27	91.35	110.96
230	40	3700	491.76	22.35	258.48	42.52	170.28	64.54	124.20	88.49	104.40	105.27
250	45	3500	553.23	21.14	290.79	40.22	191.57	61.05	139.73	83.70	117.45	99.58
270	50	3400	614.70	20.54	323.10	39.07	212.85	59.31	155.25	81.31	130.50	96.73
290	55	3350	676.17	20.23	355.41	38.50	234.14	58.44	170.78	80.12	143.55	95.31
305	60	3250	737.64	19.63	387.72	37.35	255.42	56.69	186.30	77.73	156.60	92.47
320	65	3150	799.11	19.03	420.03	36.20	276.71	54.95	201.83	75.33	169.65	89.62
335	70	3050	860.58	18.42	452.34	35.05	297.99	53.20	217.35	72.94	182.70	86.78
355	75	3000	922.05	18.12	484.65	34.47	319.28	52.33	232.88	71.75	195.75	85.35
370	80	2950	983.52	17.82	516.96	33.90	340.56	51.46	248.40	70.55	208.80	83.93
390	85	2900	1045.0	17.52	549.27	33.33	361.85	50.59	263.93	69.35	221.85	82.51
405	90	2850	1106.5	17.21	581.58	32.75	383.13	49.71	279.45	68.16	234.90	81.09
420	95	2800	1167.9	16.91	613.89	32.18	404.42	48.84	294.98	66.96	247.95	79.66
440	100	2750	1229.4	16.61	646.20	31.60	425.70	47.97	310.50	65.77	261.00	78.24

* Values computed for 1987 Ford Ranger pickup; tires = P185/75R14; revolutions/mile = 808; overall drivetrain efficiency = 0.90; DC series traction motor is Advanced DC Motors Model FBI-4001; battery pack is 120 volts; equation (12) is $T_{wheel} = T_{motor}$/(overall gear ratio × overall drivetrain efficiency); equation (13) is $Speed_{vehicle} = RPM_{motor} \times 60)/$(overall gear ratio × revolutions/mile).

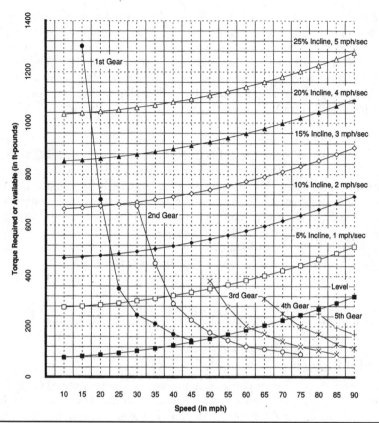

Figure 5-13 Electric vehicle torque required versus torque availlable.

curve for each gear as possible for minimum current draw and maximum economy and range. The graph confirms that second gear is probably the best overall selection. You could put the EV into second gear and leave it there because it gives you 2 mi/h per second of acceleration at startup, hill-climbing ability of up to 15 percent inclines, and enough torque to take you up to about 52.5 mi/h. For mountain climbing or quick pops off the line, first gear gives you everything you could hope for at the expense of really sucking down the amps, current-wise. At the other end of first gear, however, if you drive like there's an egg between your foot and the accelerator pedal, it actually draws only 100 A at 45 mi/h versus the 210 A required by second gear. At higher speeds, third gear lets you cruise at 60 mi/h at 270 A, and fourth gear lets you cruise at 70 mi/h at 370 A. At any speed, fifth gear appears marginal in this particular vehicle. Although it can possibly hold 78 mi/h, it requires 440 A to do so. At any other speed, other gears do it better with less current draw. While current draw is your first priority, too much for too long overheats your motor. You don't want to exceed your motor's speed rating either as you would do if you drove much above 45 mi/h in first gear. Is this a usable motor and drive train combination for this vehicle? Definitely! If you want to make minor adjustments, just raise or lower the battery voltage. This will shift the torque-available curve for each gear to the right (higher voltage) or to the left (lower voltage). A larger motor in this particular vehicle will give you better acceleration and upper-end speed performance; the torque-available curves for each gear would be shifted higher. But the penalty would be higher weight, increased current draw, and shorter range. A smaller motor would shift the torque-available curve lower while returning a small weight and current-draw advantage. But beware of underpowering your vehicle. If given the choice, always go for slightly more rather than slightly less horsepower than you need. The result will almost always be higher satisfaction with your finished EV conversion.

Tesla Model S Is Build from the Ground Up!

The Model S is engineered from the ground up as an electric car. Through meticulous and innovative engineering, Tesla created a drive experience to surpass any that have come before it. By designing a slim-line battery pack positioned below the floor and combining the motor and power electronics into a compact module between the rear wheels, the Tesla engineering team developed a car that truly integrates Tesla's proven power train. The innovative integration is the key to exploiting the advantages of an electric car and creates new opportunities in safety and performance while maximizing interior space.[8]

However, if I had the choice, I'd go with the ones who already build their own EVs. Yeah, I'd want to forget the torques and graphs and go straight to Tesla!

Buy Your EV Chassis

This is step three. Even if you go out and your EV ready-made, you still want to know what kind of a job the manufacturer has done so that you can decide whether you're getting the best model for you. In all other cases, you'll be doing the optimizing, either by the choices you've made up front in chassis selection or by other decisions you make later on during the conversion. In this section you'll be looking at key points that contribute to buying smart:

- Review why conversion is best—the pro side.
- Why conversion might not be for you—the con side.
- How to get the best deal.
- Avoid off-brand, too old, abused, dirty, or rusty chassis.
- Keep your needs list handy.
- Buy or borrow the chassis manual.
- Sell unused engine parts.

EV Conversions

In the real world, where time is money, converting an existing internal combustion engine vehicle saves money in terms of large capital investment and a large amount of labor. By starting with an existing late-model vehicle, the EV converter's bonus is a structure that comes complete with body, chassis, suspension, steering, and braking systems—all designed, developed, tested, and safety-proven to work together. And the

EV converter inherits another body bonus—its bumpers, lights, safety-glazed windows, etc. are already preapproved and tested to meet all safety requirements.

Here's a compelling argument for an existing chassis. General Motors Chief Executive Richard Wagoner back in 2001 said that global warming was not going to be solved with $100,000 (€64,000) electric cars, and his company is focusing on a vehicle more consumers can afford.

This is why I like this quote. You see that it proves the point that electric cars can come to market vis-à-vis cheap chassis and batteries. Yet batteries are getting cheaper, and chassis can be used from old cars for electric car conversions, and manufacturers can do it for car companies. So Wagoner gets it; we all get it! Just build it!

Anyway, Wagoner spoke to the Commonwealth Club of California on the topics of global warming, ethanol availability, and rising gas prices. Global warming needs a solution that can "significantly shift the fleet of vehicles to different fuels or different propulsion technologies fairly rapidly," he said after the event.[9]

The Other Side of Conversion

What's the downside? It's likely that any conversion vehicle you choose will not be streamlined like a soapbox derby racer. It will be a lot heavier than you'd like and have tires designed for traction rather than low rolling resistance. You do the best you can in these departments depending on your end-use goals: EV dragster, commuter, or highway flyer.

It's equally likely that your conversion vehicle comes with a lot of parts you no longer need: internal combustion engine and mounts and its fuel, exhaust, emission control, ignition, and cooling/heating systems. These you remove and, if possible, sell.

Then you have additional conversion vehicle components that you may wish to change or upgrade for better performance, such as the drive train, wheels/tires, brakes, steering, and battery/low-voltage accessory electrical system. On these just do what makes sense.

How to Get the Best Deal

There are a number of trends that help you to get into your EV conversion at the best possible price, but you still have to do the shopping. Shopping for your EV chassis is no different from buying any vehicle in general. Put your boots on. Grab a good book on buying new or used cars to help you. Just remember not to divulge your true intentions while bargaining so that you can entertain such scenarios as this: When the salesperson says, "Well, to be honest, only two of its four cylinders are working," you say, "No problem. How much will you knock off the price?" Or the ideal situation: "Frankly, that's the cleanest model on the lot, but its engine doesn't work," and you say, "No problem. Let me take it off your hands for $100." Your best deal will be when you find exactly the nonworking "lemon" that someone wants to sell—specifically, a lemon in the engine department. If you find a $5,000 vehicle that doesn't run because of an engine problem (but everything else works okay), you can pick it up for a fraction of its value and save nearly the entire cost of conversion. Seek these deals out; they can save you a substantial amount of money.

Specific vehicle characteristics aside, what's the best vintage vehicle for conversion in terms of cost? As a rule of thumb, used is better than new, but not too used. While older is better in terms of lower cost, you lose a lot of the more recent vehicle safety features if you go too far back. If you go back farther than 10 to 15 years, you begin getting into body deterioration and mechanical high-mileage problem territory. And if you go back to the classics—the 1960s and earlier models—the price starts going up again. But several important trends are working to the EV converter's benefit.

Late-Model Used Vehicles (Late 1990s and Onward)

These late-model vehicles make ideal EV conversions. Not only are they available at lower prices than new vehicles with all the depreciation worked out, but vehicles with 20,000 to 50,000 miles on the odometer are better for EV converters because vehicle drive trains, brakes, and wheels/tires generate less friction (burrs and ridges are worn down, shoes no longer drag, seals are seated, etc.) and roll/turn more easily. However it is always important to update your brakes because most likely the brake housing needs some work![10]

Here are some options to look for:

- **Stripped-down late-model used vehicles**—These are an even better deal. Almost everyone wants the deluxe, V-8, automatic transmission, power-everything model. You, on the other hand, are interested in a straight stick, four-cylinder, no-frills model that nobody wants.

- **Lightweight, early 1980s four-cylinder cars/trucks**—These represent problem sales for used-auto dealers and are frequently discounted just to move them.

- **Older lightweight diesel or rotary cars/trucks**—These also represent problem sales for used-auto dealers because potential buyers are unsure of engine repairability and parts. With current owners, it's more likely they have just gone out of favor. In either case, these represent a buying opportunity for you.

- **Cars/trucks with blown engines or no engines at all**—These are a real problem to move for any owner but a gift for you. It's a marriage made in heaven, and you can usually call the terms. Scan the newspapers for these deals.

Avoid the Real Junk

While late-model nonrunning bargains are great, avoid the problem situations. Avoid buying off-brand, too old, abused, dirty, or rusty chassis. The parts and labor that you have to add to them to bring them up to the level of normal used models makes them no bargain.

Dirt and rust are okay in moderation, but too much of anything is too much. If you can't tell what kind of engine it is because you can't see it in the engine compartment, pass on this choice. You don't want to spend as much time cleaning as you do converting or pay someone else to do it, only to find that essential parts you thought were present under the dirt are either in poor condition or missing.

Rust is rust, and what you can't see is the worst of it. How can you put your best EV foot forward in a rust bucket for a chassis? Don't do it; pass on this choice in favor of a rust-free vehicle or one with minimal rust.

Keep Your Needs List Handy

Regardless of which vehicle you choose for conversion, you want to feel good about your ability to convert it before you leave the lot. If it's too small and/or cramped to fit all the electrical parts—let alone the batteries—you know you have a problem. Or if it's very dirty, greasy, or rusty, you might want to think twice. Here's a short checklist to keep in mind when buying:

- **Weight**—With 120 V and a 22-hp series DC motor, 4,000 to 5,000 pounds is about the upper limit. On the other hand, the same components will give you blistering performance and substantially more range in a 2,000- to 3,000-pound vehicle. Weight is everything in EVs. Decide carefully.

- **Aerodynamic drag**—You can tweak the nose and tail of your vehicle to produce less drag and/or turbulence, but what you see before you buy it is basically what you've got. Choose wisely and aerodynamically.

- **Rolling resistance**—Special EV tires are still expensive, so look for a nice set of used radials, and pump them up hard.

- **Drive train**—You don't want an automatic transmission; a four- or five-speed manual transmission will do nicely, and front-wheel drive typically gives you more room for mounting batteries. Avoid eight- and six-cylinder cars in favor of four-cylinder cars, and choose the smallest, lightest engine–drive train combinations. Avoid heavy-duty anything or four-wheel drive.

- **Electrical system**—Pass on air conditioning, electric windows, and any power accessories.

- **Size**—Will everything you want to put in (i.e., batteries, motor, controller, and charger) have room to fit? How easy will it be to do the wiring?

- **Age and condition**—These determine whether you can get parts for it and how easy it is to restore it to a condition fit to serve as your car.

Buy or Borrow the Manuals

Manuals are invaluable. Whether you get one online or a hard copy through the mail, you must have the manual for your particular car. If possible, seek out the manual to read about any hidden problems before you buy the vehicle. After you own it, don't spend hours finding out if the red- or green-striped wire goes to dashboard terminal block number 3; just flip to the appropriate schematic in the manual and locate it in minutes. Component disassembly is easy when you learn that you always must disengage bolt number 2 in a clockwise direction before turning bolt number 1 in a counterclockwise direction, and so on. Believe me, these are labor savers.

Sell Unused Engine Parts

This is one of the most eco-friendly and green things about the EV conversion I love. I mean somebody somewhere wants that four-cylinder engine you are removing from your EV chassis. This is amazing for their car or for scrap metal.

First, make a few phone calls to place want ads; then call dealers and junk yards. If no cash consideration is offered, see if you can trade the parts for something of value. Do all this early, before you really even start your conversion, because it puts money back into your budget immediately. It's a cash-flow decision I would go with immediately. You do *not* want to be stuck holding these old parts at the middle or end of your conversion.

What Tesla Motors Came Up with from Its Calculations

Equipped with Tesla's advanced electric power train, the Model S provides instant torque and smooth acceleration from 0 to 60 mi/h in fewer than six seconds. The aluminum body is engineered for superior handling, safety, and efficiency. With the most energy-dense battery pack in the industry and best-in-class aerodynamics, the Model S has the longest range of any electric car in the world, up to 300 miles on a single charge.

The components on display demonstrate the superiority and efficiency of Tesla's EV architecture. Customers will be able to see how the flat, lithium-ion battery of the Model S improves rigidity and handling, how the rear drive unit is efficiently designed for maximum power, and how the lack of an internal combustion engine makes the Model S stronger and safer.

FedEx Helps to Launch Zero-Emission Step Vans Built on Similar Chassis—Smith's Newton Platform

If you are the likes of FedEx and want to build your own EV, you don't have to worry about the parts because you buy and build on a glider of new. However, to be really green, the company should use a chassis of old, as seen in Figure 5-14. The first Newton step vans are expected to be deployed in select U.S. markets throughout the remainder of 2012.

Smith Electric Vehicles Corp. (Smith) unveiled the Newton step van shown in Figure 5-14, an all-electric, zero-emission vehicle built on the versatile Newton platform that features a walk-in body produced by Indiana-based Utilimaster. The Newton step van, an ideal solution

FIGURE 5-14   FedEx is not reinventing the wheel—it is using the same chassis.

for thousands of delivery routes in urban environments, offers a gross vehicle weight of 14,000 to 26,000 pounds and a range of approximately 100 miles on a single charge. The vehicle incorporates Smith's proprietary Smith Drive, Smith Power, and Smith Link technologies to provide superior power train performance, battery management efficiency, and remote system monitoring.

"The leadership shown by FedEx in adopting all-electric vehicles has been instrumental to growing the industry," said Bryan Hansel, CEO and chairman of Smith. "We welcome FedEx and look forward to successful vehicle deployments that demonstrate the economic and environmental benefits of fleet electrification."

"This opportunity helps support our goal to optimize and operate our vehicle fleet in an economically and environmentally sustainable manner," said Dennis Beal, vice president of global vehicles at FedEx. "Smith's global footprint and track record of successful EV deployments make it an ideal collaborator as FedEx continues to improve its fleet efficiency, reducing emissions while providing the best possible service to our customers around the world."

The Model S suspension system was developed for the unique architecture of the Model S. It works in harmony with the rigid and light Tesla platform to provide precision handling and optimal comfort. Unencumbered by an engine, the lightweight front suspension optimizes wheel control. The rear multilink suspension is designed to seamlessly integrate with the power train, as shown in Figure 5-15.

Also, you have to love the great ride and handling package; the active air suspension combines automatic advantages with on-demand features. As the Model S accelerates, it lowers the vehicle for optimized aerodynamics and increased range. Nevertheless, the electronic power steering automatically reacts to driving conditions to stay comfortable and responsive at all speeds. The feedback is so precise that you'll feel constantly connected to the road.[11]

Figure 5-15 TurnE Conversions suspension system.

All-Over Aerodynamic Aesthetics

Chevy designers constantly talk about form and function, but when it comes to the Volt, Director of Design Bob Boniface will tell you that the company didn't compromise aesthetics to enable function. The result? A vehicle that looks sporty; is quick, smooth, and modern; yet achieves an extremely low coefficient of drag.

Countless hours in the wind tunnel influenced the styling of many exterior components, including mirrors, a closed grill, rocker panels, and the rear spoiler. In fact, the vertical blades that go from the bottom of the spoiler to the bottom of the bumper were designed specifically to trick the air into separating from the bumper more quickly, giving the Volt extraordinary aerodynamics.[12]

From the top of a new and distinctive power dome to its unique closed rear apron, the all-new BMW ActiveE design of the electric guts to the car in Figure 5-16 boasts the all-new emission-free drive system in the back to create a rear-wheel drive and a lowered center of gravity for the power-driven handling enjoyed in every BMW. Figures 5-16 and 5-17 show the lithium-ion batteries with a liquid cooling system.

FIGURE 5-16   Lithium batteries shown in BMW drawing of ActiveE.

FIGURE 5-17   Lithium-ion batteries with liquid cooling system built into the design of the car.

The thrills, exhilaration and butterflies you will get from the BMW ActiveE are all powered by innovative and high-performing lithium-ion batteries. These high voltage units feature large format storage cells equipped with a liquid cooling system. Developed exclusively for the BMW ActiveE, this active thermal management system is capable of maintaining an ideal operating temperature to ensure consistent performance and longevity, regardless of exterior temperature.

Figure 5-18 shows you the TurnE Conversions VW bug underbelly of the car. See how the bottom has been sanded, stained, and compounded to get out any rust. Also notice how well an existing updated chassis can hold the motor and controller package in the back of the van.

Figure 5-18 Structure of the VW bug refurbished for conversion by TurnE Conversions.

CHAPTER 6

Electric Motors

The superior AC system will replace the entrenched but inferior DC one.
—George Westinghouse (from *Tesla: Man Out of Time*)

lectric motors are everywhere. In your house, almost every mechanical movement
that you see around you is caused by an alternating-current (AC) or direct-
current (DC) electric motor.

A simple electric motor has six parts:

- Armature or rotor
- Commutator
- Brushes
- Axle
- Field magnet
- DC power supply of some sort[1]

By understanding how a motor works, you can learn a lot about magnets, electromagnets, and electricity in general. In this chapter you will learn what makes electric motors tick.

The heart of every electric vehicle (EV) is its electric motor. Electric motors come in all sizes, shapes, and types and are the most efficient mechanical devices on the planet. Unlike an internal combustion engine, an electric motor emits zero pollutants. Technically, there are three moving parts in an electric motor. Even with three parts, electric motors outlive internal combustion engines every day of the week.

The parts are the rotor and two end bearings. This is just one of the main reasons for the widespread adoption of EVs and the movement in general to get people to build their own EVs. Whether you get someone to build your EV or you build it yourself, the product, your EV, is a planet-saving proposition.

What this means to you is that ownership of a fun-to-drive, high-performance EV will deliver years of low-maintenance driving at minimum cost, mostly because of the inherent characteristics of its electric motor—power and economy. The objective of this chapter is to guide you toward the best candidate motor for your EV conversion today and suggest the best electric motor type for your future EV conversion or build.

To accomplish these goals, this chapter will review electric motor basics and provide you with useful equations, introduce you to the different types of electric motors and their advantages and disadvantages for EVs, introduce the best electric motor for your

EV conversion or build today and its characteristics, and show you the electric motor type that you should closely follow and investigate for future EV conversions or builds.

Why an Electric Motor?

In Chapter 1 you learned that the electric motor is ubiquitous because of its simplicity. All electric motors, by definition, have a fixed *stator*, or stationary part, and a *rotor*, or movable part. This simplicity is the secret of their dependability and why, in direct contrast to the internal combustion engine with its hundreds of moving parts, electric motors are a far superior source of propulsion.

- **Electric motors are inherently powerful**—Nearly all traction motors deliver near-peak torque at 0 rpm. This is why electric traction motors have powered our subways and diesel-electric railroad locomotives for so many years. There is no waiting, as with an internal combustion engine, while it winds up to its peak-torque rpm range. Apply electric current to it, and you've instantly got torque to spare. If any EV's performance is wimpy, it's due to poor design or electric motor selection—not the electric motor itself.

- **Electric motors are inherently efficient**—You can expect to get 90 percent or more of the electrical energy you put into an electric motor out of it in the form of mechanical torque. Few other mechanical devices even come close to this efficiency.

An electric motor is all about magnets and magnetism. A motor uses magnets to create motion. If you have ever played with magnets, you know about the fundamental law of all magnets: Opposites attract and likes repel. Thus, if you have two bar magnets with their ends marked "north" and "south," then the north end of one magnet will attract the south end of the other. On the other hand, the north end of one magnet will repel the north end of the other (and similarly, south will repel south).

Horsepower

Since electric motors are efficient, the horsepower behind them in a real EV can be shocking to the system initially (no pun intended). I remember the first time that I drove an electric car. When I stepped on that accelerator, it took off! No questions asked. No engine with excessive parts to get in the way of that.

Here are some technical points to understand when trying to find the right motor for your car:

1. Electric motors are rated at their point of maximum efficiency; they may be capable of two to four times their continuous rating, but only for a few minutes (acceleration or hill climbing). Internal combustion engines are rated at their peak horsepower. For example, the FB1-4001A motor is rated as 30 hp continuous at 144 V and 100 hp peak. The five-minute rating of the FB1-4001A motor is 48 hp at 144 V.

2. Each 1,000 pounds of vehicle weight after conversion requires 6 to 8 hp. This is the continuous rating of the motor. Thus a 3,000-pound conversion requires a motor that is rated at approximately 20 hp. More horsepower is required for higher speeds, heavier vehicles, and steeper terrains.

3. The available horsepower of a motor increases with voltage; for example, the FB1-4001A motor is rated at 18 hp continuous at 72 V but at 30 hp continuous at 144 V. As the voltage is increased, the rpms increase. Horsepower is a function of rpms × torque.

4. Although electric motors are rated as "continuous," the motor can run at less horsepower. If only 10 hp is required for the desired speed, then the motor runs at that reduced load. This is the function of the motor controller.

5. Operating continuously above the rated horsepower eventually will overheat and damage the motor. A motor that is rated at 150 A can run at 300 A for a short time (minutes), but longer periods can easily damage the motor. Do not buy an undersized motor for your EV—it will not last long. Current is what overheats components.

6. Highway speeds require greater horsepower. The horsepower required at 70 mi/h is four times the horsepower required at 35 mi/h. This means that the current required is also four times more, which means less range.

There are two types of motors: AC motors and DC motors.

DC Electric Motors

An electric motor is a mechanical device that converts electrical energy into motion and that can be further adapted to do useful work such as pulling, pushing, lifting, stirring, or oscillating. It is an ideal application of the fundamental properties of magnetism and electricity. Before looking at DC motors and their properties, let's review some fundamentals.

In Figure 6-1, there are two magnets in the WARP motor by NetGain Technologies. The armature (or rotor) is an electromagnet, whereas the field magnet is a permanent

FIGURE 6-1 Series wound DC electric motor by NetGain Technologies called the TransWarp 11 motor (11.45 inches).

magnet (the field magnet could be an electromagnet as well, but in most small motors it isn't in order to save power).

Magnetism and Electricity

Magnetism and electricity are opposite sides of the same coin. Electrical and electronics design engineers regularly use Maxwell's four laws of electromagnetism based on Faraday's and Ampere's earlier discoveries in their daily work. They might tell you, "Magnetism and electricity are inextricably intertwined in nature." In fact, you don't have one without the other. Usually, however, you only look at one or the other unless you are discussing electric motors or other devices that involve both.

Voltage is really called an *electromotive force*. You hook up a light bulb to a battery by completing the wire connections from the battery's positive and negative terminals to the light bulb, and the bulb lights. Hook up two batteries in series to double the voltage, and the bulb shines even brighter.

To tie things into the electrical realm, there is a mathematical equation that relates these parameters of force, flow, and resistance. The electrical equation, commonly known as *Ohm's law*, is

$$V = IR$$

where V is voltage in volts, I is current in amps, and R is resistance in ohms. When you double the voltage, you send twice as much current through the wire, and the light bulb becomes brighter. Alternatively, if you reduce the resistance (as, say, is done when you enlarge the hole in a bucket of water), given the same level of water, you double the flow (increase the current).

The simple bar magnet, which you probably encountered in your school science class, has two ends or poles—north and south. Either end attracts magnetizable objects to it. When two bar magnets are used together, opposite poles (north-south) attract each other, and identical poles (north-north or south-south) repel each other.

The needle of a compass is a magnetized object with its own north and south poles. Lightweight and delicately balanced, it aligns itself with the earth's magnetic field and tells you which direction is north. If you bring a bar magnet near it, it will rotate away from the earth's magnetic north pole.

You can create a magnetic field with electricity. Take an iron nail from your toolbox, wrap a few turns of insulated copper wire around it, and hook the ends up to a battery. The plain nail is transformed into a bar-style magnet that can behave just like a compass.

Conductors and Magnetic Fields

If you had a horseshoe-shaped magnet with the ends close together and you moved a wire between the poles, as shown in Figure 6-2. This relationship holds true whether the field is stationary and the wire is moving or vice versa. The faster you cut the lines of flux, the greater is the voltage, but you must do so at right angles. If the circuit is closed, the induced voltage will cause a current to flow. A handy way to remember the relationships is the *right-hand rule:* The thumb of your right hand points upward in the direction of the motion of the conductor, your index finger extends at right angles to it in the direction of the flux (from the north to the south pole), and your third finger extends in a direction at right angles to the other two, indicating the polarity of the induced voltage or the direction in which the current will flow, as shown in Figure 6-2.

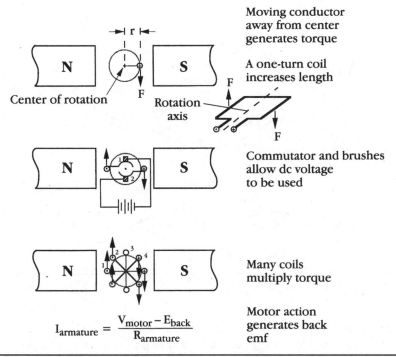

Moving conductor
away from center
generates torque

A one-turn coil
increases length

Commutator and brushes
allow dc voltage
to be used

Many coils
multiply torque

Motor action
generates back
emf

$$I_{armature} = \frac{V_{motor} - E_{back}}{R_{armature}}$$

FIGURE 6-2   DC motor basics—obtaining torque from moving conductors.

Ampere's Law or the Motor Rule

Current flowing through a conductor at a right angle to a magnetic field produces a force on the conductor. The right-hand rule is again a handy way to remember the relationships, but this time the thumb of your right hand points toward you in the direction of the current through the conductor, your extended index finger at right angles to it points in the direction of the flux (from the north to the south pole), and your extended third finger points downward in a direction at right angles to the other two, indicating the direction of the generated force, as shown in Figure 6-2. Now you're ready to talk seriously about motors.

Electromagnets and Motors

An electromagnet is the basis of an electric motor. You can understand how things work in an electric motor by imagining the following scenario: Say that you created a simple electromagnet by wrapping 100 loops of wire around a nail and connecting it to a battery. The nail would become a magnet and have a north and south pole while the battery is connected. Now say that you take your nail electromagnet, run an axle through the middle of it, and suspend it in the middle of a horseshoe magnet, as shown in Figure 6-3. If you were to attach a battery to the electromagnet so that the north end of the nail appeared as shown in the figure, the basic law of magnetism tells you what

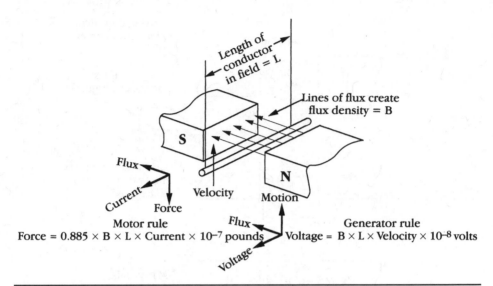

Length of conductor in field = L

Lines of flux create flux density = B

S

N

Flux

Current

Force

Velocity

Motion

Flux

Voltage

Motor rule

Force = 0.885 × B × L × Current × 10⁻⁷ pounds

Generator rule

Voltage = B × L × Velocity × 10⁻⁸ volts

FIGURE 6-3 DC motor basics—the motor and generator rules.

would happen: The north end of the electromagnet would be repelled from the north end of the horseshoe magnet and attracted to the south end of the horseshoe magnet. The south end of the electromagnet would be repelled in a similar way. The nail would move about half a turn and then stop in the position shown in the figure.

You can see that this half-turn of motion is simply due to the way magnets naturally attract and repel one another. The key to an electric motor is to then go one step further so that, at the moment that this half-turn of motion completes, the field of the electromagnet flips. The flip causes the electromagnet to complete another half-turn of motion. You flip the magnetic field just by changing the direction of the electrons flowing in the wire (you do that by flipping the battery over). If the field of the electromagnet were flipped at precisely the right moment at the end of each half-turn of motion, the electric motor would spin freely.

DC Motors in General

The current through the conductor would exert a force that would tend to rotate it in a clockwise direction. To further assist rotation, you add the commutator and brushes, shown in the middle of Figure 6-3. This arrangement allows you to power your motor from a constant supply of DC voltage. To further increase the motor's torque abilities, you add additional coils as shown at the bottom of Figure 6-2. In reality, each coil can have many windings, and you can arrange commutator segments to match the number of coils so that you have the force on each of these coils acting in unison with the force on all the other coils.

DC Motors in the Real World

Now it's time you met real-world DC motors—their construction, definitions, and efficiency. Let's start by looking at their components.

Armature

The armature is the main current-carrying part of a motor that normally rotates (brushless motors tend to blur this distinction) and produces torque via the action of current flow in its coils. It also holds the coils in place and provides a low-reluctance path to the flux. The armature usually consists of a shaft surrounded by laminated sheet steel pieces called the *armature core*. There are grooves or slots parallel to the shaft around the outside of the core; the sides of the coils are placed into these slots. The coils (each with many turns of wire) are placed so that one side is under the north pole and the other side is under the south pole; adjacent coils are placed in adjacent slots, as shown at the bottom of Figure 6-3. The end of one coil is connected to the beginning of the next coil so that the total force then becomes the sum of the forces generated on each coil.

Commutator

The commutator is the smart part of the motor that permits constant rotation by reversing the direction of current in the windings each time they reach the minimum flux point. This piece is basically a switch. It commutates the voltage from one polarity to the opposite. Since the motor's rotor is spinning and has momentum, the switching process repeats itself in a preordained manner. The alternating magnetic poles continue to provide the push to overcome losses (i.e., friction, windage, and heating) to reach a terminal speed. Under load, the motor behaves a bit differently, but the load causes more current to be drawn.

Field Poles

In the real world, electromagnets (recall your toolbox nail with a few turns of insulated copper wire wrapped around it) are customarily used instead of the permanent magnets you saw in Figures 6-2 and 6-3. (Permanent-magnet motors are, in fact, used, and you'll be formally introduced to them and their advantages later in this section.)

In a real motor, the lines of flux are produced by an electromagnet created by winding turns of wire around its poles or pole pieces. A pole is normally built up of laminated sheet-steel pieces, which reduce eddy current losses; as with armatures, steel has been replaced by more efficient metals in newer motors. The pole pieces are usually curved where they surround the armature to produce a more uniform magnetic field. The turns of copper wire around the poles are called the *field windings*.

Series Motors

How these windings are made and connected determines the motor type. A coil made up of a few turns of heavy wire connected in series with the armature is called a *series motor*. A coil of many turns of fine wire connected in parallel with the armature is called a *shunt motor*.

Brushes

Typically consisting of rectangular-shaped pieces of graphite or carbon, brushes are held in place by springs whose tension can be adjusted. The brush holder is an insulated material that electrically isolates the brush itself from the motor frame. A small, flexible copper wire embedded in the brush (called a *pigtail*) provides current to the brush. Smaller brushes can be connected together internally to support greater current flows.

Motor Case, Frame, or Yoke

In the motor's case, the magnetic path goes from the north pole through the air gap, the magnetic material of the armature, and the second air gap, to the south pole and back to the north pole again via the case, frame, or yoke. Motors operating in the real world are subject to electric power losses to the electric motor from three sources:

- **Mechanical**—All torque available inside the motor is not available outside because torque is consumed in overcoming the friction of the bearings, moving air inside the motor (known as *windage*), and brush drag.
- **Electrical**—Power is consumed as current flows through the combined resistance of the armature, field windings, and brushes.
- **Magnetic**—Additional losses are caused by the armature and the field pole cores.

In summary:

- Efficiency is simply the power out of a device relative to the power applied to the device.
- When you apply 100 W of power and get only the work equivalent to 90 W out of the motor, you have a 10-W loss. Such a device is 90 percent efficient.
- This rule of efficiency applies to motors, motor speed controllers, battery chargers, and so on.

Which Electric Motors Should I Use?

The answer to this question depends on *many* factors. The first question you should ask is: What is the intended purpose of the vehicle? Will it be used as a daily driver? Will it be used strictly for racing? Will it be a performance vehicle, or will it be designed for the greatest range between charges? In addition to knowing the answers to these questions, you should have some realistic thoughts relating to:

1. Top speed to be maintained: _____
2. Percent grade the vehicle will travel on: _____
3. Wind resistance (frontal area) of the vehicle: _____
4. Total vehicle weight (with driver/passengers/load): _____
5. Final gear ratio: _____
6. Tire diameter: _____

7. Voltage to be supplied to the motor: _____

8. Coefficient of drag: _____

9. Battery internal resistance: _____ [2]

DC Motor Types

Now that you've been introduced to DC motors in theory and in the real world, it's time to compare the different motor types (Figure 6-4). DC motors appear in the following forms:

- Series
- Shunt
- Compound
- Permanent magnet
- Brushless
- Universal

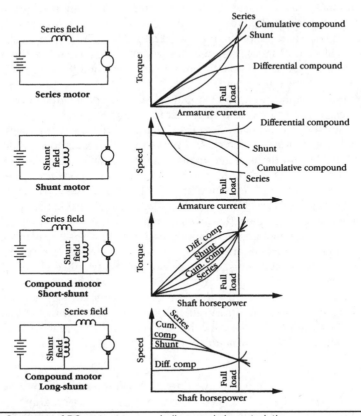

FIGURE 6-4 Summary of DC motor types—windings and characteristics.

The last three motor types are just variations of the first three, fabricated with different construction techniques. The compound motor is a combination of the series and shunt motors. For the first three motor types, we'll look at the circuit showing how the motor circuit field windings are connected and then look at the characteristics of the torque and speed versus armature current and shaft horsepower curves that describe their operation.

Each of the motor types will be examined for its torque, speed, reversal, and regenerative braking capabilities—the factors important to EV users. The motor types will all be compared at full-load shaft horsepower—the only way to compare different motor types of equal rating. Efficiency is a little harder to determine because it also depends on the external resistance of the circuit to which the motor is connected. Efficiency therefore has to be calculated for each individual case.

Series DC Motors

The most well-known DC motors, and the one that comes to mind for traction applications (such as propelling an EV), is the series-wound DC motor by NetGain Technologies. This motor has an 11.45-inch diameter. It is a series-wound DC motor with a double-ended shaft. The TransWarP 11 motor has been designed to ease the conversion process for people who want direct-drive applications. It has many unique features that set it apart from any other EV motor currently manufactured.

This motor has a "shorty" tail-shaft housing from a Chevrolet Turbo 400 transmission fitted to the drive end bell (may be ordered with or without housing). This is so because the drive end shaft is not the typical 1.125-inch single-keyed type, but rather a hefty 1.370-inch, 32-tooth involute spline that is identical to the tail-shaft spline of a Turbo 400 transmission. In other words, this motor was designed to replace a transmission and couple directly to a drive shaft.

But the manufacturer didn't stop there. The company added double-wide bearings on the drive end and grease fittings (because you now have a slip-yoke assembly). The company chose the industry standard 1350 universal, so you can easily adapt the motor to any manufacturer's drive shaft. I moved the terminals to the side of the motor to avoid road hazard damage and to allow more clearance above and below. The motor also has brush wear indicators and a temperature snap switch. It has the same high-efficiency fan and massive commutator and brushes as the WarP 11 motor. I even made the commutator end shaft the same diameter as the drive end of a typical WarP 11—just in case I wanted to connect two 11-inch motors together.

If that wasn't enough, there is a lifting ring and eight threaded mounting holes in the commutator end bell. In this way, you will find it easier to mount, as well as having a method of easily attaching a speed sensor, alternator, air conditioner, and so on.

The motor is so named because its field winding is connected in series with the armature (Figure 6-4). Because the same current must flow through this field winding as through the motor armature itself, it is wound with a few turns of heavy-gauge wire. In a series motor, torque varies with the square of the current—a fact substantiated by the actual torque versus armature current graph in Figure 6-4.[3]

Speed

Speed becomes very large as the current becomes small in a series motor—a fact substantiated by the actual speed versus armature current graph in Figure 6-4. For example, when you have high rpms at no load, it is the series motor's Achilles heel. You need to make sure that you are always in gear, have the clutch in, have a load attached, and so on because the series motor's tendency is to run away at no load. Just be aware of this and back off immediately if you hear a series motor rev up too quickly.

Field Weakening

This technique is an interesting way to control series motor speed. You place an external resistor in parallel with the series motor field winding, in effect diverting part of its current through the resistor. Keep it to 50 percent or less of the total current (resistor values equal to or greater than 1.5 times the motor's field resistance). The by-product is a speed increase of 20 to 25 percent at moderate torques without any instability in operation. Used in moderation, it's like getting something for nothing. Field weakening had its limitations, which drivers soon didn't want to contend with, given the other better options available to them.

Reversing

The same current that flows through the armature flows through the field in a series motor, so reversing the applied voltage polarity does not reverse the motor direction. To reverse motor direction, you have to reverse or transpose the direction of the field winding with respect to the armature. This characteristic also makes it possible to run series DC motors from AC (more on this later in this section).

Regenerative Braking

All motors simultaneously exhibit generator action—motors generate counter-electromotive force (EMF)—as you read earlier in this section. The reverse also holds true—generators produce countertorque. Regenerative braking allows you to slow down the speed of your EV (and save its brakes) and put energy back into its battery (thereby extending its drivable range) by harnessing its motor to work as a generator after it is up and running at speed.

Regenerative braking allows you to electronically switch the motor and turn it into a generator, thereby capturing the energy that normally would be dissipated (read lost) as heat in the brake pads while slowing down. The motor does the braking, not your brake pads.

While all motors can be used as generators, the series motor has rarely been used as a generator in practice because of its unique and relatively unstable generator properties.

Shunt Motors

Put simply, the difference in the two basic motors is that the shunt motor is capable of maintaining a constant speed throughout various current and voltage loads. The series motor, however, will drop its rounds per minute when the initial current load decreases. A third type of motor—the compound motor—includes both parallel and series windings and is used to produce heavier loads (as with the series motor) and also to regulate the resulting rounds per minute (as with the shunt motor).

Shunt Motor Troubleshooting

When a shunt motor begins to malfunction, the motor will become excessively hot. The problem is normally easy to troubleshoot. Two main malfunctions to keep in mind are the loss of a voltage supply and/or a rupture in the armature winding; these are the first things to check if a problem occurs with a shunt motor.

Shunt Motor Applications

The industrial and automotive fields use shunt motors extensively because specific levels of torque and speed are mandatory.[4]

Torque Characteristics

Because the shunt field is connected directly across the voltage source, the flux in the shunt motor remains relatively constant. Its torque is directly dependent on the armature current. In a shunt motor, torque varies directly with current. Although there is initially no counter-EMF to impede the flow of startup current in the armature of a shunt motor, the shunt motor's linear relationship is quickly established. As a result, the shunt motor does not produce nearly as much startup torque as the series motor. This translates to reduced acceleration performance for owners of shunt motor–powered EVs.

Speed

When a load is applied to any (but here specifically a shunt) motor, it will tend to slow down and, in turn, reduce the counter-EMF produced. The increase in armature current results in an increase in torque to take care of the added load. Basically, constant, counter-EMF tends to remain constant over a wide range of armature current values and produces a fixed speed curve that droops only slightly at high armature currents.

The shunt motor's linear torque and fixed speed versus armature current characteristics have two undesirable side effects for traction applications when controlled manually. First, when a heavy load (e.g., hill climbing, extended acceleration) is applied, a shunt motor does not slow down appreciably, as does a series motor, and the excessive current drawn through its armature by the continuous high torque requirement makes it more susceptible to damage by overheating. Second, in contrast to the "knee" bend in the series motor torque-speed equations, shunt motor torque-speed curves are nearly straight lines. This means that more speed control or shifting is necessary to achieve any given operating point. Series motors therefore are used where there is a wide variation in both torque and speed and/or heavy starting loads. Other than having much lower startup torque, shunt motors can perform as well as series motors in EVs when electronically controlled. The downside is that a shunt motor controller can be more complicated to design than a series motor controller.

Field Weakening

You can also achieve a higher-than-rated shunt motor speed by reducing the shunt coil current—in this case, you place an external control resistance in series with the shunt motor field winding. But here, unlike the series motor's runaway rpm region at no load, you are playing with fire because the loaded shunt motor armature has an inertia that does not permit it to respond instantly to field control changes. If you do this while your shunt motor is accelerating, you might cook your motor or have motor parts all over the highway by the time you adjust the resistance back down to where you started. Be careful with field weakening in shunt motors. Again, these motors soon lost favor owing to the inherent simplicity of the series DC motor arrangement. Heating was less of an issue and certainly not something that drivers wanted to have to worry about; the series motor was better suited to the application.

Reversing

Current flows through the field in a shunt motor in the same direction as it flows through the armature, so reversing the applied voltage polarity reverses both the direction of the current in the armature and the direction of the field-generated flux and does not reverse the motor rotation direction. To reverse motor direction, you have to reverse or transpose the direction of the shunt field winding with respect to the armature.

Regenerative Braking

A shunt motor is instantly adaptable as a shunt generator. Most generators are in fact shunt wound, or variations on this theme. The linear or nearly linear torque and speed versus current characteristics of the shunt motor manifest as nearly linear voltage versus current characteristics when used as a generator. This also translates to a high degree of stability that makes a shunt motor both useful for and adaptable to regenerative braking applications—either manually or electronically controlled.

Compound DC Motors

A compound DC motor is a combination of the series and shunt DC motors. Differentially connected compound motors are sometimes referred to as "suicide motors" because of their penchant for self-destruction. If perhaps the shunt field circuit were to suddenly open during loading, the series field then would assume control, and the polarity of all fields would reverse. This results in the motor stopping and then restarting in the opposite direction. It then operates as an unloaded series motor and will destroy itself. Differentially connected motors also can start in the opposite direction if the load is too heavy. Therefore, they are seldom used in industry.[5]

The way its windings are connected, and whether they are connected to boost (assist) or buck (oppose) one another in action, determines its type. Its basic characterization comes from whether current flowing into the motor first encounters a series field coil–short-shunt compound motor or a parallel-shunt field coil–long-shunt compound motor. If, in either one of these configurations, the coil windings are hooked up to oppose one another in action, you have a differential compound motor. If the coil windings are hooked up to assist one another in action, you have a cumulative compound motor. The beauty of the compound motor is its ability to bring the best of both the series and the shunt DC motors to users.

- **Torque**—The torque in a compound motor has to reflect the actions of both the series and shunt field coils.

- **Speed**—Similar to torque, the speed action in a compound motor also will reflect the dual variables of series and shunt field coil action. One of the initial benefits of the compound configuration is that runaway conditions at low field current levels for the shunt motor and at lightly loaded levels for the series motor can be eliminated. While the differential compound configuration is of questionable value—its curve shows a tendency to runaway speeds at high armature current values—the cumulative compound motor appears to offer benefits to EV operation. You can tailor a cumulative compound motor to your EV needs by picking one whose series winding delivers good starting torque and whose shunt winding delivers lower current draw and regenerative braking capabilities once up to speed. When you look, you might find that these characteristics already exist in an off-the-shelf model.

- **Reversing**—A short-shunt compound motor resembles a series motor, so reversing its applied voltage polarity normally does not reverse the motor direction. A long-shunt compound motor resembles a shunt motor, so reversing the voltage supply normally reverses the motor. As with speed and torque, compound motors can be tailored to do whatever you want to do in the reversing department.

- **Regenerative braking**—A compound motor is as easily adaptable as a shunt motor to regenerative braking. Its series winding gives it additional starting torque but can be bypassed during regenerative braking, and its shunt winding provides more desirable shunt generator characteristic. Controlled manually or with solid-state electronics, compound motors are usable for and adaptable to regenerative braking applications.

Permanent-Magnet DC Motors

When you were first introduced to DC motors, permanent magnets were used as an example because of their simplicity. Permanent-magnet motors are, in fact, being used increasingly today because new technology—various alloys of Alnico magnet material, ferrite-ceramic magnets, rare-earth-element magnets, etc.—enables them to be made smaller and lighter in weight than equivalent wound field coil motors of the same horsepower rating. Rare-earth-element magnets surpass the strength of Alnico magnets significantly (by 10 to 20 times) and have been used with great success in other areas, such as computer disc drives, thereby helping drive down production costs (Figure 6-5). While a commutator and brushes are still required, you save the complexity and expense of fabricating a field winding and gain in efficiency because no current is needed for the field.

Permanent-magnet motors approximately resemble shunt motors in their torque, speed, reversing, and regenerative braking characteristics; either motor type usually

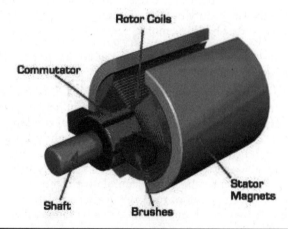

FIGURE 6-5   Permanent-magnet DC motor. (*Zero-Emission Vehicles of Australia and www.electric-cars-are-for-girls.com.*)

can be substituted for the other in control circuit designs. Permanent-magnet motors have starting torques several times that of shunt motors, and their speed versus load characteristics are more linear and easier to predict.

Brushless DC Motors

With no brushes to replace or commutator parts to maintain, brushless motors promise to be the most long-lived and maintenance-free of all motors. You can now custom tailor the motor's characteristics with electronics (because electronics now represent half the motor), and the distinctions among DC motor types blur. In fact, as seen in Figure 6-6, the brushless motor more closely resembles an AC motor (which you'll meet in the next section) in construction. Assume that brushless DC motors resemble their permanent-magnet DC motor cousins in characteristics—shunt motor plus high starting torque plus linear speed/torque—with the added kicker of even higher efficiency owing to no commutator or brushes.

There are other manufacturers of DC motors. They are UQM and AVEOX. For more information, please refer to the online resource directory: buildyourownelectricvehicle .wordpress.com.

Universal DC Motors

Universal motors generally run at high speeds, making them useful for appliances such as blenders, vacuum cleaners, and hair dryers, where high-rpm operation is desirable. They are also used commonly in portable power tools. A series DC motor is usually chosen as the starting point for universal motors that are to be run on either DC or AC. DC motors designed to run on AC typically have an improved lamination field and armature cores to minimize hysteresis and current losses (Figure 6-7). Additional compensating or interpole windings can be added to the armature to further reduce commutation problems by reducing the flux at commutator segment transitions. In general, series DC motors operating on AC perform almost the same (high starting torque, etc.) but are less efficient at any given voltage point.

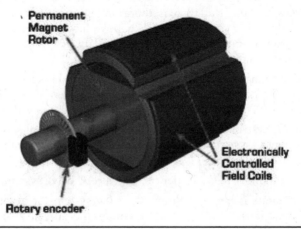

FIGURE 6-6 Brushless DC motor. (*Zero Emission Vehicles of Australia and www.electric-cars-are -for-girls.com.*)

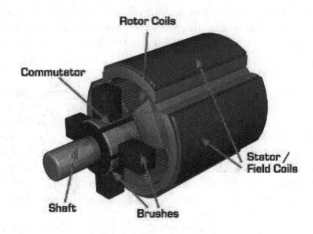

Rotor Coils

Commutator

Stator /
Field Coils

Shaft   Brushes

FIGURE 6-7   DC motor with a round stator. (*Zero Emission Vehicles of Australia and www.electric
-cars-are-for-girls.com.*)

AC Electric Motors

Now that you've met DC electric motors, it's time to meet the motor you encounter most often in your everyday life—the AC electric motor. The great majority of our homes, offices, and factories are fed by AC. Because it can easily be transformed from high voltage for transmission into low voltage for use, more AC motors are in use than all the other motor types put together. *AC motor* doesn't mean that it runs off the AC power at the house, plugged it into a wall socket with a very long extension cord. No. It uses batteries as the DC motor does. An inverter takes the place of the controller you see in DC drive systems. The system overall is a bit more expensive primarily because of these sophisticated inverters—but you get your money's worth.

- An AC motor just keeps on accelerating—the torque curve is more like that of a gas-guzzler than of a golf cart. An EV with a DC motor will reach the top of its Rpms, and that's all there is, there ain't no more. For example, 55 mi/h may be the top end of your DC motor's happy place. You can cruise at 55 mi/h all day, no problem, but don't expect to reach 60 mi/h so that you can get out of the way of that Hummer. It ain't gonna happen.

- Because the terrain around my house is hills, hills and then a couple more hills after that, as I said earlier, an AC drive system does a little better job in this terrain, all other things being equal.

- An AC system will treat your poor gas-guzzler's transmission a lot more gently than a DC system. If you're not drag racing, you probably don't have to worry about this one.

- You can move a lot heavier car with an AC drive system.

- The range is better with an AC drive system for a couple of reasons—first, because it uses lead-acid batteries more efficiently and, second, because of better regenerative braking, which acts just like a generator to put energy back into your batteries. If you're using lithium, AC/DC range difference is less of an issue.

All these things add up to an EV whose performance is considerably more like the familiar gas-guzzler's performance.[6]

Metric Mind's FAQ has a lot more to say about the advantages of AC drive systems over the typical series-wound DC electric motors and controllers you often see in EV conversions. Before looking at AC motors and their properties, let's look at transformers.

Transformers

In its simplest and most familiar form, a transformer consists of two copper coils wound on a ferromagnetic core (Figure 6-8). The primary is normally connected to a source of alternating electric current. The secondary is normally connected to the load. The other

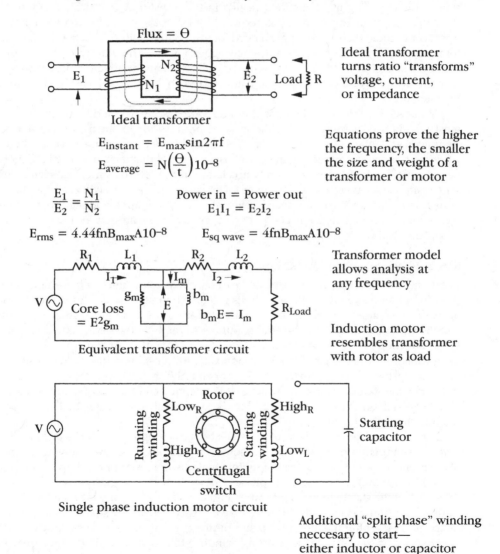

$$E_{instant} = E_{max}\sin 2\pi f$$

$$E_{average} = N\left(\frac{\Theta}{t}\right)10^{-8}$$

$$\frac{E_1}{E_2} = \frac{N_1}{N_2}$$

Power in = Power out

$$E_1 I_1 = E_2 I_2$$

$$E_{rms} = 4.44 f n B_{max} A 10^{-8} \qquad E_{sq\ wave} = 4 f n B_{max} A 10^{-8}$$

Ideal transformer turns ratio "transforms" voltage, current, or impedance

Equations prove the higher the frequency, the smaller the size and weight of a transformer or motor

Transformer model allows analysis at any frequency

Induction motor resembles transformer with rotor as load

Starting capacitor

Additional "split phase" winding neccesary to start— either inductor or capacitor

FIGURE 6-8 AC motors resemble transformers.

aspect of transformers that is useful to you (highlighted in Figure 6-8) is that an equivalent circuit of a transformer can be drawn for any frequency, and you can study what is going on. This is useful and directly applicable to AC induction motors.

AC Induction Motors

The AC induction motor, patented by Nikola Tesla back in 1888, is basically a rotating transformer. Think of it as a transformer whose secondary load has been replaced by a rotating part. In simplest form, this rotating part (rotor) only requires its conductors to be rigidly held in place by some conducting end plates attached to the motor's shaft.

When a changing current is applied to the primary coil (the stationary part or stator), the changing magnetic field results in the transfer of electrical energy to the rotor via induction. Since energy is received by the rotor via induction without any direct connection, there is no longer a need for any commutator or brushes. Because the rotor itself is simple to make yet extremely rugged in construction (typically a copper bar or conductor embedded in an iron frame), induction motors are far more economical than their equally rated DC motor counterparts in both initial cost and ongoing maintenance.

While AC motors come in all shapes and varieties, the AC induction motor—the most widely used variety—holds the greatest promise for EV owners because of its significant advantages over DC motors. These solid-state components have resulted in AC induction motors appearing in variable-speed drives that meet or exceed DC motor performance—a trend that will surely accelerate in the future as more efficient solid-state components are introduced at ever-lower costs. This section will examine the AC induction motor for its torque, speed, reversal, and regenerative braking capabilities—the factors important to EV users. Other AC motor types are not as suitable for EV propulsion and will not be covered.

Single-Phase AC Induction Motors

Recall the universal DC motor discussed earlier. When you connect it to a single-phase AC source, you have little difference in its motor action because changing the polarity of the line voltage reverses both the current in the armature and the direction of the flux, and the motor starts up normally and continues to rotate in the same direction. Not so in an induction motor driven from a single-phase AC source. At startup, you have no net torque (or more correctly, balanced opposing torques) operating on its motionless rotor conductors. Once you manually twist or spin the shaft, however, the rotating flux created by the stator currents now cuts past the moving conductors of the rotor, creates a voltage in them via induction, and builds up current in the rotor that follows the rotating flux of the stator.

How do you overcome the problem? The bottom of Figure 6-8 shows one key. This *split-phase* induction motor design—or some variation of it—is the one you are most likely to encounter on typical smaller motors that power fans, pumps, shop motors, and so on. To maximize the electrical phase difference between the two windings, the resistance of the starting winding is much higher and its induction is much lower than those of the running winding. To minimize excessive power dissipation and possible temperature rise after the motor is up and running, a shaft-mounted centrifugal switch is connected in series with the starting winding that opens at about three-quarters synchronous speed.

Polyphase or Split-Phase AC Induction Motors

Polyphase means "more than one phase." AC is the prevailing mode of electrical distribution. Single-phase 208 V to neutral from a three-phase transformer on the pole is the most prevalent form found in your home and office. These are widely available in nearly every city in the industrialized world.

Stationary three-phase electric induction motors are inherently self-starting and highly efficient, and electricity is conveniently available. If an induction motor is started at no load, it quickly comes up to a speed that might only be a fraction of 1 percent less than its synchronous speed. When a load is applied, speed decreases, thereby increasing slip; an increased torque is generated to meet the load up to the area of full-load torque and far beyond it up to the maximum torque point (a maximum torque of 350 percent–rated torque is typical).

Speed and torque are relatively easy to handle and determine in an induction motor. Metric Mind publishes AC Induction Motor Specifications, shown in Table 6-1. It seems that Metric Mind is using a vendor to get this AC induction motor out of Switzerland. Anyone have a plane ticket for me?

So are reversing and regenerative braking relatively easy to handle and determine in an induction motor. If you reverse the phase sequence of the stator supply (i.e., reverse one of the windings), the rotating magnetic field of the stator is reversed, and the motor develops negative torque and goes into generator action, quickly bringing the motor to a stop and reversing direction (Figure 6-9).

Regenerative braking action—pumping power back into the source—is readily accomplished with induction motors. How much regenerative braking you apply creates braking (moves the steady-state induction motor operating point down the torque-slip curve) and generates power (instantaneously runs with negative slip in the generator region, supplying power back to the source).

Technical Data		Motor Type								
		200-75W	200-100W	200-125W	200-150W	200-175W	200-200W	200-225W	200-250W	200-330W
Nominal power (S1)	kW	9	12	15	18	21	24	27	30	40
Nominal torque	Nm	30	40	50	60	70	80	90	100	130
Dimension "A"	mm	216	241	266	291	216	341	366	391	471
Dimension "B"	mm	18	18	18	18	18	20	20	20	20
Weight	kg	34	38	41.5	49	54.7	60.5	66.7	73	79.5
Nominal speed	rpm	2,850	2,850	2,850	2,850	2,850	2,850	2,850	2,850	2,850
Max. speed	rpm	9,000	9,000	9,000	9,000	9,000	9,000	9,000	9,000	9,000
Cooling		water	water	water	water	water	water	water	water	water
Reduction and differential gearbox		avail.	avail.	avail.	avail.	*	*	*	*	*
Bracket no. of holes		3	3	3	3	4	4	4	4	4

TABLE 6-1 Metric Mind Conversion AC Induction Motors

* Development in progress.

Source: www.metricmind.com/wp-content/uploads/2011/09/200_series_motors1.jpg.

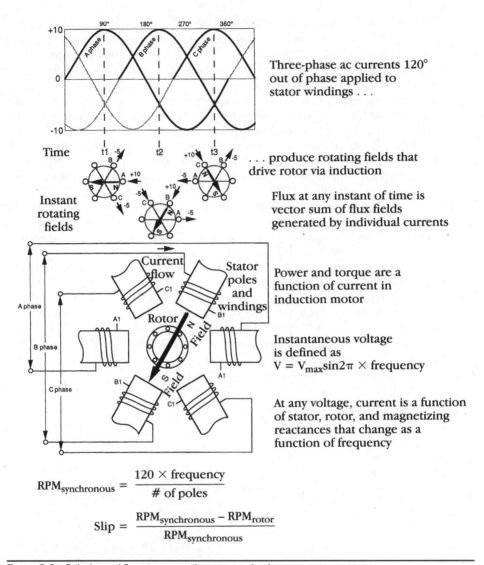

Three-phase ac currents 120°
out of phase applied to
stator windings . . .

. . . produce rotating fields that
drive rotor via induction

Flux at any instant of time is
vector sum of flux fields
generated by individual currents

Power and torque are a
function of current in
induction motor

Instantaneous voltage
is defined as
$V = V_{max}\sin2\pi \times frequency$

At any voltage, current is a function
of stator, rotor, and magnetizing
reactances that change as a
function of frequency

$$RPM_{synchronous} = \frac{120 \times frequency}{\# \ of \ poles}$$

$$Slip = \frac{RPM_{synchronous} - RPM_{rotor}}{RPM_{synchronous}}$$

Figure 6-9 Polyphase AC motor operation summarized.

Wound-Rotor Induction Motors

A wound rotor's windings are brought out through slip rings. The advantage of the wound-rotor induction motor over the squirrel-cage induction motor is that resistance control can be used to vary both the motor's speed and its torque characteristics. Increasing the resistance causes maximum torque to be developed at successively higher values. Along the way, the wound rotor has better starting characteristics and more flexible speed control. What you give up is efficiency at an increased complexity and cost.

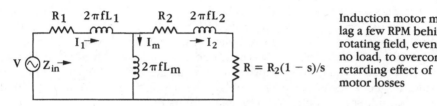

Induction motor must lag a few RPM behind rotating field, even at no load, to overcome retarding effect of motor losses

Equivalent circuit—one phase of induction motor

$$Z_{in} = \left(R_1 + \frac{R_2}{s}\right) + j2\pi f(L_1 + L_2)$$

Controller output impedance must match motor input impedance for maximum power transfer

$$Torque_{out} = \frac{3R_2 \times \# \text{ of poles}}{4\pi fs} \quad \frac{V^2}{\left(R_1 + \frac{R_2}{s}\right)^2 \times (2\pi f)^2 (L_1 + L_2)^2}$$

Torque varies with slip at any given voltage and frequency

$$Torque_{max} = \frac{3 \times \# \text{ of poles} \times V^2}{4(2\pi f)^2 \times (L_1 + L_2)}$$

Maximum torque can be maintained if voltage to frequency ratio is held constant

FIGURE 6-10 Polyphase AC motor's unique speed, torque, and slip characteristics versus voltage and frequency.

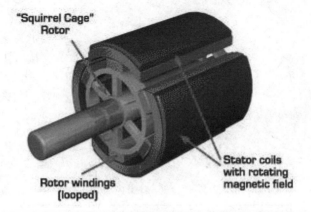

FIGURE 6-11 AC electric motor. (*Zero Emission Vehicles of Australia.*)

As I researched for this edition of this book, I noticed that the EV called the Tango from Commuter Cars Corporation used the FB1-4001 motor. When I spoke with Rick Woodbury, creator of the Tango and president of the company, we talked for over an hour about the state of electric cars. One of the greatest things that I heard from him was that the first edition of this book helped him to create the vehicle. To paraphrase, "If it wasn't for this book, the Tango would not have been built." I am glad the first edition of this book made such a contribution to the EV industry.

One of the more popular riders of the Tango is George Clooney, as shown with his Tango in Figure 6-12. Figure 6-13 shows the Advance series DC motor cutaway view. Hopefully, as I said in the second edition, Clooney's or even Leonardo DiCaprio's involvement and the wonderful aspects of the Tango or any electric car will allow Rick and other great electric car companies to build more of his own EVs. Keep it up, Rick! Way to go!

Another advantage to buying a new AC or DC motor from a reputable dealer is that you have the curves and data you need to optimize your EV conversion. There are better solutions, but regenerative braking was not important for the conversion I detail in Chapter 10, and there was already a matching controller available. I wanted good middle-of-the-road performance at a good price, as well as a product that any potential

FIGURE 6-12   George Clooney in a Tango. (*Tango.*)

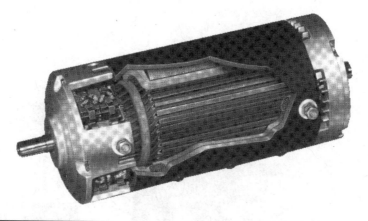

FIGURE 6-13   Advance series DC motor cutaway view.

EV converter reading this book could use and get working the first time up to bat. This motor delivers all that and more.

Tomorrow's Best EV Motor Solution

Improvements in solid-state AC controller technology clearly put AC motors on the fast track for EV conversions of the future. AC motors are inherently more efficient, more rugged, and less expensive than their DC counterparts; the reason why they are not in more widespread use today has to do with controllers, as you'll learn in Chapter 7.

"What's the best EV solution for tomorrow?" Figure 6-14 tells the story at a glance. The ideal EV drive for tomorrow has

- Low weight
- Streamlined design
- A simple drive train (one or two speeds)
- DC to get started
- AC to run above 30 mi/h
- High-frequency components (≥400 Hz)
- DC motor that gets 96 V
- Three-phase AC motor that gets 400 V
- Matching controller and motor impedance

Let's look more closely at some of the pieces.

AC Motors: HiPer Drive by ElectroAutomotive

The Seymour family has been building AC motors for nearly 50 years. They progressed from appliance motors to golf cart motors and then expanded up into high-performance

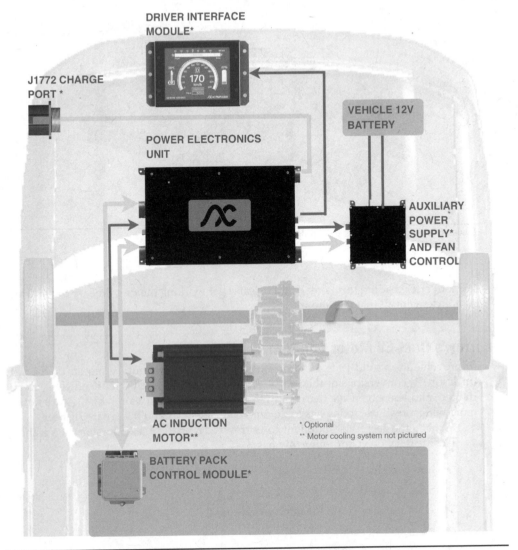

DRIVER INTERFACE MODULE*

J1772 CHARGE PORT *

POWER ELECTRONICS UNIT

VEHICLE 12V BATTERY

AUXILIARY POWER SUPPLY* AND FAN CONTROL

AC INDUCTION MOTOR**

* Optional
** Motor cooling system not pictured

BATTERY PACK CONTROL MODULE*

FIGURE 6-14 AC Propulsion LCM-150 (liquid-cooled motor). (*AC Propulsion.*)

AC motors for passenger cars. The business is now known as High Performance Electric Vehicle Systems (HPEVS), and the company refers to its system as the HiPer Drive. The systems are different from conventional AC vehicle drive systems in that they do not require high voltage. This saves the cost of extra batteries, the space needed to fit extra batteries, and the weight of extra batteries. The lighter vehicle means more peppy performance.

HiPer Drive motors feature

- High-efficiency brushless design
- Regenerative braking

- Compact, lightweight construction
- Low rotating losses
- Low electrical resistance
- High reliability
- A 1⅛-inch keyed shaft

These are integrated drive systems, with the motor, controller, and charger all custom configured at the manufacturer to fit *your* conversion. For this reason, you need very specific information about your system to place your order:

- Motor rotation direction (CCW is standard)
- Total battery pack voltage
- Battery brand and model

If you are using unusual batteries, the company will ask for some specific technical specs.

Motor Ratings

It is not accurate to refer to a "10-hp motor" or a "15-hp motor" because horsepower will vary with volts and amperes, and the peak horsepower will be much higher than the continuous rating. It is also confusing to compare electric motors with gas-powered engines because electric motors are given a continuous rating under load, and gas-powered engines are rated at their peak horsepower with no parasitic loads, such as water pumps or alternators. For accurate identification, a motor should be identified by name or model number. The HiPer Drive AC50 motor is suitable for vehicles in the 2,000 to 2,500 pounds factory curb weight range (not gross vehicle weight).

Specifications

Length	18.16 inches
Diameter	8.98 inches
Weight	115 pounds
Peak efficiency	90 percent
Battery pack nominal	96–108 V
Torque: peak (ft-lb)	90 @ 3,000 rpm
Maximum rpms	6,500
Peak horsepower	52 @ 96 V
	60 @ 108 V

The AC 35 motor, as seen in Figure 6-14, from High Performance Electric Vehicle Systems is designed for use in utility and light automotive applications (up to 2,250 pounds). It is capable of producing 60 hp and 130 ft-lb of torque, as shown in Figure 6-14. It is available in several different configurations depending on your needs. The weight of this motor is around 85 pounds.

Here is a list of the typical applications for this motor:

- Utility vehicles
- Light automotive vehicles

- Motorcycles
- Mining
- Pumps

This motor can be manufactured as vented or nonvented (sealed).

Tuning

Electrical and mechanical systems deliver better performance when balanced and tuned up. Race car mechanics set up their cars for the average case and then tune and adjust them for maximum performance. Should you do anything less with your EV? Match the impedance of the motor to the controller for maximum power transfer, and go through everything else with a fine-tooth comb.

Keep It Simple

What's the simplest controller that can implement this approach? Do you even need to use solid-state designs? How few batteries can you get by with? Which are the smallest motors you can use? Can you do something innovative with their placement? How can you simplify the drive train? Can you do something better with the tires? You get the idea.

Conclusion

At present, AC induction and permanent-magnet brushless DC motors are the best technologies available, with efficiencies up to 98 percent, silent operation, and almost never requiring any servicing. They each have various advantages and disadvantages over one another. It will be interesting to see which one becomes the new standard in the years to come.

The AC motor and controller system used in Figure 6-15 features the Azure AC24 motor with an optional AT1200 gearbox with internal differential. This drive system is designed with front-wheel compact sedans in mind. Vehicle conversions are made fast and easy because this motor/gearbox assembly will easily replace the engine and transmission of many compact front-wheel-drive vehicles. When you contact Azure, if you are interested in this product, ask for Beth Silverman. She can help you out.

Tesla Motors and Toyota Motor Corporation Intend to Work Jointly on EV Development, TMC to Invest in Tesla

Tesla Motors and Toyota announced that they intend to cooperate on the development of EV, parts, and production system and engineering support. The two companies intend to form a specialist team to further those efforts. TMC has agreed to purchase $50 million of Tesla's common stock issued in a private placement to close immediately subsequent to the closing of Tesla's currently planned initial public offering.

"I've felt an infinite possibility about Tesla's technology and its dedication to *monozukuri* [Toyota's approach to manufacturing]," said TMC President Akio Toyoda. "Through this partnership, by working together with a venture business such as Tesla, Toyota would like to learn from the challenging spirit, quick decision-making, and flexibility that Tesla has. Decades ago, Toyota was also born as a venture business. By partnering with Tesla, my hope is that all

FIGURE 6-15 Azure AC24 motor with an optional AT1200 gearbox with internal differential. (*Azure Dynamics.*)

Toyota employees will recall that 'venture business spirit' and take on the challenges of the future," he continued.

"Toyota is a company founded on innovation, quality, and commitment to sustainable mobility. It is an honor and a powerful endorsement of our technology that Toyota would choose to invest in and partner with Tesla," said Tesla CEO and cofounder Elon Musk. "We look forward to learning and benefiting from Toyota's legendary engineering, manufacturing, and production expertise."

TMC has, since its foundation in 1937, operated under the philosophy of "contributing to society through the manufacture of automobiles" and has made cars that satisfy its many customers around the world. TMC introduced the first-generation Prius hybrid vehicle in 1997 and produced approximately 2.5 million hybrids in the 12 years since. Late last year, TMC started leasing of Prius plug-in hybrid electric vehicles (PHEV),[8] which can be charged using an external power source such as a household electrical outlet. The company also plans to introduce EVs into the market by 2012.

Tesla's goal is to produce increasingly affordable electric cars to mainstream buyers—relentlessly driving down the cost of EVs. Palo Alto, CA–based Tesla has delivered more than 1,000 Roadsters to customers in North America, Europe, and Asia. Tesla designs and manufactures EVs and EV power train components. It is currently the only automaker in the United States that builds and sells highway-capable EVs in serial production. The Tesla Roadster accelerates faster than most sports cars yet produces no emissions. Tesla service rangers make house calls to service Roadsters.[7]

Roadsters Are Faster than Porsche 911 from the *Wall Street Journal*

As the car market anticipates the arrival of the Tesla Model S electric-powered sedan, new information about the car continues to trickle forth. According to numerous reports, Tesla boss Elon Musk said at a weekend event that the company will also sell a higher-performance version of the car.

The souped-up Model S reportedly will accelerate to 60 mi/h from a standstill in 4.5 seconds, which is two-tenths of a second quicker than the 2011 Porsche 911 Carrera with a manual transmission. If you order the Porsche with the company's quick-shifting PDK automatic transmission, it will match the Tesla at 4.5 seconds in the 0-to-60 mi/h sprint.

There are versions of the 911, including the Carrera S, GT3, and Turbo, that could more easily outrun the Tesla, but what is important is that the big sedan will be able to run with Porsche, Corvette, Audi R8, and other muscular sports cars. It would also establish electric power as a performance feature, not just a gasoline-saving one. Even the basic Model S is quick at 5.6 seconds from 0 to 60 mi/h.

The car is to roll out next year as a 2013 model with prices starting at $50,000. Tesla says that the Model S will have a range of up to 300 miles depending on which battery option the buyer chooses.[9]

These products are very similar to the AC Propulsion Motor called the AC-75, the AC-150, or the LCM-150 (liquid-cooled motor). It provides peak power output of 75, 150, and 150 kW with constant power for the cooled motor of 100 kW. These drive systems are designed for lightweight passenger and commercial vehicles.

BMW ActiveE Motors

The electric motor of the BMW i3 Concept is designed primarily for operation in an urban environment, developing 125 kW/170 hp, with peak torque of 250 Nm (184 lb-ft). Typically, an electric motor develops maximum torque from standstill in contrast to an internal combustion engine, in which torque increases with engine rpms. This makes the BMW i3 Concept highly agile and provides impressive acceleration. The BMW i3 Concept accomplishes 0 to 37 mi/h in under four seconds and 0 to 62 mi/h in under eight seconds.

At the same time, the abundant torque is delivered over a very large rpm range, resulting in very smooth power delivery. The single-speed gearbox provides optimal power transmission to the rear wheels and accelerates the BMW i3 Concept to an electronically governed 93 mi/h without loss of power. The electric drive also allows for deceleration by means of the accelerator pedal. After the driver eases up on the accelerator, the electric motor acts as a generator, converting the kinetic energy into electricity, which is then fed back into the battery. Energy recuperation generates a braking effect that makes a significant contribution to vehicle deceleration.

A coasting mode makes this unique "single-pedal control" of acceleration and braking using only the accelerator even more user-friendly. When the driver eases off the pedal, the electric motor's zero torque control keeps the drive train disconnected as long as the pedal is in this position. The vehicle now coasts without consuming power, driven by its own kinetic energy.[10]

Motors Everywhere!

Look around your house and you will find that it is filled with electric motors. Below is an interesting experiment for you to try. Walk through your house and count all the motors you find. Starting in the kitchen, there are motors in:

- The fan over the stove and in the microwave oven
- The dispose-all under the sink
- The blender
- The can opener
- The refrigerator (two or three in fact, one for the compressor, one for the fan inside the refrigerator, and one for the icemaker)
- The mixer
- The tape player in the answering machine
- Probably even the clock in the oven

In the utility room, there is an electric motor in:

- The washer
- The dryer
- The electric screwdriver
- The vacuum cleaner and the Dustbuster minivac
- The electric saw
- The electric drill
- The furnace blower

Even in the bathroom, there's a motor in:

- The fan
- The electric toothbrush
- The hair dryer
- The electric razor

Your car is loaded with electric motors:

- Power windows (a motor in each window)
- Power seats (up to seven motors per seat)
- Fans for the heater and the radiator
- Windshield wipers
- The starter motor
- Electric radio antennas

In addition, there are motors in all sorts of other places:

- Several in the VCR
- Several in a CD player or tape deck
- Many in a computer (each disk drive has two or three, plus there's a fan or two)
- Most toys that move have at least one motor (including Tickle-Me-Elmo for its vibrations)
- Electric clocks
- The garage door opener
- Aquarium pumps

In walking around my house, I counted over 50 electric motors hidden in all sorts of devices. Everything that moves uses an electric motor to accomplish its movement.[11]

CHAPTER 7

The Controller

If cars are so much better now with all the electronics and technological innovation onboard in the past few decades, shouldn't they be going a million miles per hour, cost only pennies, and last for decades? A great controller can solve that problem!

The controller is another pillar of every electric vehicle (EV). If one area could take credit for renewed interest in EVs, the controller would be it. You can buy a controller, plug it in, and be up and running in no time—something earlier EV enthusiasts could only dream about. In the future, there will be further reductions in size and improvements in efficiency of the motor control electronics. While the motor may benefit from only small improvements owing to technology changes, future motors may become distributed and located in the wheels themselves.

The controller takes power from the batteries and delivers it to the motor. The accelerator pedal hooks to a pair of *potentiometers* (variable resistors), and these potentiometers provide the signal that tells the controller how much power it is supposed to deliver. The controller can deliver zero power (when the car is stopped), full power (when the driver floors the accelerator pedal), or any power level in between.[1]

Unlike the plethora of choices with electric motors, your controller choices are rather simple and are dictated by the electric motor you use, your desire to make or buy your EV, and the size of your wallet or purse. In this chapter you'll learn what the different types of controllers are, how they work, and their advantages and disadvantages. Then you'll encounter the best type of controller to choose for your EV conversion today (the type used in the conversion in Chapter 10) and the electric motor controller you're likely to be seeing a lot more of in the future.

Controller Overview

The heart of an EV is the combination of:

- The *electric motor*
- The motor's *controller*
- The *batteries*

The controller normally dominates the scene when you open the hood, as you can see in Figure 7-1. In this car, the controller takes in 300 V direct current (DC) from the battery pack. It converts it into a maximum of 240 V alternating current (AC), three-

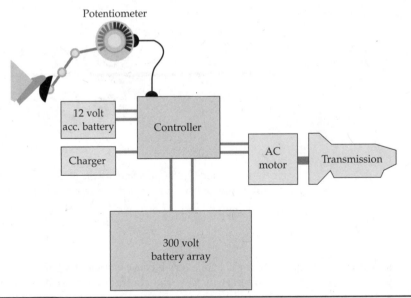

Potentiometer

Figure 7-1   Electric vehicle connections to controller. (*HowStuffWorks.com.*)

phase, to send to the motor. By using large transistors, the batteries can very quickly turn the voltage on and off. This creates a sine wave. When you push on the accelerator pedal, a cable from the pedal connects to two potentiometers. The signal from the potentiometers tells the controller how much power to deliver to the motor. There are two potentiometers for safety's sake. The controller reads both potentiometers and makes sure that their signals are equal. If they are not, then the controller does not operate. This arrangement guards against a situation where a potentiometer fails in the full-on position.[2]

Using six sets of power transistors, the controller takes in 300 V DC and produces 240 V AC, three-phase. The controller additionally provides a charging system for the batteries, and a DC-to-DC converter to recharge the 12-V accessory battery.[3]

Most controllers pulse the power more than 15,000 times per second in order to keep the pulsation outside the range of human hearing. The pulsed current causes the motor housing to vibrate at that frequency, so by pulsing at more than 15,000 cycles per second, the controller and motor are silent to human ears.

In an AC controller, the job is a little more complicated, but it is based on the same idea. The controller creates three pseudo–sine waves. It does this by taking the DC voltage from the batteries and pulsing it on and off. In an AC controller there is the additional need to *reverse the polarity* of the voltage 60 times a second. Therefore, you actually need six sets of transistors in an AC controller, whereas you need only one set in a DC controller. In an AC controller, for each phase, you need one set of transistors to pulse the voltage and another set to reverse the polarity. You replicate this three times for the three phases—six total sets of transistors.

You can relax now. The useful but largely theoretical equations of Chapter 6 are behind you. Controllers can generate even more equations, but we'll save those for the engineers who build them. The objective here is to give you a brief background on controllers and introduce you to a working controller for your EV with minimum fuss.

A controller basically is the brain or computer of an EV. This computer "controls" the performance of the electric motor. The controller integrates the motor speeds and expected battery range/speed through its energy density.

Coordinating between the controller and the motor can help an EV accelerate from 0 to 60 mi/h in six to seven seconds (or less), which can determine the range of the car and the top speed. The controller controls manual or automatic drive systems for starting and stopping, going forward or backward, governing the speed, regulating or limiting the torque, and protecting against overloads and faults.

Off-the-shelf sources mentioned in the *Build Your Own Electric Vehicle* website located at BuildYourOwnElectricVehicle.WordPress.com are really the only way to include a controller into your EV. A controller from a reputable company that can provide safety and reliability is the number one priority.

The controller correctly matched to the motor will give it the right voltage. The weight of the motor magnets and size of its brushes determine the power and torque. You can buy a reliable controller from a number of sources today at a good price.

A motor speed controller uses a microprocessor to drive a preamplifier and then a power-stage amplifier that controls the flow of power from the battery to the motor, with various feedback sensors to monitor the system's operation.

Solid-State Controllers

Solid-state controllers have been used for decades in all types of EVs and are widely available off the shelf. Basically, they use solid-state switches or a *silicon-controlled rectifier* (or *semiconductor-controlled rectifier*), a four-layer solid state current controlling device. The name "silicon controlled rectifier" or *SCR* is General Electric's trade name. The SCR was developed by a team of power engineers led by Robert N. Hall and commercialized by Frank W. "Bill" Gutzwiller in 1957.[4] Transistor, metal oxide semiconductor field-effect transistor (MOSFET), and insulated-gate bipolar transistor (IGBT) replace the mechanical switches in a switch controller. These are more common among lower-voltage controllers because low-voltage MOSFETs tend to have very low on resistance (and hence power loss). MOSFET controllers are very efficient at low power levels because power loss in a MOSFET is proportional to the square of the current. Their resistance also increases as they heat up, so when used in parallel, they tend to automatically balance the load. IGBTs are somewhat like a hybrid between bipolar transistors and field-effect transistors. IGBTs have a constant voltage drop, which makes them more efficient than MOSFETs at high power levels, although often less efficient for low-power applications. One disadvantage of IGBTs is that, like all bipolar transistors, they are prone to thermal runaway and imbalances when used in parallel. As such, IGBT controllers require good cooling systems (liquid cooling is common) and/or matched transistors to avoid imbalance.[5]

Electronic Controllers

Electronic switches can be switched on/off much faster than mechanical switches and don't wear out. The controller switches them on/off so fast (thousands of times per second) that the motor gets the average current rather than the peak or zero. This is called *pulse-width modulation* (PWM). PWM controllers are modestly priced, readily available in a wide variety of sizes, and provide smooth, stepless motor control.

AC Controllers

The AC motor with solid-state inverter is the most sophisticated motor available. Such motors are used in high-end EVs, where meeting performance objectives is more important than cost. An AC induction motor or AC PWM motor (often called a *brushless DC motor*) is driven by an inverter that converts battery DC voltage into variable-frequency three-phase AC. New EV sports cars such as the Tesla Roadster Sport and the Porsche 911 Conversion from EV Propulsion or Café Electric and all the auto companies that have recently produced EVs all used AC drive systems. Like modern internal combustion engines (ICEs), these AC controllers are very complex and expensive, but they offer the most advanced features (such as cruise control and regenerative braking) and provide the best overall range.

Controller Choice

High reliability, high performance, and minimum cost are the key factors that determine your EV controller choice. You can pick one or maybe two, but you can't have all three. For example:

- High reliability and high performance mean high cost (Café Electric Zilla controller; no know failures; 300 V, 2,000 A, $3,500).
- High performance and minimum cost mean low reliability (Logicsytems rebuilt Curtis controller; high failure rate; 156 V, 1,000 A, $1,500).
- High reliability and low cost might mean low performance.

If you've driven an EV, you'll find that most electronic controllers do *not* imitate the accelerator pedal response of a normal ICE vehicle! It is undesirable to do so. The throttle response of most ICE vehicles is unpredictable and very "lumpy." Some cars leap forward at the slightest touch; others gently creep ahead. As the pedal is pushed farther, the amount of speed increase for a given amount of pedal movement varies considerably. On some cars, there is hardly any difference between half throttle and full throttle; on others, it's the difference between 30 and 120 mi/h. There is always a noticeable delay. The car doesn't coast, and letting up the pedal produces variable amounts of engine braking. People just get used to these huge variations.

The throttle response of EV controllers is completely different. It is entirely predictable, repeatable, and instantaneous. Most people have to relearn the response to expect but find it *better* than an ICE vehicles. The motor controller normally controls motor current and voltage, from zero to maximum, as a direct function of throttle position, zero to maximum (i.e., half throttle means half voltage and half current). Electric motors don't need to idle, so releasing the pedal means the motor is off (no creep). Pressing the pedal to a certain point sets a certain motor voltage, which means a certain speed. Pressing hard gives you a constant rate of acceleration. Release the pedal, and the EV simply coasts (unless you have regenerative braking, in which case you can configure what it does).

An Off-the-Shelf Curtis PWM DC Motor Controller

You can do something today that EV converters of a decade ago could only dream about—pick up the telephone and order a brand-new DC or AC motor controller from any one of a number of sources. You can have it in your hands a few days later, mount it, hook up your EV's electrical wiring and throttle control to it, and be up and running with virtually a 99 percent chance of everything working the first time.

As with series DC motors, today's DC controllers are readily available from many sources, they work great, most of them are of the PWM variety, they are easily installed in different vehicles, and the price is right. A modern off-the-shelf PWM DC motor controller is not the ultimate, but it's pretty close to the best current solution. More important, it's one that most EV converters, manufacturers, or auto shops repairing EVs will have no trouble implementing today. After you do your first conversion or build, become the acknowledged genius in your neighborhood; once you know what you really like and don't like, you can get fancy and exotic.

The DC PWM controller recommended here is from Curtis PMC, a division of Curtis Instruments, Inc., of Mt. Kisco, New York. As with electric motors, don't read anything into the controller's appearance here. Curtis is only one of a large number of controller manufacturers from the list on the online source directory, and the recommended controller model is only one of a number the company manufactures.

The Curtis PMC Model 1221B-7401 DC motor controller, shown in Figure 7-2, is already very familiar to you. It is a PWM-type controller with the following features:

- MOSFET-based technology
- A constant switching frequency of 15 kHz
- An external 5-kΩ throttle potentiometer
- Automatic motor current limiting
- Thermal cutback at 75 to 95°C
- High pedal lockout (prevents accidental startup at full throttle)
- Intermittent-duty plug braking
- Overvoltage and undervoltage protection
- User-accessible adjustments for motor current limit, plug braking current limit, and acceleration
- A waterproof heat-sink case

The controller is also well matched in characteristics to the Advance Model FB14001 series DC motor, particularly in the impedance area (earlier you read about the

FIGURE 7-2 Curtis PMC 1221B DC motor controller.

importance of this for peak power transfer). If the Curtis PMC DC controller characteristics sound familiar, it's because they use the PWM integrated circuit (IC) technology you've been reading about throughout this chapter and bring all these benefits to you with additional features in a rugged, preassembled, guaranteed-to-work package at a fine price. The same reputable vendor comments from the Chapter 6 also apply here.

Installation and hookup are a breeze. If you look closely at the controller terminals, you'll notice the markings M–, B–, B+, and A2 appear (listed clockwise from the lower left when facing the terminals). You already know that the first three correspond to the –Mot, –Bat, and +Bat markings on the terminal bars. The A2 marking means Armature 2, the opposite end of the armature from the Armature 1 winding that is normally connected to the +Bat or, in this case, the B+ terminal. Anything else you might want to know is covered in the Curtis PMC manual that accompanies the controller.

AC Controllers

AC overwhelmingly has benefits that make it a winner despite the complications involved. In general, AC controllers require more protective devices to isolate against noise, yet DC motors make a lot more noise than AC motors.

Chapter 6 showed that the speed-torque relationship of a three-phase AC induction motor is governed by the amplitude and frequency of the voltage applied to its stator windings. The best way to change the speed of an AC induction motor is to change the frequency of its stator voltage. A change in frequency results in a direct change in speed, and if you change the frequency in proportion to the voltage (both at ¼, ½, ¾, etc.), you get the speed-torque curves.

Knowing the voltage and frequency ratio that you want to maintain allows you to calculate the voltage, current, and output torque relationship for any values of input voltage and frequency using vector math and lookup tables. In simpler terms, if you feed the speed and torque values you want into some sort of "smart box," it can provide the voltage and frequency necessary to generate the proper motor control signals.

Today's Best Controller Solutions

If you've read Chapter 6 on motors, you already know that this book recommends series DC and AC motors as the best motor solutions for today's EV converters to choose from, and you have to figure out which one to use. The same words of advice apply in the controller area as in the motor area: If someone tells you there is only one controller solution for a given application, ask another person. As with motors, you probably can find three or four good controllers for any application because there are only shades-of-gray controller solutions. And, as with the recommended motor, the controller recommended here is not *the* solution—it is just *a* solution that happens to work best in this case.

While the series DC motor and PWM controller are unquestionably best for today's first-time EV converters, the bias of this book is toward AC controllers and motors. However, one of the best EV technologies today is a DC controller. On one side, AC induction motors are inherently more efficient, more rugged, and less expensive than their DC counterparts. This translates to more driving range from a given set of batteries and less probability of failure and the possibility of graceful degradation when a failure does happen, but not less expense! The package isn't cheap, and the motor alone isn't either. These benefits come at a price. This is why nearly every newly designed commercial EV today uses one or more AC induction motors or their closely related cousin, the brushless DC motor.

No new electric car uses DC drive. However, some of the many conversions out there use DC controllers. DC is cheap and very suitable for a budget conversion. Again, for a conversion that is less expensive, a DC motor works.

What's in the laboratories today will be available to you in the not-too-distant future, and beyond that, continued improvements in solid-state AC controller technology could put AC motors in every EV conversion in the future. Let's look at developments in two areas—systems and components—that virtually guarantee this outcome.

Now, let's look at the future—today! Café Electric is a small manufacturing company located in Corvallis, Oregon. Its mission includes enabling the hobbyist and small manufacturer to build on-road, freeway-capable EVs of ever-increasing quality and integration. The nature of the company's products and expertise is not appropriate for low-speed EVs such as bicycles, neighborhood EVs, go-carts, etc.

By enabling the hobbyist, we promote EVs in general, especially at times when the "Big 3" maintain that such things cannot be practically done. We do our best to enable our customers by designing, developing, manufacturing, and selling electronic devices that improve the safety and performance of EVs while still being affordable and available to common persons converting their own cars. At this time, our range of electronic devices covers only motor controllers and the interfaces for them, but we have several more projects in the works. Enabled by our ongoing obsession with overall business efficiency, we strive to balance the ratio of results to cost. We feel that by doing this we can offer these low-volume manufactured components at a much lower cost than could be possible in a traditional company that is concerned with getting return on their R&D investments and large overhead.[6]

Zilla Controller (One of the Best DC Controllers for Conversions)

The Zilla controller, shown in Figure 7-3, is by far one of the most popular and powerful motor controllers available for EVs. The Zilla has exceptionally high power density. According to conversion specialists and people in the EV community today, the Zilla is ranked as one of the best controllers on the market. This controller can give your EV the most amazing pickup and torque like no other EV on the market (or even on your block).

I like that the producer of the Zilla cared about safety in its literature because it shows that a reasonably priced, efficient, and off-the-shelf controller can still be safe. From carefully monitoring as the controller comes up to voltage, that it communicates properly, and that the integrity of the output stage is okay before engaging the main contractor to the dual microprocessors that cross-check and have independent means of shutting the system off, there is no other DC controller that approaches this level of security.

It is no surprise that all the world's quickest EVs use Zilla controllers, but the safety features allow them to also excel in street applications. The Z1K controller in particular has become very popular for street conversions owing to its superior feature set and low price. Table 7-1 shows the specifications of this popular controller.

FIGURE 7-3   Zilla Controller Model Z2K-HV. (*Café Electric, LLC.*)

Zilla Models

Every Zilla controller comes with a Hairball 2 interface box, which is packed with features. The Hairball is the central connection hub for all of the smaller ancillary circuits and it handles communication with the controller as well as programmable parameter adjustment. The unit has a built-in automatic pre-charge function to help protect the main contactor from arcing. There is also active monitoring of the main contactor for voltage drop or a stuck condition. Other features include programmable motor voltage and current limits, programmable battery voltage and current limits, adjustable low battery voltage protection and an additional low battery indicator output. There are two

Maximum nominal input voltage range for lead-acid batteries: 72 to 348 volts
Absolute maximum fully loaded input voltage range: 36 to 400 volts*
Maximum motor current at 50°C heat sink temperature: 2000 Amps for Z2K, 1000 Amps for Z1K
Maximum battery current at 200V: 1900 Amps for Z2K, 950 Amps for Z1K
Maximum battery current at 300V: 1770 Amps for Z2K, 885 Amps for Z1K
Maximum battery current at 400V: 1600 Amps for Z2K, 800 Amps for Z1K
Continuous motor current @ 50°C coolant temp and 100% duty cycle: over 600 Amps for Z2K, 300 Amps for Z1K
Peak power: 640,000 Watts for Z2K, 320,000 Watts for Z1K
PWM frequency: 15.7 kHz
Power devices: IGBT
Voltage drop: <1.9 volts at maximum current
* At this time we are suggesting not exceeding 375 volts on the EHV models; we hope to bring that back up to 400 volts with further testing.

TABLE 7-1   Zilla Specifications

speed sensor inputs for motor over speed limiting of up to two motors and there is a pulse output that can interface with most 4 or 6 cylinder tachometers to show instantaneous motor RPM. All Zillas require a Hairball to run. You can select options for your Hairball if desired. The Zilla package also includes a 3-ft Cat-5 cable with ferrite to connect the Zilla and the Hairball, a serial cable and adapter for connecting a laptop to the Hairball, a contactor snubber diode, a shorting plug, two high-voltage stickers, and a Zilla screwdriver. The computer that you will need for setting the Hairball settings must have an RS-232 serial port with a 9-pin connection. In case your computer is not appropriate or you just want a portable way to adjust settings, the company can sell an appropriate Palm Pilot handheld terminal in the accessories section.

Zilla controllers are so powerful that the batteries have trouble supplying their high currents, and motors can get damaged by them as well. This is why it helps to have appropriate engineers assist you with the formulas needed to calculate proper voltage limits. While this book is a great resource, be *sure* to set the current limits to reasonable values for your setup before testing. Check out the information in Table 7-1.

The highly capable batteries available today, such as the 32 Series automotive-class lithium-ion cell from A123 Systems, are more energy dense and able to deliver speed and range. Combine an A123 battery and the Zilla controller and you will have a full-performance ride of your life.[7]

Manzitta Micro sells Zilla controllers for Café Electric. Manzanita Micro was formed by Rich Rudman and Joe Smalley. The two of them were roommates in college at the University of Idaho and became friends. Both had a love of racing and tinkering with high-performance automobiles as well as an affinity for things electronic. Manzanita Micro began as a floppy disk drive service company and grew into a small business that built timing systems for autocross and road racing. By the mid-1990s, Rudman and Smalley found a hobby that melded their love of cars and electronics—the EV.

After converting a Ford Fiesta and building his own 2,500-A controller, Joe went to work for the U.S. Navy in Bremerton, Oregon. Rich converted his own 1978 Ford Fiesta, "Goldie," and then began doing contract circuit-board design, laying out plans for inverters and other medical and marine electronics. He also worked for Cruising Equipment in Seattle on the original E-Meter EV gauge.

As his experience with electric vehicles grew, Rich worked on the powerful Raptor and T-Rex controllers. He also designed and built a working prototype drive system.

In his off time, Rich and his old friend Joe drag raced their Fiestas and also got into racing electric hydroplanes, setting a few EV records. Rudman and Smalley saw the need for a powerful and flexible charger for EVs as well as a good battery management system (BMS) for the ever-popular new AGM batteries. DCP moved to Oregon and later became Alltrax, and Rich moved back to the Kitsap Peninsula, where he had grown up. He and Joe finalized plans for their BMS system, the flexible power-factor-corrected chargers, and Rudman Regulator, and his line of chargers was brought to market.

Rich began working full time out of his garage, building and marketing the new EV battery components.

Today, Manzanita Micro has grown, and Rich oversees the operation of three shops and a host of employees. He and Joe still meet weekly to improve existing products and work on the next big thing. The years 2007 and 2008 saw tremendous growth for the company, and Manzanita Micro has delivered high-performance liquid-cooled chargers to companies such as Commuter Cars, Inc. (the Tango), and Phoenix Motorcars. Manzanita Micro also has delivered numerous rapid-charge-capable plug-in Prius kits to customers, including the cities of Seattle, Tacoma, and Wenatchee, Washington. Two enormous 75-kW "monster chargers" have been tested that can charge EV battery packs in just a few minutes. From the fast chargers, to battery evaluation and testing, to high-strength aluminum welding and metal fabrication, Manzanita Micro is ready to meet the needs of the next generation of EV.

The name Manzanita comes from the Manzanita Bay, just a few miles from Kingston, Washington, where Rich Rudman grew up. Now with headquarters in Kingston and over 1,000 chargers sold and in use in over nine countries, Manzanita Micro has established itself as one of the foremost EV component suppliers in the world.

Let's Go WarP-Drive Controllers

The WarP-Drive Classic (aka WarP-Drive 1) is from NetGain Controls, Inc. It has many great features that make it an attractive motor speed controller choice for lightweight to medium high-power DC EV conversions. The base unit comes programmed for 1,000 A rated output at 160 V and is user-upgradeable to 1,200 and 1,400 A and 260 and 360 V. The WarP-Drive comes from the factory with CANbus connectivity, which gives it endless capabilities to operate in conjunction with already-in-development third-party hardware solutions.[8]

The WarP-Drive Industrial (aka WarP-Drive 2), as shown in Figure 7-4, is the company's newest controller. It has many great features that make it an attractive motor speed controller choice for all classes of high-power DC EV conversions. The base unit comes programmed for 1,000 A rated output at 160 V and is user-upgradeable to 1,200 and 1,400 A and 260 and 360 V. The WarP-Drive comes from the factory with CANbus connectivity, which gives it endless capabilities to operate in conjunction with already-in-development third-party hardware solutions.

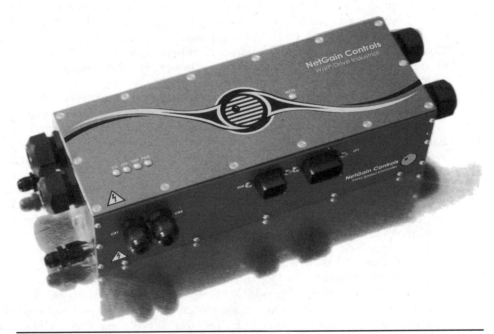

FIGURE 7-4   NetGain controls controller. (*NetGain Technologies.*)

Variable DC Voltage Input/Output.   This industrial strength controller can be purchased with 160-, 260-, or 360-V nominal limits. Customers can purchase upgrades at any time that allow the end-user to perform the upgrade. This versatility allows customers to start with a feature-packed controller and scale to higher peak voltages as budget allows.

The output voltage (motor voltage) can be configured using our interface module to limit the voltage that the motor sees. This is especially useful for higher peak voltages used with motors that are limited (such as the NetGain WarP, ImPulse, and TransWarP, which have a limit of 160 to 170 V).

Liquid Cooling.   The best controllers on the market use liquid cooling to wick away the heat produced. The WarP-Drive uses precision machined liquid chill plates. Connecting to the chill plate is extremely easy. The company also carries two custom cooling kits for the WarP-Drive. These kits are designed to be installed in minimal time.

Temperature Safety.   Designed for any temperature situation, the WarP-Drive internally protects itself from overheating. When using the available cooling kits, temperatures usually can be limited to below 55°C. However, some situations can arise that cause the temperature to rise above this (e.g., a pump failure). At 55°C, maximum power output capability begins to decrease. At 95°C, the ampere output capability hits zero.

ZAPI

ZAPI has a family of controllers that are designed to perform with standard DC traction motors. Microprocessor-based, these controllers use the latest in solid-state MOSFET technology, including high-frequency operation, microprocessor logic, digital adjustment, diagnostics, and fault-code storage. This controller is very similar to the Curtis controller but with many more features, programming options, and regenerative braking. I personally found this controller to be very reliable. For the ZAPI controllers, a handheld microprocessor-based tool is used for adjustment, testing, and diagnosis. The tester feature provides a visual display of critical operating parameters such as battery voltage, motor voltage, motor current, controller temperature, accelerator voltage, and hours of operation.[9]

EV Controllers Help to Dispel All Myths About EVs Today

EV conversion specialists today are making the greatest EVs on the market. Each is a new challenge, a new dimension, but the reality of it is that from the smaller conversion companies such as Cool Green Sheds or Electric Vehicles of America to AC Propulsion, Nissan, Ford, and Tesla, today we can have what we want in an EV. Here are some companies that say it all for the controllers of today and tomorrow.

Sevcon Controllers Built in the USA Going into Chinese Electric Trucks

Chinese firm AUCMA Electric Vehicle Co., Ltd., has selected Sevcon as the motor controller supplier for its new light truck, the A-2. Aimed at municipal fleets in both domestic and overseas markets, the new A-2 light-duty truck is powered by a 72-V, 16-kW AC motor, and Sevcon will supply its Gen4 digital motor controller with regenerative braking technology to the company based in Qingdao, Shandong Province.

AUCMA EV CEO Raymond Zhu said: "The Sevcon controller is perfectly suited for the A-2 truck application with adjustable parameters that allow us to tailor the performance to any specific truck type, from flatbed to cargo box vehicles. The customer support we have received from Sevcon has been exceptional, and the company has worked hard to ensure that the controller will work effectively across our applications."[10]

The electric drive train gives the A-2 a range of up to 150 km and a top speed of over 50 km/h. The A-2 can be fully recharged via a 220-V, 30-A socket in 8 hours, with the onboard charger safely switching off when the battery is fully charged.

"We were delighted to work with AUCMA EV on its new A-2 electric truck, which marks another customer application for our Gen4 controllers," said Sevcon President and CEO Matt Boyle. "We believe that there are huge opportunities for electrification in the commercial vehicle segment and are looking forward to seeing the A-2 in action both in China and overseas markets, such as the USA."

The A-2 is expected to go into service in municipal fleets operated by local authorities in China but also will be exported. AUCMA EV will ship part-assembled A-2 trucks to the United States, where they will be completed and sold from a base in California. For more on this controller, go to www.sevcon.com or the online source directory at: BuildYourOwnElectricVehicle .WordPress.com.

AC Propulsion, Inc., to the Rescue—Today

While others have only dreamed or talked about it, AC Propulsion has done it— designed an integrated AC induction motor and controller that has been installed into numerous prototype EVs.

AC Propulsion aims to be the leading developer and producer of high-performance, high-efficiency, high-value drive trains with integrated charging systems for the EV and HEV markets. We are driven by a commitment to advancing technology, product quality, and customer satisfaction. We envision a world in which our technology leads the way toward a green, efficient, and convenient way to drive that does not sacrifice expectations for vehicle performance.[11]

AC-150 Gen 2 System: A Reliable Proven Performer for Passenger Car Needs

The AC-150 drive system, as shown in Figure 7-5, includes a power electronics unit (PEU) and the AC induction traction motor that provide high performance, high efficiency, and rapid, convenient charging for electric and hybrid vehicles. An optional BMS and driver interface are also available. This system is used extensively in EVs all across the world, including the iconic BMW Mini-E and Taiwans's Yulon MPV.

Push the lever all the way in the other direction, and you have no regeneration. Take your foot off the accelerator pedal, and the vehicle coasts and coasts. After Bob Brant returned to earth after driving an AC Propulsion vehicle in 1992, he gave me a walk-around tour and talked about the batteries (AC Propulsion's Honda CRX uses 28 conventional 12-V, deep-discharge lead-acid batteries that produce 336 V), controller, motor, charging philosophy, and temporary instrumentation (to monitor performance). Look at the differences in technology and batteries today versus 1992.

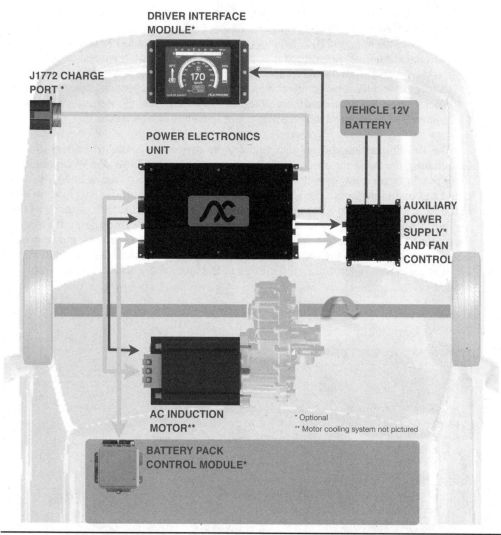

FIGURE 7-5 AC-150 electric propulsion systema and components. (*AC Propulsion, Inc.*)

AC-150 Gen 3 System: For Next-Generation EVs

Scheduled for production in 2012, this new drive system is smaller in size and has many new features. The company is confident that the Gen 3 system will surpass customer expectations. To see what's new in the third-generation AC-150 system, see the datasheet in Figure 7-5. AC Propulsion's own literature says it best about the newer AC-150 Gen2:

> The AC-150 drive train has consistently received praise the world over for its innovative design and jaw-dropping performance. First available in 1994, the AC-150 drive system retains the first generation's 150-kW (200-hp) rating but has fewer parts, is

30 percent smaller, 8 pounds lighter, and packages more functions inside the electronics enclosure than the original AC-150. By far one of its most attractive features is the integral 20-kW bidirectional grid power interface. The integrated grid interface was originally developed to serve as a high-power battery charger for battery electric vehicles. With the bidirectional capability, many new applications are opened up for electric drive vehicles of all types, including distributed generation, selling grid ancillary services, automated battery diagnostic discharge testing, and using vehicles to provide multi-uninterruptible backup power to homes or businesses.[12]

The AC Propulsion design also contains its own state-of-the-art battery charger (eliminating the need for an additional external charger), whose 20-kW capability at unity power factor allows you to fully recharge in only one hour from a 240-V, 40-A AC outlet and overnight from a conventional 120-V AC source. Figure 7-6 is a close-up view of the AC Propulsion 120-kW power electronics drive train inside the company's eBox. The figure shows how the connections are made to the 120-kW amplifier in the eBox to the motor fan, the chassis, the 260-V AC power, the motor signals, and the motor power. The company has made the controller setup with easy connections, as shown in Figure 7-7, that are centrally located in the middle of the area where an ICE engine would be located.

FIGURE 7-6   AC Propulsion 120-kW power electronics drive train in the eBox.

FIGURE 7-7   Close-up of some of the connections to the 120-kW amplifier on the eBox.

The AC-150 Gen3 drive system, shown in Figure 7-5, includes a power electronics unit (PEU) and an AC induction traction motor that provide high performance, high efficiency, and rapid, convenient charging for electric and hybrid vehicles. A flexible digital interface allows customers to use their own batteries, BMS, and driver interface or those developed by AC Propulsion.

Today, some of the best high-performance EVs use the AC Propulsion systems—from the Wrightspeed ×1 electric sports car, to the Venturi Fetish high-performance car out of Italy, to the Tesla and soon the Toyota RAV EV. In October 2011, AC Propulsion announced that its sister company, e.motor Advance Corp., of Beijing had entered into a supplier contract with Beiqi Foton Motor Co. of Beijing. The contract between the two companies provides that e.motor Advance will deliver 100 sets of its drive train and battery pack systems to Foton for their Midi-E electric car. One hundred Midi-E vehicles will be in service as taxis in the Yanqing area of Beijing in 2012. Beijing Vice Mayor Gou; the CEO of Foton, Mr. Wang Jinyu; and Mr. William Chang, chairman of e.motor Advance and the majority shareholder of AC Propulsion, attended the signing ceremony.

This is the first deal where a U.S.-based company will supply the major EV components for a China-built EV. AC Propulsion began work with Foton Motor Co. in early 2010, developing an EV conversion using AC Propulsion's advanced drive and recharge system, along with an AC Propulsion battery pack. e.motor Advance established a new facility in 2011 located in Beijing's Yizhuang economic zone and plans to start construction of a new manufacturing facility in 2013 with a capacity to build 30,000 units per year. Sales are expected to reach this level within three years. In 2005, AC Propulsion established a subsidiary manufacturing operation in Shanghai that provided components used in the BMW Mini-E program.

Present drive train production is a joint operation by AC Propulsion San Dimas, AC Propulsion Shanghai, and e.motor Advance. San Dimas remains the company's headquarters and engineering center, as well as contributing limited production.

AC Propulsion and AutoPort, Inc., Partner to Develop EV for U.S. Postal Service Feasibility Study

In October 2010, AC Propulsion and AutoPort announced that they will partner in engineering, development, and conversion to provide an EV conversion prototype and report for the U.S. Postal Service (USPS). The USPS chose this as one of five solutions in a feasibility study for the possible conversion of its 142,000 long-life vehicles (LLVs) to plug-in battery electric vehicles.

"We are thrilled to partner with AutoPort to present a long-term solution to the U.S. Postal Service," AC Propulsion CEO Tom Gage said. "Our solution provides the safety and performance required by the USPS, and it will reduce cost, increase efficiency, and improve drivability for the mail carriers."

With the AC Propulsion and AutoPort solution, the current USPS LLV design will be stripped of the gasoline engine, transmission, and other components and refitted with the AC Propulsion AC-150 drive system, an integrated power system that includes an AC induction motor, inverter, charger, and 12-V power supply. AutoPort will convert the vehicle onsite at its factory in New Castle, Delaware. The converted vehicle will comply with *Guidelines for Electric Vehicle Safety* (SAE J2344) and all applicable federal motor vehicle safety standards.

AC Propulsion and AutoPort's solution offers the low cost of ownership and revenue-producing potential of vehicle-to-grid (V2G) technology integrated into its drive system. The AC-150 is the only V2G-capable drive train worldwide. "For AutoPort, this is a landmark day in our history, to be one of the companies selected by the USPS to participate in their demonstration project involving the conversion of an LLV to an all-electric vehicle," said Roy Kirchner, president of AutoPort. "We believe that electricity is the right fuel for the USPS delivery fleet, and by including V2G capabilities, our solution will give the lowest total cost of operation"

AC Propulsion and AutoPort's solution will be tested against the USPS's requirements at AutoPort's factory and then will be placed into service in the DC metro area for at least one year, where the vehicles will be monitored for carrier satisfaction, cost of operation, and maintenance.[13]

Vehicle or Battery Management System

by David Andrea

The vehicle management system (VMS) coordinates battery management, recharging, energy monitoring, and AC grid interface functions. Features of the VMS include:

· Advanced drive control circuitry

- "Glass smooth" torque under all load and speed conditions
- Natural and transparent driving feel
- Driver-adjustable regeneration
- Traction control
- Speed control
- Interface and control for the battery management system (BMS)
- Displays for battery state of charge, ampere-hours, watt-hours, regeneration ampere-hours and watt-hours, and net watt-hours per mile
- RS232 serial port for laptop computer displays
- Global Positioning System (GPS) option for location-specific grid services
- Wireless bidirectional Internet communications for remote dispatch of grid power services and remote diagnostics
- Energy management algorithms to deploy grid power functions subject to driver range requirements
- Battery module-level upper and lower voltage limits for regeneration and motoring

Battery Management Systems

A BMS Monitors and Protects Cells in a Battery

The primary job of a BMS is to protect the battery (by preventing operation of any cell outside its safe operating area). A secondary job may be to maximize battery capacity by balancing the battery's State of Charge (SOC).

BMSs are used to manage a battery pack, such as by:

- Monitoring its state
- Calculating secondary data
- Reporting those data, protecting the BMS
- Controlling the environment
- Balancing the BMS

A BMS may monitor the state of the battery as represented by various items, such as:

- Voltage—total voltage, voltage of periodic taps, or voltages of individual cells
- Current—current in or out of the battery
- Temperature—overall pack temperature, air intake or exhaust temperature, or individual cell temperatures
- Environmental conditions—air flow in air-cooled batteries

Additionally, a BMS may calculate values based on the preceding items, such as:

- State of charge (SOC) or depth of discharge (DOD)—to indicate the charge level of the battery

- State of health (SOH)—a variously defined measurement of the overall condition of the battery
- Maximum charge current as a charge current limit (CCL)
- Maximum discharge current as a discharge current limit (DCL)
- Resistance—dynamic resistance for entire pack or individual cells
- Total energy delivered since manufacture
- Total operating time since manufacture

A BMS may report all the preceding data to an external device, using communication links such as:

- Can bus (typical of automotive environments)
- Direct wiring
- Serial communications
- Wireless communications

A BMS may protect its battery by preventing it from operating outside its safe operating area, such as:

- Overcurrent
- Overvoltage (during charging)
- Undervoltage (during discharging), especially important for lead-acid and lithium-ion cells
- Overtemperature or undertemperature
- Overpressure (typical of NiMH batteries)

The BMS may prevent operation outside the battery's safe operating area by:

- Requesting the devices to which the battery is connected to reduce their use of the battery or even terminate it
- Including an internal switch (such as a relay or solid-state device) that is opened if the battery is operated outside its safe operating area
- Actively controlling the environment, such as through heaters, fans, or even air conditioning

In order to maximize the battery's capacity and to prevent localized undercharging or overcharging, the BMS may actively ensure that all the cells that compose the battery are kept at the same state of charge. It may do so by:

- Wasting energy from the most charged cells through a dummy load (regulators)
- Shuffling energy from the most charged cells to the least charged cells (balancers)
- Reducing the charging current to a sufficiently low level that will not damage fully charged cells while less charged cells may continue to charge

BMS technology range in complexity and performance:

- Simple dissipative ("passive") regulators across cells bypass charging current when their cell's voltage reaches a certain level to achieve balancing.

- Active regulators intelligently turn on a load when appropriate, again to achieve balancing.
- A full BMS reports the state of the battery to a display and protects the battery.

BMS topologies mostly fall into three categories:

- Centralized—A single controller is connected to the battery cells through a multitude of wires.
- Distributed—A cell board is installed at each cell, with just a single communication cable between the battery and the controller.
- Modular—A few controllers, each handing a certain number of cells, communicate with each other.

Centralized BMSs are most economical, least expandable, and are plagued by a multitude of wires ("spaghetti"). Distributed BMSs are the most expensive, simplest to install, and offer the cleanest assembly. Modular BMSs offer a compromise of the features and problems of the other two topologies.

How Much Balancing Current Do You Need?

Is 100 mA Sufficient to Balance a Lithium-Ion Pack? What About 1 A? 10 A?
One of the functions of a BMS is to balance a battery. The cells in a battery may be unbalanced in multiple ways, including:

- Actual SOC (state of charge)
- Leakage (self-discharge current)
- Internal resistance
- Capacity

Only cells from better manufacturers are closely matched, and batteries that use them require very little balancing. Unfortunately, cells from many manufacturers have significant cell-to-cell variations.

Balancing takes care of equalizing the SOCs of the various cells in a battery. In so doing, it compensates also for the second one: cell-to-cell variations in leakage. Its job may be somewhat hindered by the third one: variations in cell resistance. What balancing does not do is take care of the fourth one. That's done by a different technique.

Redistribution

Redistribution allows use of all the energy in the battery; it requires significantly higher currents than balancing. The point of balancing is to maximize the charge that the battery can deliver, limited only by the cell with the lowest capacity.

- Without balancing, the capacity of a battery is limited at one end when a cell becomes fully charged and at the other end when a cell (same or different cell) becomes fully discharged.
- After balancing, the capacity of a battery is limited at both ends by the cell with the lowest capacity (or, in extreme cases, by the cell with the highest internal resistance).

A balanced battery is one in which, at some SOC, all the cells are exactly at the same SOC. This can be done at any SOC level. In batteries that are fully charged regularly, it is usually done at the 100 percent level.

A BMS balances a pack by removing extra charge from the most charged cells and/or by adding charge to the least charged cells. Balancing can be dissipative or nondissipative (dissipative—energy is wasted in heat; nondissipative—energy is transferred and therefore is not wasted). Dissipative balancing is often called *passive balancing*; nondissipative balancing is often called *active balancing*.

On a first order, how much current is required to balance a battery depends on why the battery is out of balance:

- Gross balancing—to remedy a gross imbalance right after manufacture or repair of a pack that was built using mismatched cells
- Maintenance balancing—keeping the pack in balance during normal use

Gross Balancing

The proper way to manufacture or repair a pack is to do so in a way that results in a balanced pack so that the BMS is not required to provide an initial gross balancing.

If, on the other side, the pack is built or repaired with no consideration for initial SOC of individual cells, the BMS may be expected to do gross balancing. In this case, the maximum length of time required to balance the pack will depend on the size of the pack and the balancing current.

Therefore, if the BMS is expected to balance a large, grossly unbalanced pack in a reasonable time, it will have to provide a relatively high balance current.

Maintenance Balance

If a pack starts balanced, keeping it in balance is a far easier job: All the BMS has to do is to compensate for the variation in self-discharge leakage in the cells.

In many applications, the BMS is not able to balance 24/7. Yet leakage discharges a cell 24/7. In such cases, the balance current has to be higher, in inverse proportion to how much time is available for the BMS to balance the pack. For example:

- If the BMS can balance constantly, the balance current can be just 1 mA.
- If the BMS is only able to balance for one hour every day, the balance current must be 24 mA to achieve an average of 1 mA.

Of course, it is okay if the BMS is able to run more balance current than the required minimum.

There is no reason to specify a BMS that can handle more balancing current than is required by the pack in the worst case of leakage current delta. The balancing current required is proportional to the difference in the leakage current and to what percent of the time it is available for balancing:

$$\text{Balance current (A)} = [\text{max leakage (A)} - \text{min leakage (A)}]/$$
$$[\text{daily balancing time (hours)}]/24 \text{ (hours)}$$

Typically, the self-discharge (leakage) current of lithium-ion cells is specified in terms of allowed months in storage at room temperature. (Actually, very few manufacturers specify it.) The value is on the order of many months. From that we can estimate the value of leakage current. For example, in a 100-Ah pack, if in the worst case the cells are discharged in 18 months, the leakage is

$$100 \text{ Ah}/(18 \text{ months} \times 30 \text{ days} \times 24 \text{ hours}) = 7.7 \text{ mA}$$

Let's assume that some cells have that maximum leakage of 7.7 mA, and all cells have at least 0.7 mA of leakage; then the delta leakage is 7 mA. From that, and from how long the BMS can balance, we can see how much balance current is required. For example, if the BMS is in a vehicle that is plugged in for 12 hours every night, charging takes 8 hours, and balancing occurs after charging is completed (4 hours). Then the balance current is

$$\text{Balance current (A)} = 7 \text{ mA}/(4 \text{ hours}/24 \text{ hours}) = 42 \text{ mA}$$

Thus, in this example, a BMS with a maximum balance current of 100 mA is sufficient to keep the pack in balance.

One way to increase the balance current is to increase the maximum current that the BMS can handle (say, from 100 mA to 1 A). But another way is to increase the time available for balancing. In the preceding example, balancing occurred only after charging was done, leaving only 4 hours for balancing. But what if the BMS were smart enough to know a priori which cells are likely to need balancing and do balancing whenever plugged in (12 hours in the example)? This is the approach taken by the more sophisticated BMSs (e.g., Texas Instruments).

In this case, in the preceding example, the available balancing time would be 12 hours, and the balance current would be

$$\text{Balance current (A)} = 35 \text{ mA}/(12 \text{ hours}/24 \text{ hours}) = 70 \text{ mA}$$

Thus a smart algorithm allowed the same BMS hardware (limited to 100 mA) to handle a pack with five times the leakage.

With a smarter algorithm, a BMS can handle a pack with somewhat more leakage, but only to a point (in practice, a factor of 3 at most). Beyond that, the hardware itself has to be able to handle more current.

So how much balance current is required for a lithium-ion pack during normal operation? Here are the rules of thumb that Elithion has derived to date:

- 10 mA is sufficient for small backup supply applications (10 kWh), 100 mA for large applications (100 kWh).
- 100 mA is sufficient to handle any automotive application (10 kWh, plugged in nightly).
- 1 A is sufficient for large pack applications other than backup (>100 kWh, cycled daily).

Conclusion on BMSs

- Balancing compensates for the SOC of individual cells. It does not compensate for capacity imbalance (which is what redistribution does).
- A balanced pack is able to supply the maximum charge, limited only by the cell with the least capacity (to supply all the charge, unlimited by any cell, redistribution is required).
- The single best thing for the sake of a battery pack's balance is that the pack is built balanced so that the BMS is not requir/ed to do gross balancing.
- It makes little economic and engineering sense to specify a BMS that can balance a grossly unbalanced pack just once in its lifetime. Such a BMS will be vastly more costly, bulkier, and produce far more heat than a BMS that is designed just for the job that it must do 99 percent of the time—keep a previously balanced pack in balanced. Instead, it makes more sense to build packs that are already balanced so that there's no need for a BMS that can perform gross balancing.

- If the pack is balanced at the factory, the BMS only needs to be able to handle a balancing current sufficient to compensate for cell-to-cell variations in self-discharge current during normal operation.

- A BMS that provides 100 mA of balancing current is sufficient for most lithium-ion applications.

- A smart algorithm that allows balancing all the time can increase a given BMS balancing capability by a factor of about 3.[14]

Tesla Controllers Use AC Propulsion Technologies

What can I say but that there is Tesla Motors. This new electric car on the market has an amazing AC drive of a new design with a new controller, new motor, and new battery subsystem. The Tesla Roadster delivers full availability of performance every moment you are in the car, even while at a stoplight. Its peak torque begins at 0 rpm and stays powerful at 13,000 rpms. This makes the Tesla Roadster six times as efficient as the best sports cars while producing one-tenth the pollution. Figure 7-8 shows you that

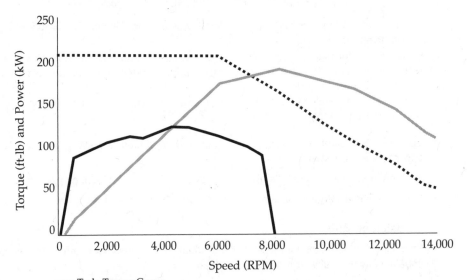

---- Tesla Torque Curve
——— Tesla Power Curve
—— 4-Cylinder High-Performance Engine Torque Curve

The black line represents the torque capabilities of a gasoline engine, which has little torque at low rpm and can only deliver reasonable horsepower within a narrow rpm range. In contrast, the dotted line demonstrates how the Tesla Roadster's electric motor produces high torque at 0 rpm, delivers constant acceleration up to 6,000 rpm, and continues to provide high power up to 13,000 rpm. The gray line shows the shaft power from the Tesla Roadster's electric motor as it builds steadily with increasing speed to a peak of 189 kW at around 8,000 rpm.

FIGURE 7-8 Performance and torque versus an internal combustion engine. (*Tesla Motors.*)

performance versus an internal combustion engine, and the numbers speak for themselves. The Tesla is a great "build your own" EV sports car for the masses today. The connection with the motor and batteries is being made in an intelligent and efficient manner.

Summary

In simple terms, computers think in binary logic: 1's and 0's, on and off, yes and no. Rather than complex feedback systems that hone in on the results you want (engines), you can implement a simple-logic approach that moves you there directly (controller). Also, as a result of improving technology, you can save on components while not compromising on safety and dollars. Cell phone and camera makers, with whom miniaturization has been raised to an artform, were ecstatic with the technology and have been improving on every new generation. What this means to EV converters is smaller and less expensive electronics. So there is a great future for EVs and EV conversions today. Going forward with each new generation of technology improves our chances for all-electric drive.

There are more controllers than ever before available: AC and DC high-quality industrial and cheap Chinese throwaways; new, used, and surplus. There are vast quantities of surplus parts and controllers readily available. You can get all the parts you need at pennies on the dollar. The Internet, search engines, and e-mail discussion lists have made information on EVs far easier to find.

The skills and tools needed to build your own EV controller have never been easier to get. There are manufacturer's application notes with complete working circuits, inexpensive motor controller demo boards that only need to be scaled up, power transistor modules that handle all the tricky high power wiring for you, and so on. The perfect EV controller design has not been created yet. There is no open-source controller project equivalent to Linux, for example—but there are people working on it!

The main thing stopping most people is simply their own apathy. Whether you believe you can do it or not—you're right!

CHAPTER 8

Batteries

For those who want some proof that physicists are human,
the proof is in the idiocy of all the different units
which they use for measuring energy.
—Richard Feynman, *The Character of Physical Law*

If you're planning an electric vehicle (EV) conversion and you're debating whether to use the traditional lead-acid batteries, as seen in Figure 8-1, or lithium, forget the lead. I've got some good news for you: Now you can now have lithium for the same price per mile as lead[1] (or even *less* than lead by the time you read this). Lithium EV batteries have become widely available in the United States over the past 10 years, the quality is great, and the performance is outstanding. I hope by the end of this chapter that you'll be persuaded to put the golf cart batteries back where they belong—in the golf cart.

The lithium batteries we've grown to love are lithium–iron phosphate ($LiFePO_4$), and the most widely available of these are the prismatic cells from China. Instead of the 30 miles on a charge of lead-acid batteries—if you're lucky—these $LiFePO_4$ batteries routinely take our EV conversions 80+ *actual* miles on a single charge (if you design your EV appropriately, of course). In the U.S. Trojan is considered a reputable lead-acid battery (Figure 8-1).

What about safety? $LiFePO_4$ is the safest lithium battery yet! Later in this chapter we'll discuss thermal runaway and why $LiFePO_4$ is so stable compared with other lithium-ion batteries, and we'll get to know our batteries in more molecular detail. We'll also talk about the ways that materials scientists have tinkered with the basic $LiFePO_4$ crystal recipe and how each alteration affects battery performance.

We'll highlight the differences between lithium and lead—especially what those differences mean to you, the EV converter, in real life. For example, with lead batteries, you may have heard that you'd be better off with an alternating-current (AC) drive system because of the Peukert effect. With $LiFePO_4$, the Peukert effect (or "lead-foot tax," as I like to call it) is so slight that it doesn't really matter which drive system you use. The performance and range are excellent either way.

We'll gossip a little about the patent-grabbing scandal of the troubled childhood of $LiFePO_4$ batteries, and see why the original equipment manufacturer (OEM) "big dogs" might have left the best EV battery possible to us while chasing their Holy Grail—energy density.[2]

And finally, we'll be talking with some friends at Electric Cars Are for Girls about some best practices for using these prismatic $LiFePO_4$ cells in your EV. This group has

FIGURE 8-1   Typical lead-acid battery for cars from Trojan. (*Seth Leitman.*)

been using these batteries in EVs for years and is familiar with all the dos and don'ts and potential pitfalls of practical use. Follow the advice of this group, and you'll get 3,000 cycles of happy EV grinning out of your new battery pack. The AA-shaped cells that A123 makes are an example of the *cylindrical* battery, as shown in Figure 8-2.

Metals, Salts, and Ions: Nature's Sad Love Story

You may be wondering what metals, salts, and ions have to do with each other—or what they have to do with batteries. (If you've studied chemistry and/or physics, go ahead and skip this part. I'll wake you up when it's over.)

A metal, you may already have noticed, is shiny and hard and a good conductor of both heat and electricity. At the level invisible to the naked eye, it is an orderly crowd of charged particles, both positive and negative, in balance with each other so that the total number of positively charged particles and negatively charged particles is exactly the same. Yeah, believe it or not, your stainless steel fork and your copper kettle are made up of nothing more than a crowd of charged particles.

Let's call the charged particles *ions*. Faraday did, and what's good enough for Michael Faraday is good enough for me. Oh, a little terminology shorthand here: When I say *ions* and refer to *lithium ions*, I am talking about a *cation* (pronounced "CAT-ion"), a positively charged Li$^+$ ion. In lithium battery discussions, this is always what people

FIGURE 8-2   Filled with A123 cells, these frames will be 72-Ah modules for an electric car, connected both in series and in parallel. Cylindrial A123 batteries. (*TurnE Conversions.*)

mean when they say *ions* and don't explain further. Lithium's stable, low-energy ion is 1+. (There are negative ions, too, but lithium doesn't want to go there at standard temperature and pressure.) Chloride, carbonate, oxide, phosphate—these are examples of negative ions, or *anions*. Their stable, low-energy states are Cl^-, CO_3^{2-}, O^{2-}, and PO_4^{3-}.

Let's talk about high- and low-energy states for a minute, too. The more an ion's charge state varies from its natural, relaxed, low-energy state, the more uncomfortable it is. The more uncomfortable it is, the more anxiously it wants to get back to its happy place, and the more energy it will spend to get there. And just how uncomfortable *is* an ion anyway? The measure of this is its *oxidation state*.

Okay, so back to metals. Metals just can't seem to help splitting up into ions—that's part of their charm! Give them half a chance and a little water, and they'll be splitting down the seams and conducting electricity like mad through that water . . . which is then dubbed an *electrolyte* solution. This is all an electrolyte solution is, really—some sort of solvent with ions in it. The trouble is, all this splitting up into negative and positive charge factions is supposed to be temporary because it upsets their energetic balance to be apart, and when the balance is upset, Nature goes looking for ways to restore it.

The positively charged ions go looking for an *equally* negatively charged particle, which we'll call an *electron*. This tendency to look for a mate is how salts are formed in nature. A positively charged ion such as sodium, for example, might meet up with a negatively charged ion such as chloride and form a salt. (This is why we nickname sodium chloride *salt*, by the way. But it's not the only salt; there are all sorts of salts. Lithium carbonate is a salt.)

The tendency for a charged particle to look for a mate is also the quality we exploit to create a battery. To make a lithium-ion battery, we simply take lithium salt (it's usually found as carbonate in nature) and deprive it of its carbonate anion. Now we've got a positively charged lithium ion, forever searching for its lost soul mate. I guess ions are easier to enslave and exploit when they're heartbroken.

And now that we've captured a bunch of miserable, desperate lithium ions? Let's build an EV battery.

Building a Battery

First, we're going to need some electrodes. An *electrode* is a location where charged particles like to go. It's like a singles bar for lonely lithium ions (okay, sorry, I'll quit). Batteries have two electrodes—a positive electrode called a *cathode* and a negative electrode called an *anode*. In a lithium-ion battery, the cathode gets all the attention and makes all the difference between one classification of lithium-ion battery and another.

$LiCoO_2$, the layered metal oxide in a classic lithium-ion battery, is named for its cathode, cobalt oxide. $LiMnO_2$, a newer type of metal oxide in a lithium-ion battery, has a cathode made of type of manganese oxide known as spinel. Spinel is a crystal that sort of looks like a three-dimensional tic-tac-toe board at the molecular level. $LiFePO_4$ is just another type of metal oxide in a lithium-ion battery, only this one has a cathode made of iron phosphate, and the crystal is an olivine (peridot) type. We'll see this more clearly a little later.

And the anode? It's usually made of carbon—a special kind of graphite. Most types of lithium-ion batteries use a graphite anode.

In a lithium-ion battery, both the cathode and the anode have a special type of structure that allows *intercalation* of the lithium ions. I liked Seth Fletcher's description in a National Public Radio interview discussing his lithium battery book entitled, *Bottled Lightning*. I'm going to borrow his analogy and embellish it a little.[3]

Imagine a huge hotel such as the Westin Hotel with multiple floors, and then imagine that the guests can only enter and leave their rooms through the windows, and that will give you an idea of how lithium inserts itself into the iron phosphate cathode. The rooms fill and empty very quickly when you don't have to use the elevator!

Lead batteries, by contrast, have actual metallic lead electrodes that are consumed and restored with each charge. Lithium batteries aren't like that. A lithium cathode has no lithium metal in it, and it isn't degraded when the battery is discharged. It's just a holding tank.

This cathode "hotel" structure allows lithium to come in from the electrolyte quickly and gather around in a somewhat orderly fashion, while at the same time electrons are leaving out the back door of the hotel (where you will capture and put them to work) just as fast. This *must* happen because Nature will not tolerate a lopsided charge. The electron movement around the circuit is called *current*, and when you have current flow, your EV will move.

The anode also employs lithium-ion intercalation. While the battery is charging, the lithium ions burrow into the anode and wait. The electrons? They never reach the electrolyte or their long-lost ions. They are trapped in the anode, the cathode, and the outside circuit—going back and forth with each charge and discharge.

The reason I mention this is because you will see that two things must happen in order for a battery to work:

- The anode and the cathode must accept and release lithium ions from the electrolyte, wasting as little energy as possible in the process.

- The anode and the cathode must both move electrons readily around the circuit.

These are called *ionic conductivity* and *electronic conductivity*, respectively. If either of these things fails, the circuit stops, and so does your EV.

The Circuit

Rechargeable batteries all do the same thing, whether they power your EV or your iPod, no matter if they're lead or lithium: They convert chemical energy into electric potential.

Every time a lithium ion inserts itself into the cathode, an electron goes into the workload circuit toward your iPod from the anode, and the result is that "Bohemian Rhapsody" flows into your ears. As I mentioned before, this ion-electron movement is kept in strict equilibrium by Nature. Ions flow one way, and electrons (and thus current) flow in the opposite direction. One lithium ion, positively charged, enters the cathode, and one electron, negatively charged, exits. Actually, more like a gazillion electrons per second. In this way, electron movement does the work of the electric circuit.

Energy acts a little like boulders rolling down a hill. You may have heard it described this way before. In your EV's system, *potential energy* is a fully charged battery pack. *Kinetic energy* is what moves your EV's wheels. As the potential energy in your battery pack is converted into kinetic energy when your EV begins rolling down the driveway, you look at the *state of charge* (SOC) meter on the dash and notice that you have less and less charge in your pack. Your potential energy is not really disappearing; it is simply doing work by being converted to movement. The amount of energy remains the same.

I should note here also that your iPod doesn't eat the electrons when it plays "Bohemian Rhapsody"; it only exploits them for a little while and sends them on their way down the energetic "hill" to the anode. When lithium ions roll down the energetic hill from the anode and crowd into the cathode, electrons are squirted uphill, not having a choice about it, to do your bidding.

Thus, at the end of the day, when the anode has collected all the electrons inside, and all its ions outside in its half of the electrolyte lake, and when the cathode has all the ions tucked in to their beds and has spit every last electron out into the circuit, your EV stops rolling because your battery is dead. (Okay, you already know not to run your EV batteries down to 0 percent SOC. I was just making a point.)

It's time to plug in, then, and reverse the energetic hill: The electricity from the charger pushes the ions back up the hill to the anode, which drags the electrons back down to the cathode, where they wait for you to hit the accelerator and run the batteries down all over again.

The battery is part of a *circuit*, so if the flow of electrons gets choked up somehow, the flow of ions slows down to match, and the whole thing gets bottlenecked. You can't have ion flow without electron flow, and vice versa. This point will be important later when we're discussing lithium-iron–phosphate (or $LiFePO_4$) batteries in more detail.

Amperes Versus Volts

You may already know that increasing your pack voltage gives your EV more acceleration. What's going on there, you might ask? Everybody knows acceleration is all about "sucking amperes."[4] Well, the higher your pack voltage, the steeper is that energetic hill. The boulders—electrons—always want to roll *faster*—that's current, measured in amperes—down a steeper hill. I should point out that the higher the current, the more *resistance* (heat) there is in the circuit as well, so you'll want your wiring to be sized appropriately for the voltage of your pack.

Battery Anatomy

In a lithium-ion battery, the cathode and the anode are actually sheets of what looks like foil. When I think of a cathode being made of crystals such as spinel and olivine, I imagine a jewel like the diamond in a necklace, but the reality is instead sort of wide and flat and covered in goopy black powder.[5] (Sorry!) The anode is pretty much the same to the naked eye. There's a separator between them, and I've heard that it kind of looks like a piece of plastic trash bag.[6]

The electrolyte? That's the goopy stuff the lithium ions swim in. In a lithium-ion battery, the electrolyte is liquid, it's an organic solvent, and it's about as toxic as you can get.[7] Here is A123's list of electrolyte nominees: "Exemplary materials include a polyvinylidene fluoride (PVDF)–based polymers, such as poly(PVDF) and its co- and terpolymers with hexafluoroethylene, tetrafluoroethylene, chlorotrifluoroethylene, poly(vinyl fluoride), polytetraethylene (PTFE), ethylene-tetrafluoroethylene copolymers (ETFE), polybutadiene, cyanoethyl cellulose, carboxymethyl cellulose and its blends with styrene-butadiene rubber, polyacrylonitrile, ethylene propylene diene terpolymers (EPDM), styrene-butadiene rubbers (SBR), polyimides, [and] ethylene-vinyl acetate copolymers."[8]

Now, to be fair to A123, the scientists didn't wake up one day on the wrong side of the bed and say, "Let's see, what toxic sludge can we put into an EV battery?" All lithium-ion batteries from any manufacturer use organic solvents taken from this same list.

Safety Note on Lithium-Ion Battery Fires and Water

The electrolyte is the real reason why we don't use water to extinguish lithium-ion battery fires. From the National Highway Transportation Safety Administration (NHTSA) website:

> In case of extinguishing fire with water, large amounts of water from a fire hydrant (if possible) must be used. **Do not** extinguish fire with a small amount of water. Small amounts of water will make toxic gas produced by a chemical between the Li-ion battery electrolyte and water. In the event of small fire, a Type BC fire extinguisher may be used for an electrical fire caused by wiring harness, electrical components, etc. or oil fire.[9]

The lithium ions in the electrolyte are reactive in water, yes, but rather than blowing up the battery like folks imagine when they visualize lithium metal in water (yeah, I remember that chemistry lab, too), the real danger is that the lithium reaction will be atomizing those organic solvents into something you can *breathe* while you're standing there hosing out the fire.

Bottom line? If you're going to use water to put out a lithium battery fire, *drown* it. Oh, and keep in mind that these organic solvents are *really* flammable.

All lithium-ion batteries have these anatomic features in common: cathode, anode, electrolyte, and separator. (Lithium-polymer batteries have a solid electrolyte but are chemically similar otherwise.) Now I'm going to talk about ways that lithium batteries can differ from each other, even batteries of the same cathode chemistry.

Shapes and Sizes of Lithium-Ion Batteries

Battery Formats

Lithium-ion batteries come in three basic formats:

- Cylindrical
- Prismatic
- Pouch

There's a little more to the differences than shape, as you'll see.

Cylindrical

The AA-shaped cells that A123 makes are an example of a *cylindrical* battery, as shown in Figure 8-3. This is a battery pack that was made by students in Germany with TurnE Conversions and Ruediger Hild. As part of the A123 pack, the students designed their own spacers made from 5-mm laser-cut plastic resin from Kolon Plastics called Kocetal POM for research purposes; unfortunately, it was very expensive. However, in mass production, the costs can easily come down.

The controller runs on Windows only and displays all parameters of the battery system. The test scenario involved seven cells in series.

FIGURE 8-3 A123's small LiFePO$_4$ batteries used in TurnE Conversions from Germany.

In fact, it would be more accurate to call this format *wound* because the difference between the various formats is more than skin deep. The shape of the packaging does not necessarily indicate what's going on inside. You can get a wound cell that's been packaged in a rectangular case. Laptop batteries are a good example of this. Tesla Motors' battery pack is also made up of cylindrical (lithium–cobalt oxide) cells.

With cylindrical cells, the electrodes and separator sheets are rolled up together like a long jelly roll, and each jelly roll fits into a package; the individual cells then can be metapackaged into any shape case you like.[10]

What is the drawback to the wound cylindrical design? It's not the most efficient packaging; there's a certain amount of wasted space in packing round objects next to each other, there can be heat buildup in the center of the wound cell, and the extra packaging adds to the overall weight of the battery pack. In the EV conversion world, we're always looking for ways to *lose* weight rather than gain it!

That said, though, I must admit that I've fallen in love with the various EV uses of A123's small 26650 cells. Tesla and Killacycle (and Vectrix) showed us that we could use small lithium batteries to great effect in an EV, and now a German conversion company called TurnE Conversions is using these excellent A123 LiFePO$_4$ cells with their amazing power density (8.5 times the power density of any other lithium chemistry out there) to make cheaper, safer, faster-charging lithium battery packs in 72-Ah modules. Higher power density means that you can build a lighter pack with all the acceleration of a larger pack, as you'll soon see.

Prismatic

Prismatic cells have their electrodes and separator sheets stacked up together like the pages of a book. The batteries we EV converters buy from Chinese companies such as CALB and Sinopoly are the *prismatic type*. They come in rectangular blocks that might remind you of Legos, except that they don't interlock.[11] Prismatic cells are reportedly easier to cool because of their ribbed design, which lets air flow between them.[12] This could be a pretty important factor when you're powering a high-current device such as an EV (high current flow tends to generate heat), although in practice it doesn't seem to be much of an issue.[13] At any rate, air cooling your batteries is cheaper and simpler than fluid cooling.

Pouch

Most of the pouch cells you see are lithium-polymer cells. With pouch cells, the electrodes and separator are not in a case.[14] They're in a bag, along with the electrolyte. The battery electrodes are accessible to *conductive foil tabs*. The advantage with these pouch cells is that you can fit them into spaces you wouldn't be able to fit batteries with rigid cases, and you can orient them any way you like. The disadvantage is that those foil tabs are delicate, and the automotive environment of your car is a bouncy, jostling, vibrating place that is trying very hard to destroy your battery connections.[15]

Lithium-polymer batteries have solid electrolyte rather than liquid, which helps them to retain their shape, but pouch cells are prone to swelling, so it is important to take this into account when building a framework for them in your EV.[16]

Prismatic pouch cells are made up of LiFePO$_4$, and A123 is making some of these, too.[17] Presumably, the electrodes and separator are layered like the pages of a book with liquid electrolyte in between, but instead of being placed into a "Lego block," they are placed into a flat pouch. With these A123 AMP20 cells, the power density is phenomenal compared with the Chinese prismatic LiFePO$_4$ batteries (because of the A123 "secret sauce," which I'll talk about later); however, using the AMP20s in an EV has proven to be a little problematic. The pouch is floppy, and the electrode tabs are delicate, the same as with lithium-polymer pouches. This means that the EV converter must create an external structure to hold the battery pouches and the connections firmly in place. Fine. But holding the pouches firmly in place sort of denotes *surrounding them with something firm*, and the problem with LiFePO$_4$ pouches with liquid electrolyte is that they tend to swell,[18] so the battery case has to flex with the swelling pouches or crack. The connections also need to flex with the swelling pouches or they also crack—but not too much flexing or the connections break anyway. You can see the dilemma.

Battery University says that with lithium-polymer pouch cells, swelling is usually due to a manufacturing defect,[19] and manufacturing defects should be detectable immediately. How this is managed with lithium-polymer pouches is that you charge the pouch cells up a little before you install them to see if one or two of them swell. If they do, send them back. With the A123 pouches, sending them back is not so simple because they were bought from surplus, which means no returns. Also, along with the recent bankruptcy of A123 Systems and court disputes with Johnson Controls, now a Chinese company will provide loans to pay off creditors.

Ampere-Hours and Voltage

For an EV, your system voltage generally will be somewhere between 72 and 320 V. You're familiar with the lead-acid batteries that come in 6-, 8-, and 12-V packages, right? That math is easy. Well, lithium battery math is a little different.

Chinese prismatic cells come in a package that's measured in ampere-hours (Ah). (That's amperes *times* hours, not amperes *per* hour, by the way.) You can buy these cells in units ranging from 40 to 1,000 Ah, but they're all 3.2 V (or thereabouts). So what do you do? You buy the ampere-hour category you want—100 Ah or so—and strap a whole bunch of them together, or you can tell your battery supplier what configuration you need (half of them in the trunk, the other half split between two front battery boxes, etc.), and the company will put your pack together for you.

More ampere-hours mean more weight, but lithium ampere-hours are a lot lighter than lead-acid ampere-hours! To give you an idea of the difference, the CALB 180-Ah LiFePO$_4$ battery is a popular size, and it weighs 12.35 pounds. If you're putting together a 120-V system, divide 120 by 3.2, which gives you about 38. That's 470 pounds of batteries. Not bad. Those old Trojan T-105 batteries, on the other hand, weigh 1,200 pounds for the same pack voltage. *Note:* The only reason we use numbers like 120 or 144 V is because back in the day we had to use lead-acid batteries, and they only came in packages of 6, 8, or 12 V or whatever, but you can certainly have a 121.6-V system with your new lithium pack if you want. You *could* round up or down, whichever, if you're switching out a lead pack for a lithium pack. (It's a lot more satisfying to round *up* and "feel the juice" after you've spent the money and gone through the trouble, though— am I right?)

Wiring in Series or Parallel

Question: Can't I buy a bunch of 40-Ah cells and wire them in parallel to get the ampere-hours I want? For example, four of the 40-Ah cells wired in parallel will give me 160 Ah, right?

Answer: Sure. People get quite creative with the battery configurations to fit in the space available! I'll talk about this at the end of the chapter in the "Best Practices for your LiFePO$_4$ EV Batteries," section.

C-Rate

The *C-rate* is a capacity rating of current flow in a battery. The battery manufacturer (in this case, a China Aviation 180-Ah LiFePO$_4$ battery) offers a graph such as the one in Figure 8-4 to show how the battery performs at different current rates.

The 1-C rate answers the question: How many amperes at a time would I have to draw out in order to drain my battery completely in one hour? The answer, in the case of the 180-Ah battery above, is 180 A. It's just a math problem after that.

Do you want the 3-C rate? Multiply your ampere-hours (in this case 180) by 3, and you've got 540 A. The 10-C rate? 1,800 A. These are examples of fast current flow. In the chart in Figure 8-4, it's discharge current that is being referred to.

Slow current flow, like the top two lines in the graph, is shown by fractions of C. The top line is 0.1 C, or a tenth of the 1-C rate, that is, 18 A. The middle line is 0.3 C, 54 A. When you see C/4, that's the same sort of thing—it's the 1-C rate divided by 4, or how fast we could drain the whole battery in four hours. There is a recommended charging rate measured in C-rate, too—usually 0.3 C.

Flat Discharge Curve

You may have heard the term *flat discharge curve* when speaking of LiFePO$_4$ batteries. It's not perfectly flat, but there's a nice wide region in there before about 80 percent depth of charge (DOD), or 144 Ah, where the voltage does not drop much during the

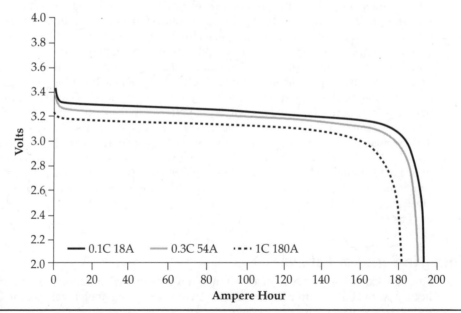

FIGURE 8-4 Typical LiFePO$_4$ discharge curve. (*www.lithiumstorage.com/ChinaAviation.*)

discharge cycle. With a lead-acid battery, the voltage drops as the discharge progresses, and the line would have a noticeable slope downward and to the right. With higher current flow, the slope would just get steeper and steeper! This is the *Peukert effect*, and while it was a major consideration with lead-acid batteries, with LiFePO$_4$ batteries it's not much of a problem.

The Diamond in the Discard Pile: LiFePO$_4$

The Scandal

I never could resist a good scandal, and the story of how we wound up with a better lithium-ion battery than the OEMs is a really juicy story.[20] A while back, some nice folks at John Goodenough's laboratory at the University of Texas had a good idea: Hey, why don't we use iron phosphate as a lithium-ion battery cathode? So Dr. Goodenough and a minion or two gave that a whirl, but the results were underwhelming. The problem? The iron phosphate cathode made for a remarkably stable lithium battery, and that part was nice, and the lithium ions seemed to move like lightning—but the electrons got bogged down in the cathode, bringing the whole battery to a crawl, if not a halt. This is not what anybody likes to see in a battery!

Maybe I should mention the minions because they'll be important later in the story. Minion one was a postdoc (newly minted scientist) named Padhi, Dr. G's right-hand man. Minion two was a visiting scientist from Nippon Telegraph and Telephone (NTT) in Japan named Okada. These two minions became friends, as people will do when they work together.

So, as I was saying, Dr. Goodenough wasn't terribly excited about the possibilities of LiFePO$_4$ from the initial research results, but an associate from Hydro-Quebec (H-Q) liked it well enough to get exclusive rights from Dr. G to produce LiFePO$_4$ batteries in North America. That guy's name was Michel Armand, if you're keeping score.

So someone from Armand's lab at H-Q got the idea of *coating* the LiFePO$_4$ cathode with a thin layer of carbon—and suddenly, a so-so battery was a pretty darn good battery.

H-Q then patented the coating process. From then on, as far as the folks at H-Q were concerned, any LiFePO$_4$ battery made in North America must be made by them because they obtained the exclusive license on LiFePO$_4$ from Dr. G and the University of Texas, and on top of that, H-Q also held its own patents on the cathode coating.

Now let's go back to Dr. G's minions for a minute. Padhi and Okada had become friends, or maybe Okada played Padhi like a fiddle, but whatever the case, after Okada went home to Japan, he and Padhi were catching up on old times, and Padhi happened to mention the research he was doing on LiFePO$_4$. Then Padhi *happened* to send Okada details of the research via e-mail, and shortly thereafter, NTT *happened* to patent LiFePO$_4$ cathodes in Japan.

H-Q was not pleased about this, understandably, and pretty soon all the parties involved were throwing lawyers at each other. Meanwhile, back in Boston, Yet-Ming Chiang (of A123 fame) was fiddling around with a LiFePO$_4$ cathode in his laboratory, too, and Chiang's own minion discovered that doping the olivine crystal with supervalent niobium or zirconium made the cathode a better conductor of electrons. *Doping*, by the way, is just the process of adding friendly impurities, very judiciously, to the cathode material—well just a little, just in some key spots.

Unfriendly impurities in the cathode are the leading cause of battery failure, but a friendly impurity stuck in just the right place makes the cathode material work better. In this case, it made the cathode a *lot* better. So it was off to the patent office for A123, too.

Poor Armand wasn't pleased about this development either. He was pretty sure that Chiang had just accidentally coated his cathodes with carbon—Armand's own invention—and *that* was responsible for the spectacular results Chiang was seeing. After a lot of public hurling of insults and threats, it turned out that Armand was mistaken about his conclusion. In retrospect, I suppose that Armand probably should have gotten a clue that Chiang was onto something completely different when Chiang's cathodes were getting *seven times* the power density of other lithium-ion batteries—but we see what we want to see, don't we?

Anyway, after the patent litigation dust settled—if it has, yet—the OEMs decided that LiFePO$_4$ was not their cup of tea and moved on. I don't know, maybe the profit density was already becoming too diluted.

That Beautiful LiFePO$_4$ Cathode

What makes the LiFePO$_4$ cathode so special? It's got three very desirable qualities for an electric vehicle.

- **Power density**—There is no other lithium-ion battery available that has the ability to deliver horsepower like LiFePO$_4$ can[21]—although not every LiFePO$_4$ battery is optimized for power density. The cathodes can be tinkered with in the laboratory in a number of ways, which I'll talk about later on. The same chemical foundation that makes this power density possible also makes LiFePO$_4$ the leading candidate for lightning-fast charge and discharge, second only to supercapacitors.[22]

- **Cycle life**—LiFePO$_4$ has less impedance than other lithium-ion chemistries, which translates into less lithium metal plating on the anode in the form of dendrites and "mossy lithium." This means that LiFePO$_4$ batteries last longer than other lithium batteries.[23] CALB (China Aviation) is claiming that its batteries are good for 3,000 cycles at 80 percent (if you keep the batteries at 80 percent DOD). That sounds to me like 10 years of service if you take reasonable care of them.[24]

- **Safety**—You have to actually *work* at getting this battery to go into *thermal* runaway; it doesn't want to go there on its own. Its runaway temperature is 518°F, considerably higher than other lithium chemistries.[25] Why? Because of what I like to call Mother Nature's battery management system, which operates at the molecular level.

Molecular LiFePO$_4$

Let's drop down to the molecular level and talk about LiFePO$_4$ for a minute, and you'll see why it's got some of the qualities it does. For purposes of the cathode, we can discard the lithium from the discussion because it's just a visitor—it doesn't actually live there. Figure 8-5 is here to help you visualize all the options today and in the future for battery chemistry, even though this chapter focus on lithium-based technologies.

FIGURE 8-5   Molecular view of lithium ion LiFePO$_4$.

In the figure, the big dark gray blocks—octahedrons, two pyramids back-to-back—are FeO$_6$. The light gray blocks are phosphate. The little dots are lithium ions. The oxygen molecules of the FeO$_6$ have the iron surrounded, so its electrons are insulated from neighboring iron's electrons, which is what makes this form of iron *nonconductive*.

Now why on earth would you want a battery cathode material stuffed full of molecular insulation? I wondered that too. I believe that it's because that insulation in the crystal structure, if you can get it lined up correctly, can be used to build little express lanes for ions and electrons to flow from hither to yon in the cathode very quickly.

But first the scientists had to get some electrons moving *at all*. It took Michel Armand about six months to figure out that a thin *coat of carbon* on the outside of the lithium–ion phosphate crystal seemed to improve conductivity quite a bit.[26] Exactly *why* carbon gets the conductivity ball rolling is still a subject of debate, but *that* it works is not in question.[27]

A student in Chiang's laboratory at MIT discovered a little later that *doping* iron phosphate very lightly with *supervalent ions*—that is, (positively) charged and high-energy ions—improved the electron conductivity by a *lot*.[28] Dope it too much, and it slows down conductivity. Dope it just enough, and it delivers *seven times* the conductivity.

Doping means adding a small amount of the substance, in this case supervalent niobium or zirconium ions, when making the crystal. The ions arrange themselves into the crystal in predictable spots based on their size and charge. And what exactly does this doping do? It (probably) warps the doors/walls of the tunnels a little to create a larger opening for the ions and electrons to pass.[29]

Tinkering with the LiFePO$_4$ Recipe

Doping is an example of one of the ways that LiFePO$_4$ has been altered through nonstoichiometry to make it a better battery cathode material. *Nonstoichiometry* is just chemist speak for "tinkering with the recipe," like a baker who adds a little more of this or a little less of that to make a cake come out differently. In the case of LiFePO$_4$, nonstoichiometry has been used a few different ways to improve it. The first was the supervalent doping we just discussed, done by Chiang and his team at MIT. In order to get those niobium ions in there, you have to add them to the LiFePO$_4$ crystal recipe before you put it in the oven, so to speak.

Another way nonstoichiometry has been used is to substitute other transition metals, such as manganese or nickel, for some of the iron ions, called *isovalent doping* (the charge is the same as the iron or the same as the lithium depending on which side of the express lane you want to put the dopant). There are other possibilities for this type of doping, and the point of isovalent doping—depending on the dopant—seems to be to increase either energy density or cycle lives.

The third way that nonstoichiometry has been used with LiFePO$_4$ is to add a coating to the outside. There is more than one way to add a coating to a cathode, but Gerb Ceder's laboratory at MIT took advantage of the fact that some substances have different solubilities and will naturally separate into layers in order to bake the outer *charge-collecting layer* right into the cake. I'll talk more about this contribution later.

A fourth nonstoichiometry trick is just to create some holes in the crystal, called *antisite defects*. A lithium ion will get stuck in the hole, and since it is a positively charged ion, it closes off that particular lane for lithium traffic and opens up a different avenue for concerted cross-channel diffusion—like closing off a surface street and opening up a freeway.[30]

Stumbling onto Something Good

It seems like a lot of LiFePO$_4$ cathode research goes this way: Scientists do something that improves conductivity; then, in trying to explain why it worked, they discover that they didn't understand it at all and now have a whole new set of questions. Or they create a computer model that works perfectly on paper and falls apart entirely in practice.

Here's a perfect example of what I mean: The cathode seems to have a slowdown in the lithium loading process, and it looks like a problem intrinsic to LiFePO$_4$. It goes something like this: There are two phases of lithium insertion; let's call the phases *loaded* and *unloaded*. You can see it clearly with x-ray diffraction because the loaded layer is a different color than the unloaded layer. The outer layer of the cathode gets loaded first, and then the loaded phase steadily moves inward toward the middle of the cathode until all the lithium is on board. (It's like getting on a bus. The lithium ions that are already on the bus don't seem to want to "move on back" and let the other lithium ions on the bus. It sort of seems like the lithium ions have to push their way in, one molecule at a time. Scientists call this *diffusion-limited lithiation*.)

Thus Chiang's team at MIT came up with a couple of ideas to overcome this problem. One idea was to make a very thin cathode so that there was a lot of surface area compared with the total volume of the cathode. Let's make the whole *bus* the front of the bus! But alas, no. It made for a very fragile cathode that didn't last long before it died. Chiang's next idea was to build more floors in the cathode hotel. Instead of the 35 floors of the Westin, with each lithium ion going to its own window, let's make 300 floors or 3,000 floors in the same space. And that worked *great*, much better than the preceding idea. This is how *nanophosphate* was born. *Reducing particle size*[31] to nanoscale makes the bulk diffusion of lithium ions go much faster.[32]

Going with the Grain

Even with those improvements, it looked like diffusion-limited lithiation was the Achilles heel of $LiFePO_4$. On paper and in computer modeling, the lithium ions should have been screaming through the cathode, and they weren't! What was the holdup?

Another materials scientist at MIT, Gerb Ceder, went to work on that puzzle. And it turns out that with $LiFePO_4$, there are two areas where lithium diffusion can get bottlenecked. The first is on the surface of the crystal particle, and the second is through the body of the crystal particle—called *bulk diffusion*.

Ceder discovered that the ions were not getting bogged down *inside* the cathode, although it looked that way; they were actually getting slowed down at the cathode's surface. Why? It turns out that diffusion of lithium into the $LiFePO_4$ cathode doesn't happen equally around the surface. $FePO_4$ has a grain to it, like wood—it is *anisotropic:* The lithium ions can only load *with* the grain.

The lithium ions could get to the cathode quickly enough, but they might bump into it on any side, and then they had to change directions and wander around a while until they found a door, which took up valuable time. Ceder found that he could give the ions some help in the form of what he described as a "racetrack" that they could use to get quickly from wherever they happened to land on the cathode directly to the door on the correct side of the grain. The "racetrack" is his *charge-collecting coating*, the one I told you he baked right into the cake. Pretty cool, right?

Once he drew the ions a map, the lithium could load and unload from Ceder's coated crystals at unbelievable rates, and humble $LiFePO_4$—the battery material the OEMs rejected because of its moderately lower energy density—became the newly crowned king of the C-rate. All other things being done correctly, Ceder said, a future version of this battery should be able to charge as fast as you can deliver the juice. (Give it four to five years for production, he said in 2009.) Something like a cell phone or a laptop, he said, will be able to charge in one minute or less. For an EV? It becomes a question of how fast can you deliver the power, not how fast you can receive it. He speculated that you'd be able to charge a plug-in hybrid-sized pack in 12 minutes.[33] If you can fill up in 12 minutes, energy density becomes a much less interesting quality.

Impedance and Fast Charging

The impedance of a cell determines how fast you can charge it before it starts plating lithium on the anode. (You want impedance to be as low as possible if you want a fast-charging battery.) Traditional $LiCoO_2$ has a relatively high impedance at 40 Ω-cm^2, and $LiFePO_4$ has quite a low impedance at 12 Ω-cm^2. A123 is working on getting its $LiFePO_4$

cells' total *area-specific impedance* down to 8 Ω-cm^2.[34] This is explained nicely in A123's most recent patent application for LiFePO$_4$ batteries, so I'll share it with you. I'll include the reference website address so that you can have a look at it yourself.

A123 says this on the implications of impedance: When you charge a battery at high rates of current, the cell voltage increases. The higher the impedance, the higher is the voltage you need to keep up the current flow. So how does this contribute to lithium plating, you ask? "As the battery is charged at higher rates, the positive electrode potential is pushed to a higher potential, and the negative electrode drops to a lower potential. At high rates, the potential at the negative electrode drops to below 0 V versus Li/Li$^+$, and plating of lithium metal at the negative electrode occurs."[35]

Lithium plating, you'll recall, is how lithium batteries die, and it's a safety hazard to have lithium metal in your battery. Fast charging ordinarily is the enemy of lithium-ion battery cycle lives, but LiFePO$_4$ is naturally pretty resistant to lithium plating.

In a conventional 4.2-V lithium-ion battery, such as one that contains LiCoO$_2$, the maximum charging current is also limited by the potential at the positive electrode. A high potential at the positive electrode will cause electrolyte oxidation, which greatly decreases the lifetime of the battery. Lithium–iron phosphate has a lower average voltage during charge. Thus a positive electrode incorporating lithium–iron phosphate as the active material can be polarized to a greater extent before reaching the electrolyte oxidation potential.[36]

A word or two about *impedance* and *internal resistance* is in order here. Impedance can be imagined as the little hassles the lithium ions encounter while they try to get to their resting spots in the cathode. These hassles slow the ions down and waste energy. It's not quite the same as resistance, which I always associate with heat generation. The cathode isn't heating up at all; it's just that there are all these energetic barriers that delay the loading or unloading process so that it can't intercalate at theoretical maximum speed. I think there's a mathematical equivalence in some electrical systems between impedance and internal resistance, but it doesn't have the same implications.

Energy Density Versus Power Density

You've probably heard that LiFePO$_4$ has been *outgrown* (shall we say?)—by the OEMs—because its energy density is too low, right? Let's look at this question and see if we agree or not.

First, what exactly is energy density? *Energy density* is measured in watt-hours per liter (Wh/L), so it's a question of run time. (How many miles can I go with batteries that take up x amount of space?) Another measure of energy density that EV converters might be very interested in is watt-hours per kilogram (Wh/kg), or *specific energy*. (How far can I go on x weight of batteries?)

Power Density

The *power density* of a battery is measured in watts per kilogram (W/kg). LiFePO$_4$ has better power density than any other lithium-ion chemistry out there, including the latest and greatest lithium technology being developed for the OEMs using your taxpayer dollars. Power density, in a nutshell, means that same structural and chemical assets that allow ultrafast charging also allow for serious horsepower delivery.

Yes, I used the "H" word—*horsepower*. Are you surprised to hear that your EV's horsepower actually comes from your batteries? It's true. In your old gas guzzler, the engine produced the horsepower, and the gas tank provided the range. In your new EV, the horsepower and range both come from the battery pack. When you stomp on the accelerator, your electric motor can either rise to the battery pack's challenge, or it can roll over and play dead. Otherwise, it doesn't have much to do with producing the EV's horsepower.

A123's AMP-20s deliver an incredible 2,400 W/kg. Your Chinese prismatic LiFePO$_4$ batteries aren't optimized for power density like the A123 batteries are, but you can still get liftoff with them, don't worry.

To get an idea of your pack's horsepower, look for "peak discharge current" in the specs. You want to know how many amperes your pack can deliver in a pulse, which is somewhere between instantaneously and 10 to 15 seconds depending on who you ask. HiPower's 200-Ah LiFePO$_4$ batteries will give you 1,000 A for a "burst" of 10 to 15 seconds—the 5-C rate.[37] That should be plenty to get around that Hummer on the freeway, even uphill. The same batteries can run continuously at the 3-C rate, 600 A.

The beauty of LiFePO$_4$, compared with other lithium-ion chemistries, is that your LiFePO$_4$ battery pack will forgive you for using the pulse-rate discharge—it doesn't shorten cycle life to do this. As we discussed in the preceding section on lithium plating, however, high current flow is *exactly* what shortens cycle life with a layered metal oxide or spinel cathode.

The beauty of LiFePO$_4$ compared with lead-acid batteries is the relative lack of a Peukert effect. There *is* a "lead-foot tax" with LiFePO$_4$, but it is slight and becomes noticeable only at around 80 percent discharged. Translation? Your battery pack is less accommodating of your lead foot at the end of the day. If you need to stretch your EV's range occasionally—and LiFePO$_4$ is quite forgiving of this, too—you'll want to baby it along after the 80-mile mark.

Mother Nature's Battery Management System

Let's talk about thermal runaway for a minute, the taboo subject of lithium batteries. No doubt you've seen the "flaming laptop" videos on YouTube, and if you're like me, you felt some hesitation about buckling yourself and your family into a car with 500 pounds of "thermal runaway" in the trunk. This is a reasonable concern. I'm happy to report that LiFePO$_4$ is the safest of all the lithium-ion batteries by a wide margin. Why? Because each molecule has a chemical safety chain on it, so the whole battery tends toward stability.

It goes something like this: Have you driven over the mountains and seen the runaway-truck turnouts? If not, I'll try to describe them. Truck turnouts are basically deep gravel-filled bogs that are designed to allow a truck that has lost its brakes going downhill to pull off and let the deep gravel and a slight uphill grade slow it safely to a stop. LiFePO$_4$ has the equivalent of a runaway truck turnout on every molecule that operates automatically, whether you are involved with the process or not. In chemist speak, the oxidation product of lithium–iron phosphate is ferric phosphate, which is a stable molecule. (This is the "gravel bog" I'm talking about—in a battery, oxidation is the downhill process, and stability is the state at which the chemical reaction wants to stop.) Additionally, complete oxidation (the bottom of the energetic hill) of lithium–iron phosphate occurs at a relatively low voltage (3.4 V), and phosphates, because of their strong chemical bonds, are very heat-resistant molecules.[38]

FIGURE 8-6   Battery management on A123's LiFePO$_4$ battery packs from TurnE Conversions.

On the other hand, other lithium-ion chemistries rely heavily on your battery's chemical brakes *not* failing and must employ elaborate and fail-safe battery management systems to ensure that thermal runaway doesn't get a chance to get started. Fail-safe lithium battery management, in practice, is a lot harder than it sounds.[39]

Since it's very hard to improve on things that Mother Nature does well, we've discovered with our prismatic LiFePO$_4$ batteries that battery management seems to go better when we just stay out of her way.[40] Good LiFePO$_4$ battery management as seen in Figure 8-6 by TurnE Conversions is more about observing than meddling—and I'll talk more about that in the "Best Practices for your LiFePO$_4$ EV Batteries" section.

Thermal Runaway Explained

What is *thermal runaway*? In a nutshell, it's the tendency for lithium to set itself on fire. When we humans first figured out that lithium excelled at energy storage, naturally, we tried to make a battery out of it. Having experience with lead batteries and their lead metal electrodes, it seemed logical to make a lithium battery with lithium metal electrodes.

Those had *awesome* energy density. Tesla Motors would be proud!

Unfortunately, they also had a propensity for bursting into flame. On a molecular level, this is so because the lithium oxidation reaction is quite *exothermic*, meaning that it releases heat as it reacts. One molecule reacting creates heat, which increases the reaction rate of its neighboring molecules, resulting in sort of a vicious cycle of uncontrolled reaction. This is the basis for the term *thermal runaway*.

Thermal runaway can happen within one battery, and it can happen in a group of batteries the same way. If you had one lithium-ion battery on fire sitting next to other lithium batteries that were not on fire . . . well, it wouldn't be long before *all* the lithium batteries in the pack burst into flame.

In the years that followed the first lithium batteries, researchers figured out that if you *dissolved* the lithium first, the positively charged lithium ions were a lot tamer than lithium metal—although the energy density was less impressive. Scientists discovered that if they took reasonable care (especially during charging) with batteries made with lithium ions rather than lithium metal, the batteries would behave themselves in a civilized fashion.

The lithium-ion battery *most* naturally prone to thermal runaway is also the most energy dense—*lithium–cobalt oxide*. It reaches thermal instability temperature at 302°F. The lithium-ion battery least naturally prone to thermal runaway is also the least energy dense—*lithium–iron phosphate*. It is thermally stable to 518°F. The other lithium-ion batteries fall somewhere in the middle, with manganese spinel being the safest $LiFePO_4$ alternative, reaching thermal runaway temperature at 482°F. There is not a natural law that I'm aware of that says that energy density and thermal instability *must* go hand in hand with lithium-ion batteries—it's just working out that way at the moment.

What causes lithium-ion batteries to go rogue, anyway? With a perfectly functioning battery, free of manufacturing defects, *overcharging* can set it off. We take care during the charging process to ensure that this doesn't happen. A *short circuit* can generate enough heat to set a good battery off. A *crash*, where the battery is breached and the lithium is exposed to the air, can set an otherwise functional battery off into thermal runaway. Batteries can have manufacturing defects, too, and that's what set those famous Sony laptops ablaze in 2006. Remember that? That was the year *thermal runaway* became a household term!

What happened in that case, according to Battery University, is that microscopic metal dust accidentally got into the batteries during manufacturing, and 1 in 200,000 batteries developed an internal short circuit.[41] With internal battery shorting like that, an external battery management system can't prevent thermal runaway—although a good, functional battery management system can keep your batteries safe from overcharging and external short circuits.

Best Practices for Your $LiFePO_4$ EV Batteries

After everything I've said up to now, the question remains: How can you, the EV converter, get the longest life and the most satisfaction out of your $LiFePO_4$ battery pack? I think the best way to answer this question is to ask people with years of real EV conversion experience.

I'd like to introduce you now to some friends whose EV conversion perspective I've come to respect and trust over the years: Mike Collier, owner of Lithium Storage in Salt Lake City, who has been selling $LiFePO_4$ battery packs to the close-knit EV conversion crowd now for several years; Randy Holmquist from Canadian Electric Vehicles, who has years of practical experience with both lithium and lead EV systems; and Jack Rickard from EVTV, who has been backyard experimenting on these $LiFePO_4$ EV batteries for the past few years, often pushing them to failure just to see what they can do. In the process of his experimentation, which you can watch online at EVTV, Jack has learned things about these batteries that I'll bet even the manufacturers would be surprised to hear.

I asked these men about pack balancing, wiring, battery box placement, battery management systems (BMSs), and cooling, and I was surprised to learn that keeping your pack warm enough for charging and discharging in cold climates is more of an issue than cooling. And there's something you should know about first startup for safety reasons.

Balancing

Question: I've heard of top balancing and bottom balancing your EV's battery pack. Can you tell me what's involved in balancing your pack, and why people do it?

Mike: Bottom balancing entails bleeding all the cells down to a base voltage. Top balancing involves shunting all the cells to a uniform peak voltage. The key idea behind balancing is that these lithium cells are not created equal. A 100-Ah cell might actually be 105 or 109 Ah. So within a pack there will be variation in the capacity of individual cells, and the pack will have issues in use. If you top-balance the pack, all the cells will be off-balance at the bottom, and vice versa. Some folks hate top balancing because they know a user will drive an EV until the pack is "drained" and flat-line some cells. Under a bottom-balanced pack, you'd simply stop charging once the first cell in the pack reaches peak. Balancing is all about establishing a common voltage reference point and staying reasonably within it.

Jack: We advocate balancing the cells at the bottom *once* on installation and slightly undercharging them thereafter. It appears to work on multiple cars, many cells, and many thousands of miles.[42] We do not use shunts and haven't since some very early experiments with Thundersky cells. . . . they are unnecessary and in most cases dangerous. Cell shunts increase the time any cell spends at max voltage. They basically hold it there bypassing the cell and feeding energy to the remaining cells. They are kept there until the others catch up.

Mike: Lithium cells have a really flat discharge curve, and they work well within an 80 percent DOD max. [There's a new line of thinking among those of us in the EV conversion business.] Why not simply restrict the operating voltage range to only take advantage of the flat part of the curve? So, for a CALB cell, instead of using 2.5 to 3.6 V, restrict the use to something closer to 2.9 to 3.45 V (cutting off charging just as the voltage picks up and cutting off use just as the voltage begins to drop). The steep tail of the curve doesn't provide useful energy, and the steep head requires extra energy and time to charge. Maybe this only uses 70 percent of the pack, but cycle life will increase, and balancing wouldn't be such a sensitive issue.

Battery Management

Question: What are the properties of good BMSs? How can I tell the difference between a good BMS and a bad BMS?

Mike: [Right now] it's the wild west of BMS. How to tell a good BMS? Ask around what distributors resell. Anyone can make a BMS, but a distributor isn't going to resell a BMS that brings a lot of support issues. A good BMS will do what you need it to; [that is] really subjective. My company requires a BMS that protects a pack against overcharging and signals an alert when the pack is reasonably depleted. A BMS can be as exquisitely simple or as beautifully complex as you require.

Jack: We don't use BMSs at all.

Cooling

Question: How do I keep my EV's battery pack cool enough, and just how cool is cool enough, anyway?

Mike: I think it's ridiculous that the new OEM EVs have extensive cooling systems. CALB (China Aviation) likes a max charging temp of 113°F and a max discharging temp of 131°F. Sinopoly is supposedly even more temperature-tolerant. Both will be Underwriters Laboratories certified by the end of the year, so we'll see how the stated specs hold up, but 113°F *happens*. Imagine a pack installed in a hatchback EV in Arizona parked out in the sun. It probably gets hotter than 113°F! Folks cool their packs with fans and/or creating a water jacket to go around the pack and sometimes between cells or groups of cells. The worst that happens with our brands is [that] they will vent some of the liquid electrolyte, and the vented cells will lose some Ah capacity. Some swelling of the plastic cases may occur.

Warm Enough?

Question: If I live in a cold climate, do I need to take precautions to keep my batteries warm enough? What about during charging: Is there a minimum charging temperature?

Mike: Again referencing CALB (China Aviation) for its conservative specs: 32°F minimum temperature for charging and –4°F minimum temperature for discharging. Best idea for a cold climate is to keep the EV in a garage with some heating/insulation for charging. [Some converters] . . . have built water jackets for keeping the cell pack above freezing.

Figure 8-7 shows the battery case designed for the Smart Electric Car at 60 Ah on a computer aided design (CAD) software program in Germany. TurnE Conversions is using 70-Ah cells now in 4 rows, 16 cells across, 2 cells in parallel. Some get by with insulating the pack superwell. The worst that happens is the lithium plates out, and the pack will begin to fail.

Charging

Question: When people are switching out lead batteries and putting prismatic LiFePO$_4$ batteries in, do they need a different charger? And do they need to learn different charging habits?

Randy: Most chargers can be used to charge lithium as long as a BMS is used; the BMS will shut down the charger if the charge gets too high. Even an older charger can be used with a simple external relay to cut charging.

Battery Box Placement

Question: Where do you put the battery boxes for best performance?

Figure 8-7   CAD drawing of battery box for Smart car. (*TurnE Conversions.*)

Randy: Batteries should be as low and centered in the car as possible. In a car they should be sunk into the floor rather than installed above the floor. In an accident, a battery box that is above the floor can become a missile; if it's below the floor, it can't go very far. In Figure 8-7 you can see the Smart battery case designed in a computer-aided design (CAD) application. This is sent to be laser cut, bent, and welded—aluminum at 2.5 mm. Most cars have the fuel tank under the rear seat, and this makes an ideal location for some or all of the batteries. We cut a hole in the floor and weld in a battery box that is at least half its height below the floor. The box should be welded in to retain the structural integrity of the car body, as shown in part in Figure 8-8 with the cardboard and tape at the bottom of the Smart car. We usually install the remainder of the batteries in the engine compartment to retain a good front-to-back weight ratio. It's not a good idea to mount batteries behind the rear axle because this can have a negative effect on the handling of the vehicle.

Battery Wiring

Question: I believe the wiring gauge needs to be lower for higher pack voltages, right? Can you sort of give me a rule of thumb for that?

Randy: We use 2/0 on all our vehicles, [although] it's a bit small for low-voltage vehicles in the 96- to 120-V range and will get warm under heavy acceleration or hill climbing. On higher-voltage systems, it will have less resistance than smaller cables and run cool. I don't think the small weight reduction is worth using slightly smaller cable on higher-voltage systems.

FIGURE 8-8 This *is* where we put the batteries. (*TurnE Conversions.*)

Question: With LiFePO$_4$, do you ever wire in parallel? Is there ever a need for that, or do you just get the desired Ah and wire in series?

Randy: I have never used lithium in a parallel system. It is harder to get a BMS to work properly with parallel systems. Just installing the capacity in larger cells is the best solution, and you have fewer connections to give trouble as the vehicle ages.

Safety Tip—First Startup

Question: Is there anything else about the first startup we should know?

Randy: Just one safety tip on starting up your new conversion: We *always* power up a new installation with a 30-A automotive fuse in the system. We leave one battery cable off and use a small jumper wire with the 30-A fuse. The 30-A fuse is enough to power up all systems from the DC-DC converter to the motor controller. You should be able to spin the motor in neutral with a light throttle so [that] the whole system is checked out before applying full power potential. We power up both the 12-V and high-voltage systems this way.

Summary

Energy density has been the Holy Grail of the OEMs for some time, maybe because they're trying to sell cars to folks who are currently married to petroleum-fueled vehicles. I don't quite get it, but what else could it be?

As for me, as long as my battery pack's energy density is adequate or more than adequate, I'm happy. Eighty-plus real-life miles on a charge is okay with me, and LiFePO$_4$ has no problem delivering that.[43]

Better still, the LiFePO$_4$ batteries you can get *today* at a reasonable price will deliver 80+ real-life miles on a charge, accommodate your lead foot, *and* will deliver them every single day for the next 10 years, even if you stretch your range occasionally. Still think the conversion crowd's prismatic LiFePO$_4$ battery pack is second best to the OEM's battery pack? I think we got *lucky*. However, for the sake of argument, Figure 8-9 shows a battery pack being built in the United States by Ford Motor Company for the electric Focus.

FIGURE 8-9   Ford Focus electric vehicle battery pack being built for an electric vehicle in the United States.

CHAPTER 9

The Charger and Electrical System

An efficient charger is a very critical part of any electric vehicle.
—Seth Leitman

The charger is an *attached* and *inseparable* part of every electric vehicle (EV) battery system. Discharging and recharging your batteries are opposite sides of the same coin; you cannot have one without the other. As you learned in Chapter 8, how you recharge your batteries determines both their immediate efficiency and their ultimate longevity. As with motors, controllers, and batteries, technology advancements also have made today's chargers superior to their counterparts a decade ago.

Because your motor, controller, batteries, and charger are also inseparable from the electrical system that interconnects them, it too is covered in this chapter, along with the key components needed for its high-voltage, high-current power side and its low-voltage, low-current instrumentation side.

In this chapter you'll learn about how chargers work and the different types, meet the best type of charger to choose for your EV conversion today (used in the conversion in Chapter 10), and look at likely future charging developments. You'll also look at your EV's electrical system in detail and learn about its components so that when you meet them again during the conversion process in Chapter 10, they will be familiar to you.

Charger Overview

Chapter 8 dealt with discharging and recharging; now we'll take a closer look at the recharging side. It's a wise business decision to invest a few hundred dollars in a battery charger that gets the most out of a battery pack that can cost a thousand dollars or more and might be replaced several times during your EV ownership period. An efficient charger is an indispensable part of any EV.

The objective here is to give you a brief background and get you into the recommended battery charger for your EV conversion with minimum fuss. You have three battery charger choices today: Build your own, buy an off-board charger, or buy an onboard charger. We'll look at each area in turn, and I'll give my recommendations. Let's start with a look at what goes on during the lead-acid battery discharging and charging cycle to understand what has to be done by the battery charger.

Battery Discharging and Charging Cycle

As you already know from Chapter 8, batteries behave differently during discharging and charging—two entirely different chemical processes are taking place. Batteries also behave differently at different stages of the charging cycle. Let's start with a look at an actual battery and then look at the discharging and charging cycle specifics.

What You Can Learn from a Battery Cycle-Life Test

Figure 9-1 shows cycle-life test results for the Trojan 27TMH deep-cycle lead-acid battery we looked at in Chapter 8. Two parameters are being monitored versus number of cycles: the minutes at 25-A capacity and the end-of-charge current.

There are more wiggles in a real battery's data curve, but this battery didn't actually fail at the end of 358 cycles; that's just the name assigned (by a battery test engineer) to the point at which this battery dropped below 50 percent of its rated capacity.

The *end-of-charge current* (EOCC) might be new to you. Notice that it's quite low early in the battery's cycle life (around 1 A) but rises steadily until at some point around midlife it shoots up to its limit value (around 20 A in this graph).

What does this mean to you?

- It means that a battery's charging current fluctuates widely over its lifetime.
- It means that you can quickly kill a new battery that only requires a small amount of current to kick in its charging cycle by placing an unregulated voltage source without any current control across it.

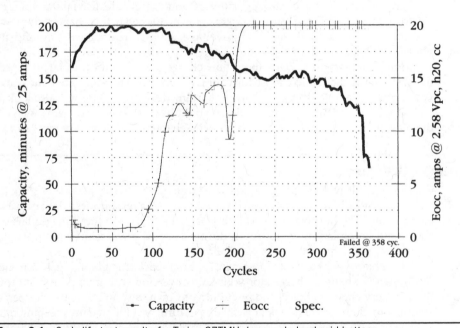

FIGURE 9-1   Cycle-life test results for Trojan 27TMH deep-cycle lead-acid battery.

- It means that you have to crank up the voltage and current when charging a more mature battery. These last two means that you cannot plug a charger into your battery, set it, and forget it because a battery's charging needs also change from cycle to cycle and with temperature and depth of discharge.

Sealed lead-acid batteries, with a small amount of calcium added to eliminate the need for rewatering, don't exhibit this characteristic; their EOCC is relatively flat, so you can be a little more tolerant with them. (They are also more expensive—you pay your maintenance costs up front.)

Battery Discharging Cycle

Let's observe the discharge cycle first, to contrast what is happening to the parameters with what goes on during charging. Capacity, cell voltage, and specific gravity all decrease with time as you discharge a battery. Figure 9-2 shows how these key parameters change (a standard temperature of 78°F is presumed):

- **Ampere-hours**—A measure of the battery's capacity and percent state of charge (the area under the line in this case). It decreases linearly versus time from its full-charge to its full-discharge value.

- **Cell voltage**—Cell voltage predictably declines from its nominal 2.1-V full-charge value to its full-discharge value of 1.75 V.

- **Specific gravity**—Specific gravity decreases linearly (directly with the battery's discharging ampere-hour rate) from its full-charge to its full-discharge value.

Battery Charging Cycle

Battery charging is the reverse of discharging. Figure 9-2 again shows you how the key parameters change.

- **Ampere-hours**—This is the opposite of the discharging case, except that you have to put back slightly more than you took out (typically 105 to 115 percent more) because of losses, heating, etc. The area under the line increases linearly versus time from its full-discharge value to its full-charge value.

- **Specific gravity**—Specific gravity increases wildly over time as a battery is charging, so making specific gravity measurements during the charging cycle is not a good idea. At the early part of the charging cycle, specific gravity increases slowly because the charging chemical reaction process is just starting. Specific gravity increases rapidly as the sulfuric acid concentration builds, and gassing near the end of the cycle contributes to its rise.

- **Cell voltage**—Voltage also increases wildly over time as a battery is charging, so making voltage measurements during the charging cycle is not a good idea either. Notice that cell voltage jumps up immediately to its natural 2.1-V value, slowly increases until 80 percent state of charge (SOC; approximately 2.35 V), increases rapidly until 90 percent SOC (approximately 2.5 V), and then builds slowly to its full-charge value of 2.58 V.

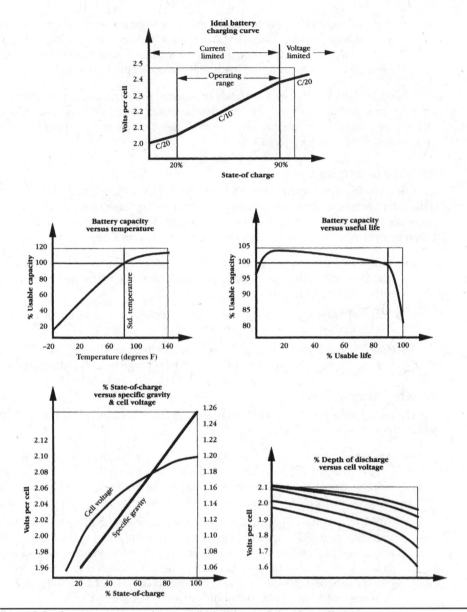

FIGURE 9-2   Graphic summary of battery discharging and charging cycles.

The Ideal Battery Charger

Battery charging is the reverse of discharging, but the rate at which you do it is critical in determining the battery's lifetime. The basic rule is this: Charge it as soon as it's empty, and fill it all the way up. The charging rate rule is this: Charge it slower at the beginning and end of the charging cycle (below 20 percent and above 90 percent).

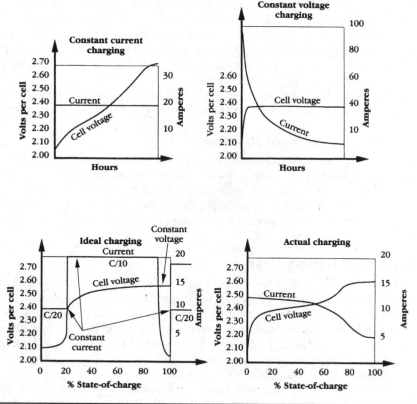

FIGURE 9-3   Ideal and actual lead-acid battery charging characteristics.

When a lead-acid battery is either almost empty or almost full, its ability to store energy is reduced owing to changes in the cell's internal resistance. Attempting to charge it too rapidly during these periods causes gassing and increased heating within the battery, greatly reducing its life. Ideally, you should limit battery current during the first 90 percent of the charging cycle and limit battery voltage during the last 10 percent of the charging cycle. Either method by itself doesn't do the job. Figure 9-3 shows why.

When people think a charger is a big deal, I say look at what AeroVironment and the car company TH!NK are doing. As I wrote on my website:

> There is no question that for mass acceptance of the electric car, there is not just a need for EVs to be introduced into the marketplace. It has to be as easy as fueling a car so [that] the transition is easier.

"TH!NK and AeroVironment are setting a new standard for extremely quick recharging—*zero to 80 percent in just 15 minutes,*" said TH!NK CEO Richard Canny. "This is a major leap forward for electric vehicles. The development and deployment of very-fast-charge stations will help speed the electrification of automobiles in the United States and globally."[1]

The graph in the upper right of Figure 9-3 shows constant-voltage charging in the ideal case. The constant voltage is usually set at a level where gassing causes a decrease in current flow through the battery with time as the battery charges. Unfortunately, with no restrictions on current, this method allows far too much current to flow into an empty battery. Feeding 100 A or more of charging current into a fully discharged battery can damage it or severely reduce its life.

Let's look at the ideal approach during all four SOC phases—0 to 20 percent, 20 to 90 percent, 90 to 100 percent, and above 100 percent. Figure 9-3 shows the results.

Charging Between 0 and 20 Percent

The first 20 percent of a fully discharged battery's charging cycle is a critical phase, and you want to treat it gently. You learned in Chapter 8 that all batteries have a standardized 20-hour capacity rating. Every battery is rated to deliver 100 percent of its rated capacity at the C-20 rate. During the first 20 percent of the charging cycle, you ideally want to charge a battery at no more than this constant-current C-20 rate.

The Real-World Battery Charger

The starting battery in your internal combustion engine vehicle is recharged by its engine-driven alternator, whose output is controlled by a voltage regulator. The starting battery is discharged less than 1 percent in its typical automotive role. The entire rated output power of the alternator is placed across it, and the voltage regulator makes sure that the voltage doesn't climb above 13.8 V—simple. Why not use the same setup to recharge your deep-cycle EV batteries?

The answer is voltage, current, and *boom*!

This approach was not used because the voltage of your EV's deep-cycle battery pack string is probably 96 V or more, but the alternator is typically set up for driving a 12-V starting battery. Assuming that you could adjust your alternator or buy a correct voltage model and had a suitable electric motor around to turn it, the full alternator output applied to a completely discharged deep-cycle battery pack would deliver too much current and charge it far too quickly. You'd damage and/or destroy your batteries in short order—that's the *boom* part. The voltage regulator not only would be useless in stopping this, but it also would prevent raising the voltage to 2.58 V per cell for the final 10 percent of the cycle and would not allow the voltage to be further raised to 2.75 V for the equalizing-charge process.

The solution is alternating-current (AC) power, a transformer, a rectifier, a regulator of either the electronically smart or manually adjustable variety, and a timer—an accurate description of today's EV battery charger.

The X pattern in the graph in the lower right of Figure 9-3 shows what most actual battery chargers deliver. Using a variation or combination of constant current, constant voltage, tapering, and end-of-charge voltage versus time methods, all battery chargers arrive at a method of current reduction during the charging cycle as the cell voltage rises. Fortunately, you can buy something off the shelf to take care of your needs. But you have to investigate before buying to make sure that a given battery charger does what you want it to do.

Battery chargers are "sized" using a formula stipulating that charging current determines the size charger you need, the 115 percent is an efficiency factor to take losses into account, and the direct-current (DC) load is whatever else is attached to the battery (this is zero, assuming that you disconnect your batteries from your EV's electrical system while recharging). You're already familiar with battery capacity and time. You can plug chargers up to 20 A into your standard household 120-V AC outlet. Higher-current-capacity chargers require a dedicated 240-V AC circuit—the kind that drives your household electric range or clothes dryer. Let's take a look at your actual options: Build your own and several different types of off-the-shelf chargers you can buy.

Also, three-phase power is found in industry and the power grid, not in the home. Homes are usually single-phase. AC electric cars usually use three-phase motors, where the voltage and frequency must be changed as driving conditions vary.

Whether to have an onboard or off-board charger is another consideration. An onboard charger gives you driving convenience (charge whenever you like), is light in weight, and is low in power consumption. Some commercial EV controllers, such as the AC Propulsion model discussed in Chapter 7, even share components between the onboard inverter-controller and the battery charger for minimum parts count and maximum efficiency (current flow decides the circuit function automatically—nifty!). An off-board charger gives you high power capability that translates into minimum charging time in a permanent charging station, which can incorporate many additional features.

Below I discuss two of the most popular chargers on the market today. They are the standard for the industry and are really accepted by the marketplace.

ChargePoint America Program

The ChargePoint America program will provide, according to the *New York Times*, 4,600 charging stations to program participants in nine regions in the United States. They are

1. Austin, TX
2. Detroit
3. Los Angeles
4. New York
5. Orlando, FL
6. Sacramento, CA
7. San Jose/San Francisco Bay Area
8. Redmond, WA
9. Washington DC

This is a strategic partnership with three leading automobile brands, including, Ford, Chevrolet, and Smart USA. Coulomb currently has the largest established base of networked charging stations worldwide, with more than 700 units shipped to more than 130 customers in 2009.

The Manzita Micro PFC-20

Some say, "More expensive than the Zivan, but worth it." A Manzita Micro is pictured in Figure 9-4. Here are some of the specifications:

FIGURE 9-4 The Manzita Micro PFC-20, an extremely versatile battery charger.

- The PFC-20 is designed to charge any battery pack from 12 to 360 V nominal (14.4 to 450 V peak). It is power factor–corrected and designed to either put out 20 A (if the battery voltage is lower than the input voltage) or draw 20 A from the line (if the line voltage is lower than the battery voltage). The buck boost charger enhancement option on the PFC-20 will enhance the output to 30 A. There is a programmable timer to shut off the charger after a period of time set by the user.

- For installation instructions, go to www.manzanitamicro.com/installpfc20rev Cnophotos.doc.

Figure 9-5 shows the Eltek Valere charger used by TurnE Conversions in Germany. It is an IP20 variant that communicates only in a standard controlled area network CAN. The charger uses 3×2.5 mm^2 cables for charging, as shown in Figure 9-6. This is a cable with CEE connectors, and inevitably, the plan is to switch to level 1 and 2 chargers, which I will get into later in this chapter. We'll all switch to type 1 or 2 chargers in the future.

The Zivan NG3

The Zivan is a very popular charger. It is used by the two notable EVs, the Corbin Sparrow from Corbin Motors and the GEM, a low-speed EV from Global Electric

FIGURE 9-5   An Eltek Valere charger used by TurnE Conversions for its electric Smart car. (*TurnE Conversions.*)

FIGURE 9-6   A simple cable with CEE connectors for the charging system for the electric Smart car. (*TurnE Conversions.*)

FIGURE 9-7   The Zivan NG3.

Motorcars, a Chrysler Company. Personally, I have used this charger before with the GEM car, and it works well (Figure 9-7).

The Zivan is a notable charger in the EV community. What I like the most about the Zivan charger is that it is microprocessor-controlled and protected against overload and short circuiting. The charger also can determine whether the batteries have problems and let you know so that you can keep a properly maintained battery pack at all times. It also has thermosensor options that allow the electric output to adapt to any type of battery. This increases the life of the batteries. The charger also operates at 85 to 90 percent efficiency for maximum savings of electric power costs, which is really efficient.

Other Battery-Charging Solutions

Against the numerous tradeoffs available to you with today's commercially available battery chargers, along with the additional options of the build-your-own approach (using today's advanced electronic components) are the still newer developments coming down the road for tomorrow and beyond. These include rapid-charging techniques, induction charging, replacement battery packs, and infrastructure development. Let's take a brief look at each.

Delphi Automotive has equipped an EV with its Delphi Wireless Charging System, a highly efficient wireless energy-transfer system featuring technology developed by WiTricity Corporation. Delphi will display its test vehicle at this year's Society of Automotive Engineers SAE World Congress.

"This is a significant advancement in our research and development efforts to offer automotive manufacturers a practical wireless charging solution we believe is superior to others being proposed," said Randy Sumner, director of global hybrid vehicle development for Delphi Packard Electrical/Electronic Architecture. According to Sumner, engineers at Delphi's Customer Technical Center in Champion, Ohio, have installed the Delphi Wireless Charging System on an all-electric TH!NK city test vehicle and have confirmed that system performance meets automotive market requirements.

A wireless charging system eliminates the need for a charging cord.

- Drivers can simply park their EV over a wireless energy source situated on the garage floor or embedded in a paved parking spot.

- Other wireless charging systems under development make use of traditional inductive charging, the revolutionary technology that offers a most efficient and convenient wireless charging option to future EV drivers and flexible installation to EV infrastructure. This is the same technology used in electric toothbrushes, which is based on principles first proposed in the mid–nineteenth century.

- These systems only work over a limited distance range, require precise parking alignment, and can be very large and heavy, making them impractical for widespread use with EVs.

"The Delphi Wireless Charging System offers more practical and flexible installation than traditional inductive systems because it uses highly resonant magnetic coupling, a modern technology that safely and efficiently transfers power over significantly larger distances and can adapt to natural misalignment often associated with vehicle positioning during parking," Sumner said.

This means that Delphi charging stations for EVs can be

- Buried in pavement
- Unaffected by environmental factors such as snow, ice, or rain
- Accommodating of a wide range of vehicle shapes and sizes and their differing ground clearances

According to Sumner, the system will automatically transfer power to the EV's battery pack at a rate of 3,300 W (equal to most chargers in the home). "We are excited by our testing and validation of the system and believe we have a valuable and unique wireless charging solution that offers the most potential for widespread use in the automotive market. With the support of automotive manufacturers, this technology can be integrated into the next generation of electric vehicles," Sumner said.[2]

ECOtality Partners with Regency Centers at 19 Locations Nationwide

ECOtality, Inc., announced a partnership with Regency Centers (NYSE: REG) to install Blink EV charging stations at 19 Regency locations nationwide. As the newest EV Project partners, ECOtality and Regency have worked together to identify the 19 sites and approximately 40 charging stations to be installed in EV Project markets in Arizona, California, Oregon, Tennessee, Texas, and Washington, DC. One of the chargers is shown in Figure 9-8.

"Our goal has been to deploy a network of chargers that are conveniently placed where consumers work, shop, and play—and aligning ourselves with major property owners is a key part of achieving that goal," said Colin Read, vice president of corporate development at ECOtality. "We are excited to work with a forward-thinking company like Regency to deploy chargers nationwide and understand the true business case and value of offering commercial EV charging access to their customers."[3]

The Blink network of charging stations provides EV drivers with the freedom to travel as they choose and conveniently charge at Blink commercial locations along the way. By becoming Blink members, consumers also may yield even greater advantages of the Blink network, such as local incentive programs, reservation systems, and enhanced Blink network capabilities.

FIGURE 9-8   ECOtality Blink DC fast charging station. (*ECOtality.*)

"The EV stations provide an added shopper convenience, demonstrate and encourage a commitment to the environment, and generate a potential future revenue stream," said Scott Prigge, senior vice president of national property operations for Regency Centers.

Regency's sustainability program, Green Genuity, uses best practices in green building design, construction, operations, and maintenance to reduce long-term operating costs and the environmental impact of shopping centers. EV charging services are now available in workplace and retail locations in San Francisco, Oakland, and Walnut Creek, as shown in Figure 9-9.[4]

CarCharging Group, Inc. (OTCBB: CCGI), announced the inauguration of EV charging services at six properties managed by Ace Parking in San Francisco, Oakland, and Walnut Creek, California. CarCharging is currently accelerating its deployment of workplace charging services in visible, convenient, and accessible locations for EV drivers. These new EV charging services are also located near popular high-traffic retail sites, including Fisherman's Wharf in San Francisco.

"The locations we selected for new EV charging stations are frequented by customers who are environmentally aware and seek out Ace Parking locations while at work, shopping, and entertainment locales in the Bay Area," said Jonathan Wicks, Ace Parking director of operations in northern California. "We pride ourselves on our green initiatives, and we're pleased to work with CarCharging Group as we continue to push the envelope as a very environmentally friendly company."[5]

Figure 9-9 EV charging station from Blink network of ECOtality in commercial parking spot. (*ECOtality.*)

"Working with Ace Parking, we've established a definitive presence for EV charging services in high-traffic locations in San Francisco, Oakland, and Walnut Creek," said Michael D. Farkas, CEO of CarCharging. "We're happy to partner with Ace Parking to deploy our services and to reinforce their environmentally friendly efforts. We share in a similar mission to attract and retain customers who are conscious about the environment."[6]

At each Ace Parking location, CarCharging's service uses a Coulomb Technologies EV fast-charging station, known as *level 2*, which provides 240 V with 32 A of power to quickly recharge an EV's battery. The EV charging stations use the standard SAE J1772 connector widely adopted by nearly all automobile manufacturers.

EV drivers can easily register and create a CarCharging account online and attach a small card to their keychain to initiate use and payment at any intelligent CarCharging station. The CarCharging keychain card also allows drivers to use charging locations on the ChargePoint Network, the largest national online network connecting EV drivers to EV charging stations. CarCharging also accepts direct payment via credit card. Users can pinpoint EV charging station locations using the CarCharging map at www.carcharging .com/.

The ChargePoint mobile application for the iPhone, Android, and BlackBerry devices also provides real-time charging station location information with turn-by-turn directions. Drivers will soon be able to reserve a time slot, guaranteeing access to EV charging stations to recharge their EVs.

EVs already on the market include the Nissan Leaf, Chevy Volt, Toyota Prius Plug-in Hybrid, BMW ActiveE, Ford Focus Electric, Coda, and Fisker Karma. Additional EV models are expected to be available later this year.

The EV Project

On August 5, 2009, ECOtality was awarded a $99.8 million grant from the U.S. Department of Energy to embark on this project. The project was officially launched on October 1, 2009. In June 2010, the project was granted an additional $15 million by the U.S. Department of Energy. With partner matches, the total value of the project is now approximately $230 million.

ECOtality is deploying chargers in major cities and metropolitan areas across the United States. The Nissan Leaf has partnered with the EV Project, and drivers who qualify to participate receive a residential charger at no cost. In addition, most, if not all, of the installation cost is paid for by the EV Project.

The EV Project collects and analyzes data to characterize vehicle use in diverse topographic and climatic conditions, evaluates the effectiveness of charge infrastructure, and conducts trials of various revenue systems for commercial and public charge infrastructures. The ultimate goal of the EV Project is to take the lessons learned from deployment of the first thousands of EVs and the charging infrastructure supporting them to enable streamlined deployment of the next generation of EVs to come.

The EV Project has qualified Leaf and Volt customers for participation based on home electrical power capabilities. Because a significant amount of vehicle charging takes place at EV driver residences, a portion of the EV Project funding supports home charging units or, more correctly, *EV supply equipment* (EVSE). In exchange for allowing the collection of vehicle and charging information, participants receive a Blink wall-mount charger at no cost (Figure 9-10) and, in select locations, up to a $1,200 credit toward the installation. This information includes data from both the vehicle and the EVSE, including energy used and time and duration of charger use. No personal information is being shared or included in the data to be analyzed.

Rapid Charging

A number of modern papers have discussed the alternatives of rapid charging. The Japanese are perhaps the most forward-thinking in this area. In short, if charging your EV's battery pack in eight hours is good, then accomplishing the same result in four hours is better. Fortunately, you can use pulsed DC, alternating charge and discharge pulses, or just plain high-level DC to accomplish the result.

Unfortunately, you need to start with at least 208 to 240 V AC (208 to 240 V three-phase is better), and you probably will overheat and shorten the life of any of today's lead-acid batteries in the process. However, if you design your lead-acid batteries to accommodate this feature (a larger number of thinner lead electrodes, special separators, and electrolyte), it becomes easy. Even nickel-cadmium batteries can be adapted to the process if your wallet is bigger. You'll hear more about this idea as time goes by (208 V will supply 50 percent more power than 480 V single-phase). Because adding one wire

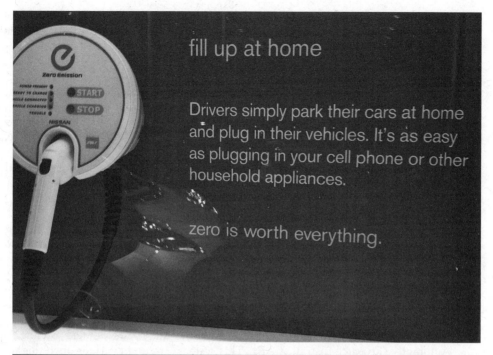

fill up at home

Drivers simply park their cars at home and plug in their vehicles. It's as easy as plugging in your cell phone or other household appliances.

zero is worth everything.

FIGURE 9-10   The Nissan Leaf charger with the EV Project at the New York Auto Show in 2012.

gives you three times the power, the world's power grid is three-phase. In a residential area, blocks are rotated between phases A, B, and C. Full-wave rectification of three-phase gives you DC with very little ripple. (Second, third, and fourth harmonics cancel.) In EVs, three-phase motors can be reversed by swapping any two phases.

Induction Charging

Nissan and Nichicon Launched the Leaf to Home Power-Supply System with EV Power Station in Japan

Nissan Motor Co., Ltd., recently launched the Leaf to Home power-supply system, which can supply electricity from batteries onboard the Nissan Leaf EV to homes when used with the EV Power Station unit developed by Nichicon Corporation. Leaf to Home, or as we use to call it when I worked for the New York State Energy Research and Development Authority (NYSEDA), Vehicle To Grid (V2G), is an industry-first backup power-supply system that can transmit the electricity stored in the large-capacity batteries of the Nissan Leaf to a residential home. Nissan will showcase this system at its Japanese dealership showrooms beginning in June 2012 to help promote efficient electricity management and demonstrate the features built into EVs. Ironically, the Nissan Leaf charges the same way it did when the first edition of this book was released.

Many alternatives have been proposed for induction charging. General Motors' Hughes Division is probably the most forward-thinking in this area. Figure 9-11, taken at the September 1992 Burbank Alternate Transportation Expo, shows the Hughes concept. You drop the rear license plate to expose the EV's inductive charging slot, drop the Hughes paddle into the slot, and turn on the juice. With other inductive floor-mounted or front-bumper-mounted approaches, you had to position your vehicle accurately. The Hughes approach makes it even easier than putting gasoline into a vehicle. There are no distances to worry about, no spills, and no risk of electric shock.

The process of putting electrical energy into your EV can be computerized with a credit-card, EV charging kiosk, or whatever shape it assumes. This will someday be as familiar as cell phones, and Hughes-style inductive charging will probably lead the way. The Toyota RAV4 and General Motors EV1 use these chargers.

Nichicon, an innovator in power-supply systems, will provide the technology to move electricity from vehicle to the home through its EV Power Station units. This power-transfer system enables electricity stored in high-capacity lithium-ion batteries onboard a Nissan Leaf to be sent to an ordinary home by connecting the car to the home's electricity distribution panel with a connector linked to the Leaf's quick-charging port. The EV Power Station system is similar in size to an external air-conditioning unit, can be installed outdoors, and conforms to the Charge de Move ("charge for moving"; CHAdeMO) Protocol for EV quick chargers. CHAdeMO is the trade name for a quick-charging method used to charge the batteries of EVs. The Nichicon system's connector complies with the standards defined by the Japan Automotive Research Institute (JARI). This system can run in various operating modes and has a timer function that can be controlled with a liquid-crystal display (LCD) touch panel. Electricity is stored or supplied automatically in accordance with a household's electricity capacity and consumption.

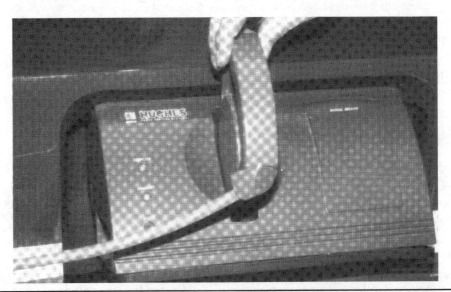

FIGURE 9-11   Early Hughes inductive charging system paddle is relatively the same as that used for the Nissan Leaf today.

The EV Power Station can fully charge a Leaf in as few as four hours, which is approximately half the time required when a normal charger is used. All current Nissan Leaf owners in Japan will be able to use the system depending on their home's installation requirements. With Japanese government subsidies taken into account; the EV Power Station is estimated to cost 330,000 yen with the consumption tax and installation charge included (300,000 yen excluding tax).

This system helps to balance the electrical supply system and can even lower a consumer's electricity bill. The Leaf to Home system will help to encourage Nissan Leaf owners to charge their cars with electricity generated during the night, when demand is low, or sourced from solar panels. This assists in balancing energy needs by supplying electricity to the grid during the daytime, when demand is highest. It also can be used as a backup power source in times of power outages and/or shortages.

The lithium-ion batteries can store up to 24 kWh of electricity, which is sufficient to supply an average Japanese household for about two days. This system underscores an additional attribute of EVs: They can be used as a storage battery whether they are moving or stationary.

Nissan and Nichicon will continue to work together to create new value in EVs as part of the way forward toward the realization of a zero-emission society.

The EV Power Station: Specifications

When charging the Nissan Leaf	
Input voltage	Single-phase AC, 200 V (±15 percent), 50 to 60 Hz (±5 percent)
Input current range	AC 0 to 36 A
Output voltage range	DC 50 to 500 V (CHAdeMO Protocol)
Peak power output	6 kW
Conversion efficiency	90 percent or more (at rated output)
Power factor	99 percent or more (at rated output)
When supplying power to a household	
Input voltage range	DC 150 to 450V
Input current range	DC 0 to 30 A (limited by cable specifications)
Output voltage	Single-phase three-wire system (AC 100 V x two-phase) AC 100 V (±6 percent), 50 to 60 Hz AC 200 V (±6 percent), 50 to 60 Hz (max. ±2 percent)
Output current range	AC 0 to 30 A
Peak power output	6 kW (single-phase AC, 100 V · 3 kW x two-phase)
Conversion efficiency	90 percent or more (at rated output)
External dimensions	650 mm (W) x 350 mm (D) x 781 mm (H) (excluding projecting parts)
Mass	Approximately 60 kg

Replacement Battery Packs

Many attractive alternatives also have been proposed in this area, and now the most popular company, Project Better Place, has countries and cities nationwide starting to implement battery-pack swapping stations. Bob Brant remembers the EV racers at the Phoenix Solar & Electric 500 had used them for years. "When they wheel into the pits, saddle-bag-style battery packs are dropped from their outboard mounting positions, and fresh battery packs are attached in their place. This same approach, with neighborhood "energy stations" replacing gasoline stations, also has a role in the future. Future EV designs could be standardized with underbody pallet-mounted battery packs."[7]

Then Bob mentioned how battery swapping in the first and second editions of the book *Your Own Electric Vehicle* was based on the same principles being used by Shai Agassi in his company Project Better Place: "You wheel into the energy station, drop the old battery pack, raise the new one into place and latch it down, pay by credit card (probably a deposit for the pack plus the energy cost for the charge), and you're on your way within a few minutes time."

Better Place Logs 1 Million Electric Kilometers

Continues to Invest in the Network as May Deliveries in Israel Reach 5 Cars per Day

Better Place has logged more than 1 million electric kilometers from the electric cars operating on its charging infrastructure in Israel and Denmark, as shown in Figure 9-12. The company reported that it invested an additional $96 million in the quarter, a majority of which was dedicated to development and deployment of the Battery Switch networks in Israel and Denmark. Better Place also announced that in May it delivered 110 cars to new customers in Israel, averaging five cars per day.

With a total of 55 Battery Switch stations in both countries in the final stages of testing and certification, following 15,000 switches of batteries in the Battery Switch stations, the company has expanded deliveries of the first Renault Fluence ZE electric cars with switchable batteries and Better Place mobility services to customers after successfully completing four months of network testing with employees, friends, and family.

Battery Switch stations that have completed their tests are starting to open to first customers in Israel, making Israel the first country in the world where a purely electric car is not limited by the range of its battery. In the coming weeks, the company expects to see additional Battery Switch stations come online as tests are completed. Nationwide coverage is expected in both markets by the end of the third quarter.

"We're excited to be delivering our services to our first set of customers," said Shai Agassi, founder and CEO of Better Place. "Step by step we're completing final tests on our system and scaling our network, the delivery of cars, and the number of kilometers driven. I'm delighted to see cars going to our members who are helping us change the world. Change takes time, and I am thankful for the extremely supportive Better Place community which has cheered us on along the way."

The company expects to continue to make additional investments in network deployment in Israel and Denmark during the second and third quarters of this year, with the majority of the investment incurred by the end of the second quarter. At the same time, revenues from

Figure 9-12   Project Better Place charging station. (*Project Better Place.*)

kilometers driven on its first two networks will begin to grow as commercial operations expand throughout this year. The company's operations in Australia will begin Battery Switch station deployment in Canberra in the second half of this year.

"Our Board believes that Better Place is uniquely positioned to unlock the mass-market potential for affordable electric driving," said Idan Ofer, chairman of the Better Place board of directors. "With the network in place, the Better Place team is delivering on the promise of bringing an unmatched experience for electric mobility. As Chairman and proud Better Place member, I've seen firsthand that this company is indeed delivering on its promise. The Board continues to enthusiastically support Better Place as they go to market."

Better Place is backed by a world-class group of investors, including HSBC which is named after its founding member, The Hongkong and Shanghai Banking Corporation Limited and Israel Corp., that are fully dedicated to the company's long-term success and view Better Place as a long-term, transformational investment in their portfolio. The company has raised more than $750 million of equity financing to date.[8]

Beyond Tomorrow

Nothing stops you from using rapid charging, induction charging, and/or replacement battery pack techniques today. Delphi Automotive is offering a wireless charging option, as shown in Figure 9-13. The other two can be done by an individual but obviously require infrastructure development for widespread adoption. Articles have even been written on how you can charge your EV from a solar source today.

DELPHI

Delphi Wireless Charging System

Delphi is developing a Wireless Charging System that will automatically transfer power to a vehicle providing a convenient, wireless energy transfer system. The system was developed in cooperation with WiTricity Corp., a wireless energy transfer technology provider. It will enable an electric vehicle's battery to be recharged without the hassle of cords or connections. This hands-free charging technology is based on highly resonant magnetic coupling which transfers electric power over short distances without physical contact, allowing for safer and more convenient charging options for consumer and commercial electric vehicles.

The wireless technology is a product differentiator for advanced technology vehicles. The high efficiency wireless energy transfer technology will require no plugs or charging cords. Instead, a magnetic field from a source resonator on the ground is aligned with a capture resonator mounted underneath a vehicle.

The Delphi Wireless Charging System eliminates the need for a charging cord. Drivers can park their electric vehicle over a wireless energy source that sits on the garage floor or is embedded in a paved parking spot.

Delphi has rights under a patented MIT-developed wireless energy transfer technology based on the following principle: Two properly designed devices with closely matched resonant frequencies can strongly couple into a single continuous magnetic field.

The wireless charging system is comprised of the following:

- Vehicle mounted capture resonator and interface electronics — fitted to the bottom of the vehicle
- Vehicle power and signal distribution systems
- Stationary source resonator pad — mounted on the ground
- Stationary charging controller

Compared to inductive systems, this highly resonant magnetic coupling technology will efficiently transfer power over significantly larger distances and will allow more parking-related vehicle misalignment. The system can fully charge an electric vehicle at a rate comparable to most residential plug-in chargers, which can be as fast as four hours.

The Delphi Wireless Charging System offers more practical and flexible installation than traditional inductive systems because it uses highly resonant magnetic coupling, a modern technology that safely and efficiently transfers power over significantly larger distances and can allow more misalignment due to parking.

FIGURE 9-13   Delphi Automotive wireless charging system specification sheet. (*Delphi Automotive.*)

CSG and Better Place Open Switchable Electric Car Experience Center in Guangzhou

China Southern Power Grid (CSG) and Better Place announced opening their Switchable Electric Car Experience Center in Guangzhou's Pearl River New Town. A switchable battery electric car today drove out from the experience center, marking Guangzhou as the starting point for EV network infrastructure in the five Southern provinces served by CSG.

As a member of the State-Owned Enterprise EV Industry Alliance, CSG is aiming to implement China's national energy conservation and low-emission strategy, as well as EV infrastructure construction, in accordance with local needs. Not unlike fuel stations for gas-powered cars, widespread EV adoption requires an extensive network of infrastructure.

In March, CSG and Better Place signed a strategic cooperation framework agreement in Guangzhou to develop EV infrastructure in China. As a global EV network provider, Better Place has extensive experience in providing EV solutions for commercial operations at scale.

After extensive research, development, and planning, CSG announced in July that it would follow a centralized, open, and government-led but enterprise-guided approach in providing energy for EVs. The company's strategy has battery switch at its core combined with centralized EV charging. CSG will promote the development of national technical standards and build a smart EV network.

CSG has signed EV infrastructure agreements with Guangdong, Guangxi, Guizhou, and other provinces. The company also has signed similar cooperative agreements with Guangzhou, Shenzhen, Nanning, Haikou, and other cities. CSG has 14 EV charging stations in its network, with 2,901 charging poles in operation and 206,000 kWh of usage over 45,000 charge cycles from January to November 2011.

During the "12th Five Year Plan" period and under the government's leadership, CSG will localize smart EV network infrastructure for the Chinese market and meet the needs of the five southern provinces to support development of the EV industry.

The Switchable Electric Car Experience Center in Guangzhou is the first fully automated battery-switch facility in the five provinces of southern China. The Switchable Electric Car Experience Center covers 1,900 square meters and is an important component of Guangzhou's efforts to promote new energy vehicles.

The semitranslucent light-green structure sits near the Guangzhou Auto Mall, the biggest auto mall in southern China, and includes a Better Place battery-switch demonstration, electric cars, auditorium, and reception center. Visitors can see the switch process and experience electric cars right next to the Guangzhou Auto Mall. Outside the facility are three Better Place charge spots, showing the charging capabilities of EVs. The customer-service center makes reservations, and knowledgeable guides will accompany visitors throughout the facility and explain more about EVs.

The entire switch process takes fewer than five minutes. The process is completely automated, and the system safely and efficiently switches out depleted batteries for fully charged ones.

The Switchable Electric Car Experience Center not only furthers knowledge about the industry but also promotes widespread EV adoption and raises public awareness and acceptance of EVs. This is an important step in promoting the smooth development of new energy vehicles in the five provinces of southern China.[9]

But serious infrastructure development is necessary for the real prize: roadway-powered EV. Numerous papers and articles have been written, all because of the mouth-watering appeal of the idea. A simple lead-acid battery–powered EV has more than enough range to carry you to the nearest interstate highway. Once there, you punch a button on the dashboard, an inductive pickup on your EV draws energy from the roadway through a "metering box" in your vehicle (so that the utility company knows who and how much to charge), and you get recharged on your way to your destination without all the pollution, noise, soot, and odors.

If the roadway and your EV are both of the "smart" design, you can even get traveler information (e.g., weather, directions, etc.) while the roadway guides your vehicle in a hands-off mode, and you read the morning newspaper or scan the evening TV news.

Nissan and Sumitomo to Collaborate in the United States on DC Quick Charger Sales

Newly Developed Unit Lower in Cost and Size

Nissan and Sumitomo Corporation (Tokyo, Japan) agreed to collaborate on sales and marketing activities in the U.S. market for the new DC quick charger. The two companies have agreed to join hands in order to accelerate the realization of a zero-emission society.

The newly developed DC quick charging unit (Figure 9-14) retains the high performance of the current quick charger manufactured by Nissan but is nearly half its size by volume. Since the new charger unit is much smaller in size, it will occupy less space and enable easier installation. Furthermore, when it is introduced to the market, the new unit is expected to cost significantly less than the current model.

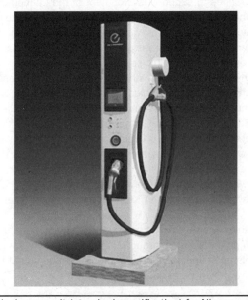

FIGURE 9-14   New quick charger unit (standard specification) for Nissan and Sumitomo DC quick charger. (*Nissan.*)

Nissan North America Co. and Sumitomo Corporation of America are jointly working on a sales and marketing plan for the product. Details of the plan will be announced in the near future.[10]

Nissan Goes to the United Kingdom with the Nissan Leaf

Nissan began deliveries of the Nissan Leaf to customers in the United Kingdom starting March 2011. Tony Whittaker, sales director at Cleaner Air Solutions in Durham, today took delivery of the first of five Nissan Leafs the company is buying at the Benfield Motor Group dealership on Portland Road, Newcastle upon Tyne. This groundbreaking moment represents the start of Nissan's vision of bringing sustainable mobility to the mass market in the United Kingdom.

Mr. Whittaker said: "We're very excited about the arrival of the Nissan Leaf. Our existing fleet is made of LPG converted vehicles as well as some hybrid vehicles, so the move to all-electric was an obvious choice. The range, performance, and running costs of the Leaf are exactly what we're looking for in a new fleet vehicle."

Kevin Fitzpatrick, vice president for manufacturing at Nissan's Sunderland Plant, was on hand to witness the handover of the first Nissan Leaf in the Northeast. He commented: "The Nissan Leaf is pioneering the electric car revolution in the United Kingdom. Since we first unveiled the car in August 2009, we've been building up to this historic moment for over 18 months. Today's deliveries mark the beginning of Leaf's introduction to customers across the country as our network of EV dealers will deliver to their first customers later this week. Currently built at our Oppama Plant in Japan, we're already preparing for the start of Leaf production at Sunderland in 2013.

Nigel McMinn, managing director of Benfield Motor Group, said: "As a Northeast-based family business, we are honored to be at the forefront of the electric car revolution and this historic occasion. Demand for the Leaf is already very high, and there is no doubt this car is going to revolutionize driving in the United Kingdom. Benfield is absolutely delighted to have been chosen as one of Nissan's first ever EV specialists.

In a similar event in Waltham Abbey, Hertfordshire, Smooth FM DJ Mark Goodier and electronics engineer Richard Todd took delivery of their Nissan LEAFS at Glyn Hopkin Nissan.

The Nissan Leaf is currently built in Japan but will be manufactured at Nissan's award-winning plant in Sunderland from 2013.

A comprehensive charging network is currently under development in the United Kingdom, and Nissan's network of EV dealers—currently 26 sites across the country—will be equipped with a quick charger, which will charge the battery from zero up to 80 percent capacity in under 30 minutes. There are programs under way to install around 1,300 charge points in the Northeast and 9,000 charge points across the United Kingdom by 2013.[11]

Crackel Barrel Building EV Charging Stations
Future Forward Fast-Charging Stations

When I heard this story off the wire, I was left with nothing to say—jaw drop! Cracker Barrel is getting into electric cars and building charging stations. Their pilot-project state is

Tennessee, where the iconic chain of restaurants is based and where there are extra incentives for electric cars.[12] As I wrote in the article on the Daily Green:

> Cracker Barrel will be rolling out fast chargers called the Blink, made by ECOtality, Inc., at select restaurant locations across Tennessee for phase 1 of the project. The company will start in "The Tennessee Triangle," the 425-mile stretch of interstate highway that connects Nashville, Knoxville, and Chattanooga. Twenty-four Cracker Barrel locations will have chargers, meaning [that] EV owners could drive the entire region, charging up at Cracker Barrels along the way.
>
> Digging starts in the spring of 2011 and will reportedly be completed within a few months. Half of the first 12 locations will feature 480-V fast chargers (referred to as "level 3" chargers), while the balance will get 240-V "level 2" chargers. Level 3 can charge an EV battery in about 20 minutes, as long as the system is able to handle it, which many early models can't. In all cases, the juice won't be free, as customers will be asked to swipe credit cards. The project is being paid for by the public-private partnership called The EV Project (www.thevproject.com).

"Our goal is to make sure there are readily available Blink public charger stations where people need them, in convenient locations," Jonathan Read, CEO of ECOtality, said in a statement. "Cracker Barrel is a place that people like to visit and is uniquely located to provide a great service and convenience to the public. Cracker Barrel's work with ECOtality will allow us to move a step closer toward creating an interconnected network of EV infrastructure," he added.

How Many and Where?

The partners plan to install 24 charging sites, although they are starting phase 1 with 12, as shown in Figure 9-15. Here are the first 12 for Tennessee:

- Athens
- Cleveland
- Cookeville
- Crossville
- East Ridge
- Farragut
- Harriman
- Kimball
- Lebanon
- Manchester
- Murfreesboro
- Nashville–Stewart's Ferry Pike

The other 12 sites will be in and around the cities of Nashville, Knoxville, and Chattanooga to support more local users. Specific locations will be announced as the EV Project progresses.

"Cracker Barrel was founded along the interstate highways with the traveler in mind and has always tried to anticipate what our guests might want and need as they stop in for some

FIGURE 9-15   Nissan Leaf soon to be charging at a Cracker Barrel near you. (*Cracker Barrel.*)

good country cookin' and to experience genuine Southern hospitality," said Cracker Barrel chairman and CEO, Michael A. Woodhouse. "Becoming a leader in the EV Project continues our tradition of striving to anticipate and meet our guests' expectations and, at the same time, allows us to participate in a meaningful way in the nation's explorations of energy independence."

Woodhouse added: "While ownership of electric cars is small compared with traditional vehicles, there's great curiosity about them, and so we expect our guests will be quite interested in seeing these charging stations when they stop in with us. We like to think that our guests will be pleased to see Cracker Barrel taking an active role in exploring energy alternatives that are aimed at protecting the environment, as well as strengthening our economy. This is a way of showing that Cracker Barrel is focused on the future even as we provide guests with a genuinely hospitable experience reminiscent of times past."

Interestingly, Cracker Barrel stores had previously maintained gas pumps, but these were removed in the 1970s during the oil embargo.

Toyota RAV4 Electric Vehicle with Tesla Motors Goes Leviton

Leviton launched its level 2 (240-V, 40-A) charging station. Designed, tested, and assembled in the United States, the next-generation Leviton charging station is able to provide 9.6 kW of output and will charge the Toyota RAV4 EV in approximately six hours. It also includes a 25-foot charging cable that offers flexibility in the mounting location.

The weatherproof, thermoplastic NEMA type 4 enclosure makes the device safe for use in both indoor and outdoor locations. The unit comes standard with Leviton's patent-pending "cord-connected" system, which is ideal for indoor applications, with the capability to be converted to hard wire for outdoor applications.

"Leviton has designed a next-generation charging station that provides unique ergonomics and aesthetics at a leading price point for Toyota consumers," said Michael Mattei, vice president and general manager for Leviton's Commercial and Industrial Business Unit. "Through our relationship with Toyota, customers know the products they purchase are designed by two companies that value safety, quality, and efficiency."

The 40-A charging station includes an industry-leading 10-year limited product warranty when installed by Leviton's s highly trained electric vehicle supply equipment (EVSE)-certified electrical contractors. As an added customer safety feature, Leviton designed its charging station with supervisory, contactor, and ground monitoring circuits, as well as built-in communications to verify proper connection before charging can commence. The device's "Auto-Restart" feature enables charging to restart following a minor fault, thereby reducing the chance of being stranded with an undercharged battery.

"Toyota has a long-standing commitment to connecting our customers with high-quality products and services such as those provided by Leviton," said Rick LoFaso, Toyota corporate manager of car marketing. "The new 40-A home charging station is yet another example of how Toyota's relationship with Leviton helps our EV customers access solutions that work best for their needs."

The Leviton level 2 40-A charging station was unveiled for the first time at the 26th World Battery, Hybrid and Fuel Cell Electric Vehicle Symposium and Exhibition (EVS26) held in Los Angeles on May 6–9, 2011. The charging station will be available for purchase through Leviton's customer website for Toyota owners (www.Leviton.com/Toyota).[13]

Your EV's Electrical System

To say that the electrical system of an EV is its most important part is not an oxymoron—far from it. The idea is to leave as much of the internal combustion vehicle instrumentation wiring as you need intact and to carefully add the high-voltage, high-current wiring required by your EV conversion or electric car build.

Do a great job on your EV conversion or electric car build high-voltage, high-current wiring, and it will reward you with years of trouble-free service. Do it in a sloppy manner, and aside from the obvious boom or poof (accompanied by smoke), you open yourself to a world of strange malfunctions.

This section will cover the electrical system that interconnects the motor, controller, batteries, and charger along with its key high-voltage, high-current power and low-voltage, low-current instrumentation components. Figure 9-16 shows you the system at a glance. We'll look at the components that go into the high- and low-current sides separately and then discuss wiring it all together.

High-Voltage, High-Current Power System

The heavier lines of Figure 9-16 denote the high-current connections. When you put the motor, controller, battery, and charger together in your EV, you need contactor(s), circuit breakers, and fuses to switch the heavy currents involved. Let's take a closer look at these high-current components.

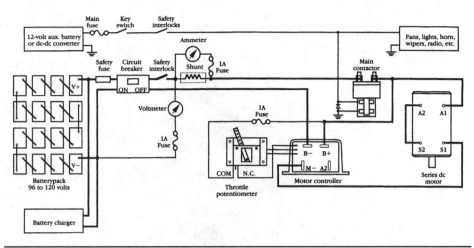

FIGURE 9-16 EV basic wiring diagram.

Main Contactor

A contactor works just like a relay. Its heavy-duty contacts (typically rated at 150 to 250 A continuous) allow you to control heavy currents with a low-level voltage. A single-pole, normally open main contactor is placed in the high-current circuit between the battery and the controller and motor. When you energize it—typically by turning the ignition switch on—high-current power is made available to the controller and motor.

Reversing Contactor

This contactor is used in EVs when electrical rather than mechanical transmission control of forward-reverse direction is desired. The changeover contacts of this double-pole contactor are used to reverse the direction of current flow in the field winding of a series DC motor. When this contactor is used, a forward-reverse center off-switch is added to the low-voltage wiring system after the ignition switch.

This enables forward-reverse or series-parallel shifting and series parallel contactor controls. Figure 9-17 is used to reverse the direction of current flow in the series DC motor. When this contactor is used, a forward-reverse-center off switch is added to the low-voltage wiring system after the ignition key switch or button.

Main Circuit Breaker

A circuit breaker is like a switch and a resetting fuse. The purpose of this heavy-duty circuit breaker (typically rated at 300 to 500 A) is to instantly interrupt main battery power in the event of a drive system malfunction and to routinely interrupt battery power when servicing and recharging. For convenience, this circuit breaker is normally located near the battery pack. The switch plate and mounting hardware are useful—the big letters immediately inform casual users of your EV of the circuit breaker's function.

The purpose of the safety fuse is to interrupt current flow in the event of an inadvertent short-circuit across the battery pack. In other words, you blow out one of these before you arc-weld your crescent wrench to the frame and lay waste to your battery pack in the process, as shown in Figure 9-18.

FIGURE 9-17   Main contactor–single pole.

FIGURE 9-18   Fuses. (*TurnE Conversions.*)

Safety Interlock

There is an additional switch that some EV converters incorporate into their high-current system, usually in the form of a big red knob or button on the dashboard—an emergency safety interlock or "kill switch." When everything else fails, punching this will pull the plug on your battery power.

Low-Voltage, Low-Current Instrumentation System

The instrumentation system includes a key switch, throttle control, and monitoring wiring. Key-switch wiring, controlled by an ignition key, routes power from the accessory battery or DC-to-DC converter circuit to everything you need to control when your EV is operating: headlights, interior lights, horn, wipers, fans, radio, etc. Throttle-control wiring is everything connected with the all-important throttle potentiometer function. Monitoring wiring is involved in remote sensing of current, voltage, temperature, and energy consumed and is routed to dashboard-mounted meters and gauges. Let's take a closer look at these low-voltage components.

Throttle Potentiometer

This is normally a 5-kΩ potentiometer, but it has a special purpose and important safety function. The Curtis model is designed to accompany and complement its controllers and to use the existing accelerator foot pedal linkage of your EV. An equivalent model is designed for replacement use or for ground-up vehicle designs not already having an accelerator pedal. With either of these, the potentiometer provides a high pedal disable option that inhibits the controller output if the pedal is depressed; that is, you cannot start your EV with your foot on the throttle—a very desirable safety feature. Since the Curtis controller also contains a fault input mode that turns the controller off in the event an open potentiometer input is detected (e.g., in the case of a broken wire)—a condition that would result in a runaway—you are covered in both instances by using the Curtis controller and throttle with high pedal disable option. Figure 9-19 shows that the throttle potentiometer wiring goes directly to the controller inputs, with the interchangeable potentiometer leads and the common and normally closed contacts wired as shown.

FIGURE 9-19 Curtis throttle potentiometer with high pedal disable switch (note the three switch contacts at left).

FIGURE 9-20   Reversing contactor—double pole, cross connected good for 48- to 96-V battery packs.

Whatever you're doing in the electrical department, a simple terminal strip such as the one shown in Figure 9-20 makes your wiring easier and neater. Using one or more of these as convenient tie-off points not only reduces error possibilities in first-time conversion wiring but also makes it simpler to track down your connections later if needed. Of course, it's only as valuable as your hand-drawn sketch of what function is on which terminal.

According to Grassroots Electric Vehicles, it is one of the best throttles on the market. I'd take the company's word for it, but you can always shop around. Price is about $180 with shipping, so I don't think you are breaking the bank.

The throttle assembly uses a high-quality original equipment manufacturer (OEM) throttle-position sensor (TPS) that is rated for 1 million full cycles. The TPS and its electrical connector are fully sealed. No stamped steel and exposed potentiometer element here.[14]

Auxiliary Relays

Figure 9-21 shows these highly useful control double-pole relays in both 20-A-rated 12-V DC. A typical use for the AC coil would be as a charger interlock. Wired in series

FIGURE 9-21   Auxiliary relays—20-A-rated 12-V DC.

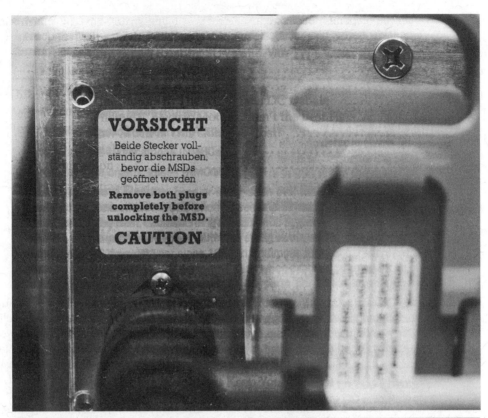

VORSICHT

Beide Stecker voll-
ständig abschrauben,
bevor die MSDs
geöffnet werden

**Remove both plugs
completely before
unlocking the MSD.**

CAUTION

Figure 9-22   Service disconnects or safety interlocks ready to install. Be careful with this part—both in German and in English. (*TurnE Conversions.*)

with the on board charger, AC voltage sensed on the charger's input terminals would immediately disable the battery pack output by interrupting the auxiliary battery key switch line, which, in turn, opens the main contactor. DC coil uses are limited only by your imagination: additional safety interlocks, as shown in Figure 9-22; voltage, current, or temperature interlocks; and controlling lights, fans, and instrumentation.

Terminal Strip
Whatever you're doing in the electrical department, a simple terminal strip like the one shown in Figure 9-23 makes your wiring easier and neater. Using one or more of these as

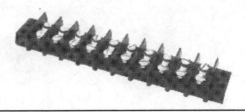

Figure 9-23   Terminal strip.

convenient tie-off points not only reduces error possibilities in first-time conversion wiring, but also makes it simpler to track down your connections later if needed. Of course, it's only as valuable as your hand-drawn sketch of what function is on which terminal.

Shunts

Shunts are precisely calibrated resistors that enable current flow in a circuit to be determined by measuring the voltage drop across them. Two varieties are shown in Figure 9-24: The left measures currents from 0 to 50 A, and the right measures currents from 0 to 500 A.

Ammeter

The most useful of all your EV onboard instruments is your ammeter. The higher range enables you to determine your motor's instantaneous current draw; it functions much like a vacuum gauge in an internal combustion engine vehicle—the less current, the higher is the range, and so on. The lower range functions the same as the ammeter that might already be on the instrument cluster of your internal combustion conversion vehicle—it tells you the amount of current your 12-V accessories are consuming.

Voltmeter

The second most useful EV onboard instrument has to be your voltmeter. The TurnE Conversions model is more for voltage testing purposes, but other conversions deliver 50- to 150-V or 9- to 14-V ranges at the flip of a switch. Expanded scale means that only the required voltmeter range is used; the entire scale is expanded to fill just the range of voltages you use. The higher range enables you to determine your battery pack's instantaneous voltage (it functions like the fuel gauge in your internal combustion engine vehicle—the less voltage, the less range is remaining). The lower range functions like the voltmeter that might already be on the instrument cluster of your internal combustion conversion vehicle—it tells you the status of your system.

Battery Indicator

Users seem either to like the battery indicator a lot or find it redundant (with the voltmeter), so investigate before you invest. The Curtis Model 900 in Figure 9-25 indicates battery pack SOC as of your last full charge-up.

The battery indicator is wired directly across the battery as if it were a voltmeter. Proprietary circuitry inside the module then integrates the voltage state into a readout of remaining energy that's displayed on one of the 10 LED bars. While this is not as useful for those employing onboard charging and certainly does not come close to being a battery energy management system, it is useful for those who charge only from a fixed site as a guide to when the next pit stop is required.

FIGURE 9-24 Ammeter current shunts—50 amp (left) and 500 amp (right).

Figure 9-25 Typical Curtis battery tester. (*TurnE Conversions.*)

Rotary Switch

Rotary switches such as the four-pole, two-position switch are ideal companions to your instrumentation meters for range, sensor, and function switching. While you can opt for only an ammeter and a voltmeter in your finished EV, you might want to check voltage and current at numerous points during the testing stage. A handful of rotary switches helps you out in either case.

Fans

Fans fall into the mandatory category for keeping temperature rise in check for engine compartment components and for keeping the battery compartment ventilated during charging. Whether DC-powered full time from the key-switch circuit, DC-powered intermittently via relay closure, or powered from an AC outlet, spark-free brushless 12-V DC motor fans and 120-V AC motor fans are the types you want to choose. As you can see in Figure 9-26, there are fans for attachment to almost every set of batteries.

Low-Voltage Protection Fuses

All your instrumentation and critical low-voltage components should be protected by 1-A fuses (the automotive variety work fine), as shown in Figure 9-27.

FIGURE 9-26 Cool those batteries! (*TurnE Conversions.*)

FIGURE 9-27 Instrumentation meters—converted electric Porsche Speedster. (*TurnE Conversions.*)

Low-Voltage Interlocks

Many EV converters prefer to implement the "kill switch" referred to earlier in this section on the low-voltage side. Often this is easier because there are a number of interlocks already there—seat, battery, impact, etc. In addition, a low-voltage implementation takes just a simple switch, possibly a relay, and some hookup wire—a few ounces of weight at the most—whereas a high-current solution takes several pounds of wire plus bending and fitting, etc.

DC-to-DC Converter

Figure 9-28 shows you a newer typical Delphi 2.2-kW DC/DC converter that creates DC voltages in hybrid and EVs necessary to power accessories and the heating, ventilation, and air-conditioning (HVAC) system. Liquid-cooled, the DC/DC converter operates at input voltages from 216 to 422 V DC with a power range to 2.2 kW. And most DC/DC converters give you a nice, stable 12-V output, even with widely varying battery pack voltage swings. By this time, you already know it's inadvisable to draw power from anything less than all the batteries in the pack's string (or risk reduced battery life, etc.), so choose your DC/DC converter accordingly—no 12-V to 12-V models, please.

Wiring It All Together

Five things are important here—wire and connector gauge, connections, routing, grounding, and checking. I'll cover them in sequence.

Wire and Connectors

This might be one of the last things you think about, but it's by no means the least important. While your wire size and connector-type choices on the instrumentation side are not as important as the connections you make with them, all these are important on the power side.

Working with American Wire Gauge (AWG) 2/0 cable-gauge wire is not my favorite pastime—think of it as involuntary aerobic exercise—but its minimal resistance guarantees you a high-efficiency EV as opposed to the world's greatest moving toaster. Minimal resistance means that how the connectors are attached to the wire cable ends is equally important to the overall result. Crimp the connectors onto the cable ends using the proper crimping tool (ask your local electrical supply house or cable provider),

FIGURE 9-28   Sevcon DC-to-DC converters—72/12 model (left) and 128/12 model (right).

or have someone do it for you. A "dinky" triangle contact crimp will cause you a hot spot that sooner or later will melt (or be arc welded) by the routine 200-A EV currents. Meanwhile, you will get poor performance. Treat each crimp with loving attention and craftsmanship, as if each were your last earthly act, and you will be in heaven when it comes to your EV's performance.

Connections

On the power side, connections are important. Check to ensure that surfaces are flat, clean, and smooth before attaching. Use two wrenches to avoid bending flat-tabbed controller and fuse lugs. Torque everything down tight, but not gorilla-tight. Check everything, and retighten battery connections at least monthly.

Routing

Aim for minimum-length routing on the power side. Leave a little slack for installation and removal and a little more slack for heat expansion, and then go for the line that's the shortest distance between the two points. On the instrumentation side, neatness and traceability count; you want it neat to show off to friends and neighbors, and you want it orderly so that you (or someone else) can figure out what you did.

Grounding

The secret of EV success is to be well grounded in all its aspects. *Well grounded* in electrical terms means three things:

- **Floating propulsion system ground**—No part of the propulsion system (batteries, controller, etc.) should be connected to any part of the vehicle frame. This minimizes the possibility of being shocked when you touch a battery terminal and the body or frame and diminishes the possibility of a short circuit occurring if any part of the wiring becomes frayed and touches the frame or body.

- **Accessory 12-V system grounded to frame**—The 12-V accessory system in most EV conversions is grounded to the frame, just like the electrical system of the internal combustion engine vehicle chassis it uses. The body and frame are not connected to the propulsion system, but the frame can and should be used as the ground point for the 12-V accessory system, just as the original vehicle chassis manufacturer did.

- **Frame grounded to AC neutral when charging**—The body and frame should be grounded to the AC neutral line (the green wire) when an onboard or off-board AC charger is attached to the vehicle. This prevents electrical shock when the batteries are being charged. To guarantee shock free performance, transformer-less chargers always should have a ground-fault interrupter, and transformer-based chargers should be of the isolation type.

Checking

This is not a paragraph about banking. It's a paragraph about partnership. Whatever system you decide to use—continuity checking, verbal outcry, color coding, matching terminal pairs to a list, etc.—at least have one other human help you. It'll make the conversion go faster, plus chances are that you'll find something that you alone might have overlooked. Speaking of chances, be sure your EV's drive wheels are elevated the first time you hook up your batteries to your newly wired-up creation so that it doesn't accidentally "wander" through your garage door.

Summary of the Parts

The following packages are suitable for conversions of cars and trucks of all sizes. They work with medium-range battery pack voltages (72 to 342 V). Your EV can achieve top speeds of 65 to 100+ mi/h depending on chassis and battery pack voltage. You can mix and match parts on these packages and switch the controller to the Soliton 1 controller on any package.[15]

No. 1: Similar in power to a four-cylinder gas vehicle.		
Impulse 9-inch motor	$1,600.00	Potential 80 hp
KDH12401A, PM 120-V, 800-A controller with Regen	$1,180.00	
Emergency disconnect	$ 115.00	
Main contactor	$ 89.00	
PC-6 accelerator pot	$ 85.00	
DC/DC converter, 55 A	$ 155.00	
Safety fuse	$ 40.60	
Analog state of the charge meter	$ 50.20	
BG 2000-360-V DC charger	$ 670.00	
Total	$3,984.80	

No. 2: Similar in power to a four-cylinder gas vehicle.		
WarP 9-inch motor	$ 1,770.00	Potential 125 hp
KDH14800D,24- to 156-V, 800-A series/PM controller	$ 1,199.00	
Emergency disconnect	$ 115.00	
Main contactor	$ 156.00	
PC-6 throttle	$ 85.00	
DC/DC converter, 55 A	$ 155.00	
Safety fuse	$ 54.60	
Cycle Analyst multimeter	$ 169.95	
BG 2,000- to 360-V DC charger	$ 670.00	
Total	$ 4,374.55	

No. 3: 96-V AC system similar in power to a 60-hp gas-powered vehicle.*

IOT 9-inch AC motor	$ 2,000.00	
Curtis AC 650-A Regen controller	$ 2,100.00	With wire harness
Curtis 840 multigauge controls	$ 295.20	
Emergency disconnect	$ 85.00	
SL3559 contactor, 24 V	$ 156.00	
PC-6 accelerator pot	$ 85.00	
DC/DC converter, 55 A	$ 155.00	
Safety fuse	$ 54.60	
BG 2,000- to 360-V DC charger	$ 670.00	
Total	$ 5,531.60	

*Batteries are not included, nor the transmission adapter plate and hub.

No. 4: Similar in power to a six-cylinder gas-powered vehicle.

WarP 9-inch motor	$ 1,770.00	Potential 185 hp
Soliton Jr. controller	$ 1,995.00	
Emergency disconnect	$ 115.00	
EVnetics throttle	$ 156.00	
DC/DC converter, 55 A	$ 155.00	
Safety fuse	$ 74.60	
Cycle Analyst multimeter	$ 169.95	
BG 2,000- to 360-V DC charger	$ 670.00	
Total	$ 5,105.55	

No. 5: Similar in power to a small-block eight-cylinder gas-powered vehicle.

WarP 9-inch motor	$ 1,770.00	Potential 350 hp
Soliton 1 DC motor controller	$ 2,995.00	
Emergency disconnect	$ 115.00	
EVnetics accelerator pot	$ 156.00	
DC/DC converter, 55 A	$ 155.00	
Safety fuse	$ 90.80	
Cycle Analyst multimeter	$ 169.95	
Manzanita PCF-20 battery charger	$ 2,050.00	
Total	$ 7,501.75	

No. 6: Similar in power to a small-block eight-cylinder gas-powered vehicle.

WarP 11-inch motor	$ 2,965.00	Potential 350 hp
Soliton 1 DC motor controller	$ 2,995.00	
Emergency disconnect	$ 115.00	
EVnetics accelerator pot	$ 156.00	
DC/DC converter, 55 A	$ 155.00	
Safety fuse	$ 209.15	
Cycle Analyst multimeter	$ 169.95	
Manzanita PCF-30 battery charger	$ 2,900.00	
Total	$ 9,664.75	

No. 7: No transmission needed.

Trans WarP 11-inch motor	$ 3,500.00	Potential 350 hp
Soliton 1 DC motor controller	$ 2,995.00	
Emergency disconnect	$ 115.00	
Reversing SW202-84 12-V DC coils	INCLUDED	
Contactor with blowouts	$ 489.95	
EVnetics accelerator pot	$ 156.00	
DC/DC converter, 55 A	$ 155.00	
Safety fuse	$ 209.15	
Cycle Analyst multimeter	$ 169.95	
Manzanita PCF-30 battery charger	$ 2,900.00	
Total	$10,690.05	

No. 8: Lightweight vehicles.

Mars 0913 brushless DC motor	$ 899.00	Potential 30 hp
KHB72701, 24- to 72-V,700-A controller with Regen	$ 999.00	
Contactor, cooling fans assembly	$ 259.00	
throttle pedal	$ 79.00	
DC/DC converter, 72 to 13.5 V, 25 A	$ 149.00	
72-V/10-A charger, 110-V AC input	$ 199.00	
Total	$ 2,584.00	

CHAPTER 10

Electric Vehicle Builds and Conversions

Any of the best-selling vehicles are also the best EV builds and conversions.
—Porsche Electric Cars Are a Favorite Electric Car Across the Globe

Whether you've read through all of this book thus far or have just picked it up and skipped ahead to this chapter, you've finally made it to the chapter that tells you how to do it. The conversion process is what ties your electric vehicle (EV) together and makes it run. You've learned that choosing the right chassis to meet your driving needs and goals is a key first step. You've also learned about motor, controller, battery, charger, and system wiring recommendations. Now it's time to put them all together. A carefully planned and executed conversion process can save you time and money during conversion and produce an efficient EV that's a pleasure to drive and own after it's completed.

This chapter goes through the conversion process step by step with the assistance of a few conversion specialists. It also recommends the type of chassis to choose for your EV conversion today—the compact pickup truck. You'll discover that after the simple act of going through the conversion process that your efforts and results will perform even better—you are now an expert.

Conversion Overview

What do you do for a fresh point of view when constructing something mechanical, even if it's an EV? You go to someone like Steve Clunn from Green Shed Conversions, who as I write this book is traveling across the country to fix people's EV conversions. With more than 20 years' experience having owned his own auto repair shops, he is able to offer a completely fresh point of view on how to build an EV.

What do you do first? While you might love to be like Steve Clunn from Green Shed Conversions and get a sport utility vehicle (SUV), which has picked up great popularity over the past few decades, if you don't have a lot of money for a battery bank, you should go with a smaller car. Why do I say this?

Everyone has his or her own choice of car. I am not going to get in the middle of that.

I just want to build electric cars. Please realize, though, that SUVs and trucks make excellent platforms for EVs because it is possible to isolate the batteries from the passenger compartment very easily; the additional battery weight presents no problems for a vehicle structure that was designed to carry weight anyway, and a pickup is far roomier in terms of engine compartment and pickup box space to do whatever you want to do with component design and layout. Keep in mind that the principles set forth here will apply equally well to any sort of EV conversion or build that you do or buy. The Ford Ranger, the Chevy S-10, and the Dodge Dakota platforms also have amazing conversion opportunities according to Electric Vehicles of America (EVA) and other proponents. In Germany, TurnE Conversions is making converted Porsche speedsters with serious battery packs, as shown in Chapter 8, as well as in this chapter, as shown in Figures 10-1 and 10-2 from Paul Liddle in Florida. Another conversion company is AMP Electric Vehicles, which is building EV vehicle conversions on Dodge and Chevrolet platforms.

Some of the greatest results were the first EV Bob Brant wrote about in the first edition of this book—a 1987 Ford Ranger. Yet places such as Green Shed Conversions and EV Porsche (not the Porsche car company), Electric Vehicles of America, TurnE Conversions, AMP Electric Cars, Electro Automotive, and countless others are also creating great conversions. AMP Electric Cars amps up, "Building your own electric Dodge Ram"—and Chevy cars too! This is an awesome company to see, and it shows that you can build your own EV from just about any car brand. Figure 10-3 shows the Nissan that Joe Porcelli and Dave Kimmins of Operation Z converted to electric.

The objective here is to get you into a working EV of your own with minimum fuss, converting from an internal combustion engine vehicle chassis. For those of you who are into building from the ground up and kit-car projects, there are other books you can

Figure 10-1 TurnE Conversions' electric Porsche Speedster.

FIGURE 10-2   Paul Liddle's electric Porsche. (*EV Porsche.*)

read, and the techniques discussed here can be adapted. The actual process for your EV build or conversion is straightforward:

- Before building or converting (planning)—who, where, what, and when?
- Building or converting (doing)—chassis, mechanical, electrical, and battery
- After building or converting (checking)—testing and finishing

Figure 10-4 shows you the entire process at a glance based on the Toyota Motor Corporation RAV4 EV. Let's get started.

FIGURE 10-3   The electric Nissan created by Joe Porcelli and Dave Kimmins of Operation Z.

FIGURE 10-4   Toyota RAV4 EV drive system. What's under their hood? (*Toyota Motor Corporation.*)

Before Building or Converting

A little effort expended before conversion, in the who-where-what-when planning stage, can pay large dividends later because you've thought out what you need beforehand and you don't have to go running around at the last minute. Let's look at the individual areas.

Arrange for Help

This is the who part. Help comes in two flavors: inside help and outside help.

Inside Help

Whether it's lifting out the engine, pulling the high-current wiring, installing the batteries, or just halving the assembly time and having someone to talk with while you work, an inside helper can work wonders. It's strongly suggested that you schedule one for all the heavy tasks and for any and all others as it suits you.

Outside Help

This involves subcontracting out entire tasks to professionals who are more competent and faster at doing their specialty. Excellent candidates for outside help would be in internal combustion engine removal and fabrication of electric motor mounting components.

Arrange for Space

This is the where part. Your work area can be owned, rented, or borrowed, but it has to be large enough, clean enough, heated or cooled enough, and with good lighting and appropriate electrical service available to take care of your needs during the length of time your EV conversion requires. Let's run down the checklist of what you should look for.

Ownership or Rental

You'll need the space for 80 to 100 hours of work time or more—one to three months of calendar time and up. If you're using your own two-car garage for the project, household rules of companionship dictate that you don't tell your wife or husband, "I just need it for a little while, honey—you don't mind parking your Porsche in the street, do you?" Plan ahead. It's going to take at least a couple of months.

An alternative solution is to rent the space you need. This can be anything from a neighbor's unused garage to an oversized public storage locker to space at a local garage. At some point, your conversion is going to look like a dinosaur with its bones strewn all about; at another, it's going to be very messy and greasy; at another, you might need to attach a winch to an overhead beam that can support the weight of the engine you are removing; and at most times, you are going to have upwards of several thousand dollars' worth of components sitting around—consider all these needs. Another alternative is to do it as a two-step process. If you have the internal combustion engine parts removed by a professional and then tow your chassis to your work area for the conversion, there is less need for a large work area and heavy ceiling joists, and it won't get as dirty.

Size

Realistically, an EV conversion will overflow the needs of any single-car garage or storage locker. While it can be done, you will be cramped, so why not do it right to begin with. Unless you opt for the two-step approach just mentioned, you need at least a two-car garage or an equivalent amount of space.

Heating, Cooling, Lighting, and Power

Suit yourself here. You're going to be working in this spot for several months. Why not do it in comfort and convenience?

Arrange for Tools

This is the what part. What kind of tools are you going to need to do the job? If you have to get into areas and tools with which you are unfamiliar, maybe you are best served by subcontracting those tasks to an outside professional. A little forethought in this area lets you quickly sort out the tasks you want to farm out from those you want to do yourself—all just by using the tools criteria.

Arrange for Purchases and Deliveries

This is the when part. Ideally, you have a just-in-time arrangement—the exact part you need comes magically floating through the door exactly when you need it. Reality might fall somewhat short of this, but nothing stops you from setting it up as your goal by thinking about what you are going to need and when in advance.

Building or Converting

Conversion planning can pay even greater dividends. As you can see in Figure 10-5, there are four parts to a conversion, or the doing stage, and each is further subdivided:

- **Chassis**—Purchase, preparation, removal of internal combustion engine parts
- **Mechanical**—Motor mount fabrication, motor installation, battery mounts, and other mechanical parts fabrication and installation
- **Electrical**—High-current, low-voltage and charging system components and wiring
- **Batteries**—Purchase and installation of batteries

A simple way of looking at the procedure is this: Buy and clean up the chassis, remove all the internal combustion engine parts, make or buy the parts to mount the electric motor and batteries, mount and wire the electrical parts, and then buy and install the batteries. Let's look at the individual areas with the help of the master, Ken Koch. In addition, Figure 10-6 shows that the song remains the same with a typical KTA Services, Inc., EV conversion kit that the company has been selling since the first edition of this book.

While the chassis and body were being readied, I looked around in earnest for a shop that could put everything together—everything except the electrical components and electronics. Having a builder far distant from my shop didn't make sense. Who wants to hop on an airplane every time you need to answer a question, give directions, or make critical decisions? So I looked within a 30-mile radius and found several hot rod builders who had experience putting together street rods with fiberglass bodies. Some were too busy and couldn't take on my project for months. Others wanted me to ransom one of my kids for a down payment. By sheer coincidence, though, I received three referrals to one builder all within the same week. I picked up his business card at a powder coating shop, another at a street rod show, and a third from my neighbor who has a beautiful '34 Ford roadster rebuilt by Brian Noppert of Ultimate Customs. Brian's shop is in Ontario, California, only eight miles from my house. All these coincidences had to mean something, so I gave him a call. "Hey Brian, we've never met before, but I feel as though I know you. Your reputation and your work precede you, and I've got a project that I think you'll find unique and interesting." The conversation went well from there on, so we cemented the deal. Brian is a multitalented young man whose portfolio of custom vehicles grows with each passing day. A recent build of his is a '50 Ford with a blown Lincoln Navigator engine, featured in the September 2007 issue of *Hot Rod Magazine*. I feel very confident with my pet project placed in Brian's hands.

To date, Ultimate Customs has accomplished about 85 percent of the work required. The company has designed and installed an adapter and motor mounts to put the transmission and two motors in place. The motors are securely mounted to the frame and are joined together in the middle with a heavy-duty chain coupler. Boy, what a job the motor installation was! Ordering the roadster body with a 6-inch recessed firewall (for a big-block engine) was a good idea. Can you imagine installing 39 inches of motor length under the hood of any vehicle, much less a '34 Ford? Nevertheless, the two motors and the transmission have been engineered into place with perfect alignment to the differential.

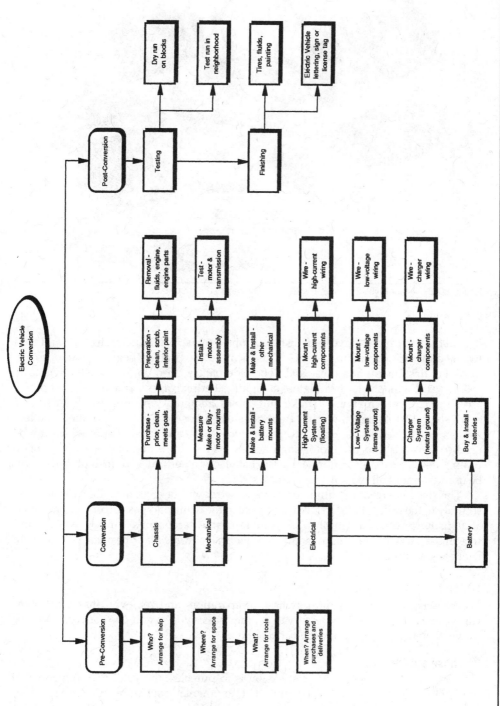

Figure 10-5 Summary of the EV conversion process.

287

FIGURE 10-6 Typical KTA Services EV kit. Yes, they are still in business.

All the body pieces have been installed, joined to the frame and each other, and then tweaked for best fit and appearance. This includes the running boards, fenders, grille, trunk lid, dashboard, hood top, and hood side panels. Also in place are the two low-profile seats, windshield, headlights, taillights, steering column and wheel, foot brake, accelerator pedal, emergency brake and cables, remotely operated trunk latch, and drive shaft. Just behind the grille a special shroud has been fabricated and installed. This shroud closes off the grille except for two 400 ft^3/min fans, each of which can be used to force air into the motors and across a small transmission cooler to provide cooling for the two Zilla controllers. Also installed is a vertical 3/16-inch plate that will be used to mount all the power electronics.

The last 15 percent of Ultimate Customs' work will be to mount the battery boxes securely to the frame, build a bulkhead between the trunk and passenger compartment, and finally, to design and install a six-point roll bar. After that, it's tear down the car and sort all the parts into four piles: paint, powder coat, chrome plate, or metal polish. More later as work progresses.[1]

Chassis

The chassis part involves the purchase, preparation, and removal phases. In other words, you do everything necessary to get the chassis you're going to convert ready for conversion. Let's take a closer look at each step.

Purchase the Chassis

The first step in a step-by-step conversion is to purchase the vehicle you're going to convert to electric. As an added incentive, take another look at Figures 10-1 through 10-4. These are the finished conversion photographs. Notice, in each case, that the vehicle is absolutely stock on the outside (Green Shed Conversions, the Smart car from TurnE Conversions, Paul Liddle's Porsche, or the Nissan Z). These companies have crews that built a great car like no other. All of them exemplify what building your own EV is all about! (*Note:* Bob Brant liked to mention that the only giveaway is the lettering. If you're going for acceleration, make sure to make the word "electric" on the back of your tailgate very large.)

The chassis purchase details were covered in Chapter 5; the bottom line is to get the most for the least. You want the stripped four-cylinder (six-cylinder in some models and years), manual transmission, least-weight version. Ideally, you don't need the engine, so you can do a tradeout with the dealer or selling party on the spot. Just make sure that everything else is as close to perfect running condition as possible—less work for you later. Most of all, make sure that there's little or no rust; clean is nice, too. Mechanical parts you can replace, but a rusted body is nearly impossible to deal with, and dirt and crud just exacerbate the condition and possibly hide additional problems.

For Tesla Motors, though, the process starts with aluminum. The sheets are placed in a hydraulic press machine and stamped into three-dimensional fenders, hood panels, doors, and roofs. Stamped aluminum saves about half the weight of steel, and the decrease in weight enables the builder to increase overall vehicle efficiency. Even before I finalized the factory location, I secured machinery and items I'd need to quickly bring the facility online. My hydraulic press line is of the finest quality, and I expect to use it for decades. The press line will produce body panels you can see and structural features that you can't. The remaining body stampings, castings, and extrusions will be produced at outside facilities and shipped to the shop for assembly.

Purchase Other Components

Once you select your chassis, you can make other parts decisions based on your overall performance goals: high mileage, quick acceleration, or general-purpose commuting. Expected vehicle range and acceleration can be calculated from the vehicle's total weight, tire rolling resistance, aerodynamics, and amount of energy and power available from the motor and batteries.

One of the easiest and quickest ways to do your build or conversion is to order all the parts you need in one kit from someone such as EVA, Green Shed Conversions, KTA Services, Manzita Micro, or NetGain. Their kits typically contain the right parts, and you get nearly everything you need to complete your build. The parts are matched to work together, and you have someone to turn to in case you need additional assistance. Buying a prepackaged kit will greatly simplify your build in terms of both what it takes to get the job done and the time it takes to run around and get all the parts you need from various vendors, make sure they match, and so on. Green Shed Conversions provides the builtup control box and motor-to-transmission adapters that really save you time and guarantee results or the entire kit for a midrange or low-end pickup truck of your choice, which includes all heavy-duty coil springs, battery brackets, cables, mounting hardware, and so on.

On the other hand in Figure 10-7, Tesla Motors built a chassis or an undercarriage from the ground up. I saw this display of the Tesla Motors Model S undercarriage display at the Westchester mall in White Plains, New York.

The other items you want to order at this time are the detailed shop maintenance and electrical manuals for your chassis. These also may be available in the library or in a technical bookstore in a larger city. Study these before starting your conversion.

Figure 10-7   Tesla Motors Model S undercarriage display at the Westchester mall in White Plains, New York getting ready to place an electric motor in a Ford Focus. (*Seth Leitman.*)

Prepare the Chassis

The next step is to clean the chassis and make a few measurements. The first chassis cleaning step is to give its engine compartment a good steam cleaning. Well-used chassis might require extra scrubbing to remove accumulated dirt and grime. This has a twofold purpose: It minimizes the grease and gunk you have to deal with in parts removal, and you can see what you need to look at and measure. You might even want to repaint the engine compartment area at this time.

The measuring step involves determining the position of the transmission/drive train or transaxle in your internal combustion engine vehicle and reproducing this position as closely as possible in your EV. Your internal combustion engine chassis will have one of three possible engine/drive train arrangements:

- Lateral-mount front engine, transmission, and driveshaft
- Transverse-mount front engine with transaxle
- Rear engine with transaxle (VW-style)

SUV or truck conversion chassis will nearly always involve one of the first two variations. Two measurements are important: vertical height from transmission housing to floor (measured on a level surface) and a vertical and horizontal reference point on transmission housing to body or frame.

Vertical clearance from the floor is important in any conversion where your motor mounts also support your transmission. You want to reproduce this as closely as possible so that your transmission is not running uphill or downhill from the way it started.

A firewall or frame horizontal and vertical alignment mark is also important. It helps in obtaining proper mechanical alignment of the mating parts later on. Scribe this mark accurately with an awl or nail, and highlight it with a dab of spray or touchup paint so that you can find it later. Tesla Motors, though, started from scratch with its builds.

Removing Chassis Parts

The next step is to drain the chassis liquids, remove all engine-compartment parts that impede engine removal, remove the engine, and then remove all other internal combustion engine–related parts. The parts removal process starts with draining all fluids first: oil, transmission fluid, radiator coolant, and gasoline. Remember to dispose of your fluids in an environmentally sound manner or recycle them. Draining gasoline from your tank is particularly dangerous and tricky. Drain as much as you can before you physically remove the tank. Discharging the air-conditioning system, if your conversion vehicle has one, is a job best left to a professional.

Next, disconnect the battery and remove all wires connected to the engine. Carefully disconnect the throttle linkage—you will need it later—and set it aside out of harm's way. Then remove everything that might interfere with the engine lift-out process. The hood is a good item to start with, followed by the radiator, the fan shroud, the fan, coolant and heater hoses, and all fuel lines. Disconnect the manifolds, and remove the exhaust system at this time, too.

If your work area isn't equipped for heavy lifting (the smallest four-cylinder engine with accessories attached can weigh around 300 pounds), you might be best served by letting a professional handle the heavy work. Engine overhaul shops are usually glad to do the job, and you might be able to cut a deal with them to buy the engine and parts that results in a net gain on the transaction. In addition, there are a number of other aspects— draining fluids, storing parts, and not breaking cables and wires—and the physical act of actually removing the engine without damage to you, your chassis, or the engine. To use an analogy, it's like changing your own motor oil at home versus at the local service station in town or a tire store such as Mavis Tire Store in Ossining, New York.

This will take you several hours, and you have to clean up the mess (yourself and your shop). The local service station or, better yet, a Mavis Tire Store and the people that do it for a living take 10 minutes, and neither they nor you get dirty.

If you do the job yourself (schedule an inside helper for this step), pick a sturdy joist to attach your lifting winch to—not a two-by-four—or rent an engine-lifting dolly. Cover the vehicle's fenders and sides with moving pads to protect them during lift-out. Attach the chain or cable to the engine at two different points. Remove the bolts from the engine mounting brackets and those attaching the engine to the transmission bell housing. The disconnected engine slides straight back away from the transmission and then up and out (in theory). In practice, since you have to both pull down on the winch cable and pull back on the engine block, another set of hands helps greatly. The objective of the whole process is to remove the 300-pound engine without banging, bending, or damaging anything inside the bell housing (transmission spline shaft, etc.) or on the now-exposed business end of the engine (clutch flywheel assembly, etc).

Then carefully set the removed engine down in an out-of-the-way yet protected part of your work area. And finally, remove everything else associated with the internal combustion engine: gas tank, gas lines, muffler, exhaust pipes, ignition, and cooling and heating systems.

Most EV conversions vehicles have been manual transmissions because they are more efficient than automatic transmissions and provide greater range, require less motor torque, require no transmission cooler, and are easier to convert. The problem with an automatic transmission is that it shifts at about 2,000 rpms; electric motors are usually designed to operate efficiently at between 4,000 and 5,000 rpms. Consequently, an automatic transmission is a poor choice that results in decreased range. If you buy a vehicle with an automatic transmission, you can replace it with a manual transmission. The additional cost is $250–$300 and up depending on the transmission and used auto parts dealer. Consider trading the automatic transmission. As stated earlier, it is helpful to use a lower gear to get the motor rpms up to improve efficiency, torque, and motor cooling. At higher speeds, it is desirable to use a higher gear to keep the motor from overrevving. So if you decide to keep your automatic transmission, you will still have to do some shifting.[2]

Mechanical

The mechanical part involves all the steps necessary to mount the motor and install the battery mounts and any other mechanical parts. In other words, you next do all the mechanical steps necessary for conversion. You follow this sequence because you want to have all the heavy drilling, banging, and welding—along with any associated metal shavings or scraps—well cleaned up and out of the way before tackling the more delicate electrical components and tasks.

Let's take a closer look at the steps.

Mounting Your Electric Motor

Your mission here is to attach the new electric motor to the remaining mechanical drive train, or in a complete ground-up build, you are connecting the motor to a new drive system. The clutch-to-flywheel interface is your contact point. Figure 10-8 gives you an overview of your task in generalized form from Ford Motor Company getting ready to mount the electric motor in its Ford Focus on the assembly line. In addition, Figure 10-9 shows a display at this year's Plug-In Day out in Long Island with my friend Carl Vogel who authored the Green Guru Guide book for me called *Build Your Own Electric Vehicle*.

Four elements are involved in mounting your electric motor:

- The critical-distance flywheel-to-clutch interface
- Rear support for the electric motor
- Front support for the motor-to-transmission adapter plate
- Flywheel-to-motor-shaft connection via the hub or coupling

I'll cover what's involved in each of these four areas in sequence. Understand that this discussion has to be generalized because there are at least a dozen good solutions for any given vehicle. So I'm going to talk in general terms here. You'll have to translate them to your own unique case. And if your skills do not include precision machining of automotive metal parts, this is another good area to enlist the services of a professional, such as KTA Services (back from what some thought extinction) and Green Shed Conversions, which has great adaptor plates, as seen in Figure 10-10, or a local machine shop.

Notice there are bolt-hole ring patterns in the mounting plates. As shown in the figure, the inner-ring pattern with its countersunk mounting holes allows flathead mounting bolts to attach the mounting plate to the motor and to be flush with the plate's surface on the inside when tightened. TurnE Conversions made the outer-ring pattern so that it allows hex-head mounting bolts to be used to attach the mounting

FIGURE 10-8   Ford Motor Company getting ready to place an electric motor in a Ford Focus. (*Ford Motor Company.*)

FIGURE 10-9   Electric motor into motor mount at Long Island's celebration of National Plug-In Day!

Figure 10-10   Electric motor mount to transmission from TurnE Conversions.

plate to the transmission bell housing and to be torqued down tight. Motor brackets help to hold the motor in place while safely attached to the car. The adaptable motor mount kit that can be used in each adapter is sized to accommodate a collar around the motor or the outside diameter of the coupling. This can vary widely from case to case. Just look at the depth to which the hub fits on the motor shaft to control the critical-distance measurement—adjust as needed.

Here's where the fun begins. While you can get two motor mounting kits from Cool GreenSheds, among many possibilities, the large plate in the background of Figure 10-11 on which the other parts are resting is the motor-to-transmission adapter plate overhang all around (to aesthetically suit your taste). Then transfer the electric motor mounting bolt hole patterns to this template using the transmission pilot shaft center as a reference. Take this completed template or pattern to the machine shop, and ask the machinist to reproduce it in metal.

If you are in a let's-get-it-done mode, Steve Clunn's company makes a complete motor-to-transmission assembly with adapter plate and precision-cast flywheel and hub/bushing parts that guarantees perfect results for Ford Ranger conversion projects.

The Critical Distance—Flywheel-to-Clutch Interface or
Using the Proper Front Support-Motor-to-Transmission Adapter Plate

After you remove the engine and before you take the flywheel off, carefully and accurately measure the distance from the front of the engine to the face of the flywheel (the part the clutch touches). This is the critical distance you want to reproduce in your electric motor mounting setup, and it will vary for different vehicles. You're going to put an adapter plate on the electric motor, put a hub or coupling on the electric motor's shaft, and attach the flywheel to this hub. When you've done this, the critical distance

FIGURE 10-11   Always make sure to measure the transmission first! (*TurnE Conversions.*)

from the front of the flywheel to the front of the motor adapter plate should measure exactly the same.

Explanation of the Importance of a Proper Chassis and Motor Mounting

Figure 10-12 shows a great compartment that Paul Liddle created to allow the motor to fit very nicely into the motor, controller, charger mount in one for his electric Porsches.

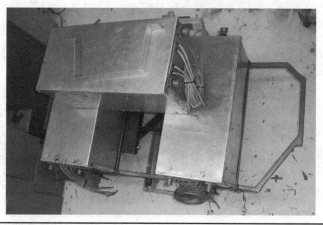

FIGURE 10-12   Electric motor mount from Paul Liddle.

Figure 10-13 shows a proper mounting of the motor by TurnE Conversions and Christian von Hösslin; however, Ken Koch's story is something to hear:

During the 7½-month body shop odyssey, it was heartening to see finished pieces of the rolling chassis trickle in for assembly while the major part of the body and its pieces were being readied for paint. The rolling chassis photo in the figure is a real "money shot" for a magazine. It shows how nicely all of the driveline, suspension, steering, and brakes are done. The four-link and Panhard bar at the rear should help the car hook up nicely. The 13-inch brakes will give lots of "whoa." The two motors, the transmission, and their mounts look exquisite. Even the brake lines are polished stainless. The idea behind all of this was that the car should look consistently nice throughout, just in case some car show judge should poke a mirror under it. No excuses required.

Also during the 7½-month period I was able to design and configure the power components to their mounting plate, plus the two pump systems to their polished firewall plates. I delivered them to the shop for installation a little bit before the body pieces were finished and painted. The photo in the figure was taken after installation into the painted car. One look under the hood and you know that it "ain't no hemi or big block." The two Zilla Z2KEHVs and their Hairballs are obvious, along with the (had to be) orange #4/0 power cables. Note the right-left symmetry. The finned "oil" coolers really are reservoirs. On the right is the power brake vacuum system and reservoir. On the left, there's a controller coolant pump and reservoir that circulates (orange) coolant through a fan-cooled transmission cooler mounted behind the grille.[3]

FIGURE 10-13   Bolting the electric motor to the transmission with an adapter plate in between. View from underneath the electric Smart conversion. (*TurnE Conversions.*)

A mounted series direct-current (DC) motor or an alternating-current (AC) motor and controller as shown by TurnE Conversions in Figure 10-13 is usually attached to the transmission (notice the earlier-mentioned torque arm between motor and frame) with its S1 power cable going up to the controller and motor and controller power cables (with standoff clip to separate them) going down the chassis frame rail *toward the rear battery pack.*

Knowing your goal—the critical distance—makes it easier to navigate toward it. Whether you have a front engine plus transmission or front or rear engine plus transaxle, this goal will be the same, although the specifics will differ.

Your original flywheel might weigh 24 pounds and have an attached ring gear (for the starter motor). Remove the ring gear that fits around the flywheel's outside edge, and have the flywheel machined down to 12 pounds or so by removing metal from its outside edge and rear (the part that faces the motor). Don't touch the flywheel's face (unless it has obvious burns or other defects). These steps don't hurt your flywheel at all yet save weight—the most important factor in your EV conversion's performance.

This is also a good time to give the clutch (the inside layer of the clutch pressure plate/clutch/flywheel "sandwich") a thorough examination. If the clutch is old (used for more than 30,000 miles) or obviously worn, now is the time to replace it. Also take a look at the transmission seals and mounts and replace them if they are excessively worn. Keep in mind that you're going to be using your clutch in an entirely new way (not using it is a better description because there's nothing to start up—you put it in gear first), so this new clutch ought to last you 200,000 miles or more.

Rear Support for the Electric Motor

This is a fairly straightforward area. You basically have two possibilities: Support the motor around its middle or support it from the end opposite its drive connection to the transmission or transaxle. The middle-style motor mounts available from Green Shed Conversions accommodate different electric motor diameters, and other conversion shops or some electric motor companies offer similar products.

The bottoms of the mounts have bolt holes that attach to your vehicle's motor mounts. The two halves of the curved steel strap go around the motor and hold it securely in place. This is the preferred and most common method of mounting electric motors, particularly in the larger sizes. End mounts are similar to motor faceplate adapter mounts; they bolt onto the motor's end face through mounting holes and are then tied off to the frame through heavy rubber (old tires) shock isolators. This approach was very popular with early VW conversions back in the day of the first edition of this book. However, today it can apply to *any* car you want!

Also needed is a torque rod to support/stabilize the motor-transmission combination and prevent excessive rotation under high acceleration loads. This adjustable-length rod normally attaches between one of the adapter-plate-to-transmission bolts and the frame.

Flywheel-to-Motor-Shaft Connection

This is the least standardized area. The coupling is press-fit onto the electric motor's shaft, and the set screw further secures it. Figure 10-10 shows the physical positioning of the six-bolt coupling in front of the electric motor to give you a better idea.

Usually, the controller and charger sit on opposite sides of the motor. The pilot bearing fits into the hub and mates with the transmission's pilot shaft (a transaxle does not require a pilot bearing). The clutch attaches to the transmission spline shaft, and the clutch pressure plate fits over it and bolts to the flywheel to complete the "sandwich."

Hubs (with or without pilot bearing, etc.) are going to differ widely in size and shape for different vehicle, transmission, and motor types, yet their function is the same. You must determine what you need in your particular case. Call Steve Clunn, Bob Batson, or your local mechanic to help you if you're not sure. This is not the time and place to be shy.

Mounting and Testing Your Electric Motor

Depending on your accumulated skill and luck, your motor installation can either go smoothly the first time or involve several cut-and-try iterations. The basic approach is to mount the hub on the motor, attach the motor to its adapter plate, attach the flywheel to the hub, measure the critical distance to make sure that it's exactly the same as it is for the internal combustion engine (add spacers as needed, etc.), and then attach this completed assembly to the transmission. The winch used to take out the internal combustion engine is probably the easiest way to install your motor assembly into your vehicle (the completed assembly weighs upward of 150 pounds with a 20-hp series DC motor), but a floor jack (elevate the front of the vehicle on jack stands first) or engine-lifting dolly also will work. The installation step is basically the reverse of the removal step: Attach the assembly at two points, and then slide the motor up and toward the transmission until the shafts are in alignment. When everything fits exactly, tighten the bolts on the transmission and the motor mounts.

For a quick test, first jack up the vehicle so that its front- or rear-drive wheels are off the ground. Then put a 2/0 cable in place across the motor's A2–S2 terminals. Attach another 2/0 cable from a battery's negative terminal to Sl—you can borrow the 12-V starter battery you just removed for this purpose—and attach another 2/0 cable to the battery's positive terminal, but don't connect it yet; as shown in Figure 10-14. With the transmission in first gear, briefly touch the positive cable from the battery to Al and do two things:

- Look to see if the rear (or front) wheels move.
- Listen for any strange or grinding noises, etc.

If the wheels move, good. If the wheels move and there is no strange grinding, this is doubly good, and you can go on to the next step. If you hear something strange or the wheels don't turn (in this case, first ensure that the battery is charged), you need to unbutton your motor assembly from the transmission and look into the problem.

Ken Koch shows the entire battery pack later in this chapter. Each battery module contains 24 A123 No. 26650 lithium–iron phosphate cells in a 4S6P configuration. Nominal module voltage is 13.2 V. Each of the two battery boxes consists of 54 modules. A series string of 27 modules for 356.2 V is paralleled with another 27 modules in each box. Maximum current per battery box is about 1,400 A. The two-box combined output is 2,800 A. Interconnections are made with 3/16- × 3/4-inch copper bars, now nickel plated. Maximum theoretical power is 917 hp. Combined total capacity is 19,673 kWh.[4]

Figure 10-15 shows the transmission component to the vehicle with the motor controller power cables. A close-up of the wired controller with the battery and motor cables going through glands that ensure 360-degree EMC connection for the shield is also shown. Even on the Porsche and Smart car conversions shown by TurnE Conversions

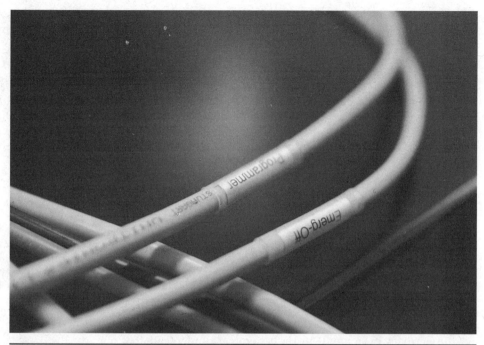

FIGURE 10-14 Cable loom for the preconfigured Smart electric car.

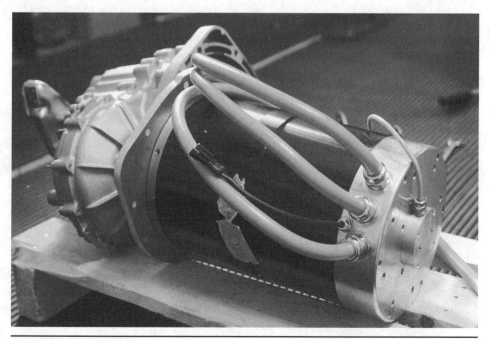

FIGURE 10-15 Motor and power cables getting ready to be mounted. Any questions? (*TurnE Conversions.*)

FIGURE 10-16   Wired-up eBox. (*TurnE Conversions.*)

in Germany in Figure 10-16, the batteries are underneath the Smart car, the Porsche, and the VW bug and far enough away from the motor controller/transmission components as well. Christian von Hösslin provided a picture of a cable loom that's preconfigured for the Smart car. Each wire comes in exactly the length needed and with a sticker to tell them apart, as shown in Figure 10-14.

Fabricating Battery Mounts

TurnE Conversions in Germany has a very creative battery pack and mounting system. The company's battery pack mount is pretty ingenious, as shown in Figure 10-17. The

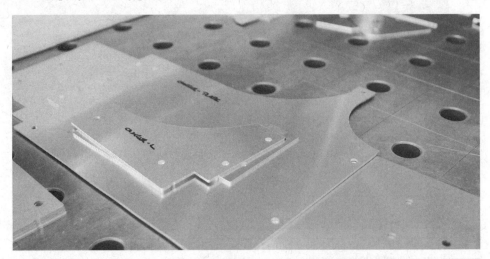

FIGURE 10-17   Clean compartment for EV Conversion. (*TurnE Conversions.*)

figure shows Paul Liddle's frame rail for his electric Porsche conversion and how it should be safely underneath the vehicle, as shown in TurnE Conversions EV conversion in Figure 10-18. Figure 10-19 shows a Tesla Motors Model S getting cleaned up and

Figure 10-18   The battery pack is safe underneath the Smart electric car conversion. (*TurnE Conversions.*)

Figure 10-19   Tesla Roadster getting cleaned up and ready for the next phase of the build. (*Tesla Motors.*)

FIGURE 10-20   See how the motor is mounted into the compartment that Paul Liddle built.

ready for the next phase of the build. The motor mount for Paul Liddle's electric Porsche in Figure 10-20 shows a great design for interconnecting the motor, controller, and battery pack in one safe place.

Additional Mechanical Components

Any metal fabricated parts inside the engine compartment (e.g., air dams, etc.) are better done at this stage because they are easier to get at before the electrical components and wiring are installed. For example, the car's shocks, coil and leaf springs, and external body parts are also best added at this time while you still have your heavy tools out (Figure 10-21).

Cleanup from Mechanical to Electrical Stage

It's a good idea to use an air hose to clean up the engine compartment and rear pickup box of your conversion after the mechanical phase is completed. Follow this with a broom sweep-up of the work area and/or a damp mop-up. The reason for this cleanup is to minimize the chance of metal strips or shavings finding their way into your electrical components or wiring during the electrical phase.

Electrical

The electrical part involves mounting the high-current, low-voltage and charger components and doing the electrical wiring to interconnect them. The electrical wiring requires knowledge of your EV's grounding plan: The high-current system is floating,

FIGURE 10-21   Completed rolling chassis assembly. (*Ken Koch, EV Consulting, Inc.*)

the low-voltage system is grounded to the frame, and the charging system AC neutral is grounded to the frame when in use. Doing the electrical wiring also involves knowledge of your EV's safety plan. Appropriate electrical interlocks must be provided in each system to ensure system shutdown in the event of a malfunction and to protect against accidental failure modes. The electrical wiring is also greatly facilitated by using a junction-box approach so that wiring is neat (no wires running every which way inside the engine compartment), and you can later trace your wiring, as shown in Figures 10-22 and 10-23. Let's take a closer look at the steps brought to us by TurnE Conversions and Christian von Hösslin.

High-Current System

First, you attach the high-current components and then pull the American Wire Gauge (AWG) 2/0 cable to connect them (another step where scheduling inside helper is appropriate). Refer back to Chapter 9 to remind yourself that there are seven components in the high-current line (in addition to batteries and charger—these I'll save for later):

- Electric motor
- Motor controller
- Circuit breaker

FIGURE 10-22 Relays for ignition, brake light (regenerative braking), and reverse light.

FIGURE 10-23 Power cables go into a small box that is attached to the battery case. It holds fuses, the battery management system, modules, and current sensors.

- Main contactor
- Safety fuse
- Ammeter shunt(s)
- Safety interlock

You've already mounted the motor. Figure 10-24 shows the Operation Z layout of the batteries, and Figure 10-25 shows the layout of the electric motor in Paul Liddle's car.

Just for additional reference points, John Wyland used 192 Dow Kokam lithium-polymer cells contained in twelve 29.6-V clear Lexan modules in his fast drag racer. A

FIGURE 10-24   Operation Z battery mounted in place.

FIGURE 10-25   Porsche electric motor power cabling in the trunk. (*Paul Liddle.*)

battery pack at 355 V is capable of 2,400-A discharges and stores 22.7 kWh of energy. This provided a system voltage of 355 V. What did that get Wyland? The best lithium-ion-powered battery had an estimated track time (ET) of 10.258 seconds at 123.79 mi/h in the standing start National Hot Rod Association (NHRA) timed quarter mile, per National Electric Drag Racing Association (NEDRA). Do it right with the wiring and battery management systems, and you can have a dragster that will go like this:

- 1.58-s 60-ft time; 0 to 60 mi/h in 1.8 seconds, 1/8-mile Estimated Time (ET) 6.443 seconds at 108.63 mi/h
- Highest quarter-mile trap speed: 126.01 mi/h
- Highest eighth-mile trap speed: 109.28 mi/h[5]

> It worked out *perfect*, and they work like a charm. The plate still has to be trimmed a little . . . but I may not. I may just leave it for show purposes. . . . the motor is an 8-inch cut just a little out of the frame so that it sits nice and level. This project is coming together so easily . . . and we are having sooo much fun![6]

Figure 10-12 showed you a great motor compartment designed by Pail Liddle at EVPorsche.com. When you flip over and place this compartment in the undercarriage, you will see how snug the motor fits in the EV Porsche WarP motor in the vehicle prior to fastening it in place in Figure 10-26.

Figure 10-26 shows the WarP motor in one of Paul Liddle's Porsches prior to fastening it in place. The circuit breaker has two inputs and two outputs. Other EV converters prefer to mount this circuit breaker within reach of the driver. The tradeoff

FIGURE 10-26 EVPorsche.com WarP motor in the vehicle prior to fastening it in place.

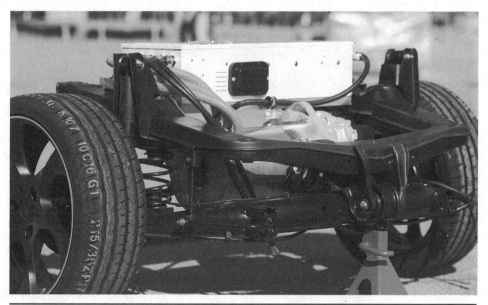

Figure 10-27 TurnE Conversions' electric motor mount to the transmission.

is that you have to pull heavy AWG 2/0 power cables behind the dash or to some other location inside the driving compartment. The benefit is the peace of mind afforded by knowing that if everything goes haywire (contactor and relay contacts weld shut, etc.), there is one switch you can reach over and touch to shut everything down. Figure 10-27 is the reverse side of the electric motor cage fastened to the back of the car. Notice there are two compartments on each side of the electric motor. One side of the electric motor mount is for the controller. The other side is for the charger with enough room for wires (on both sides) to connect the batteries. See how the motor is mounted into the compartment from Figure 10-25.

The high-current safety fuse, ammeter shunt(s), and main contactor are all located inside the junction box (which will be discussed in a moment). Steve Clunn chose to implement a safety interlock on the low-voltage rather than the high-current side. Those who insist on a dashboard-mounted high-current safety or kill switch (usually with a big red knob on it) have to endure the pain of pulling AWG 2/0 cable over, under, around, and through the rear of the dashboard—not my idea of a good time. By the way, the AWG 2/0 being discussed here is stranded (solid wire would be nearly impossible to work with in this size) insulated copper (never aluminum) wire—and it's about ¾ inch in diameter, so you have your hands full.

The high-current system has a floating ground. This means that the negative terminal of the battery pack is not connected to the frame or body at any point—it floats instead. This eliminates the possibility of an accidentally dropped tool arc welding itself to the body or chassis or, worse yet, causing your vehicle to bolt forward or backward while simultaneously causing battery pack meltdown. It also eliminates the possibility of your receiving a 120-V electric shock while casually leaning over the fender to measure battery voltage.

Low-Voltage System

On the low-voltage side, the idea is to blend the existing ignition, lighting, and accessory wiring with the new instrumentation and power wiring. There are six main components on the low-voltage side:

- Key switch
- Throttle potentiometer
- Ammeter, voltmeter, or other instrumentation
- Safety interlock(s)
- Accessory 12-V battery or DC-to-DC converter
- Safety fuse(s)

Every EV conversion should use the already-existing ignition key switch as a starting point. In an EV, the key switch serves as the main on/off switch with the convenience of a key—its starting feature is no longer needed. You should have no problem in locating and wiring to this switch.

In Paul Liddle's Porsche conversions, he mounts the throttle potentiometer on the driver's-side fender. Figure 10-28 shows the throttle potentiometer and Figure 10-29 shows it mounted in place and wired up from a 1987 Ford Ranger that Bob Brant converted. The top shows the throttle and the bottom shows the wired throttle potentiometer. Instrumentation wiring is simple; just be sure to observe meter polarity markings—the plus (+) marking on the meter goes to the positive terminal on battery. The ammeter is connected across the shunt(s) already wired into the high-current system. The voltmeter goes across the battery.

Switching potentiometer and light-emitting diodes (LEDs) show that the switch gear is in place as shown in Figure 10-30 by Christian von Hösslin.

Figure 10-31 shows the battery management system (BMS) wires and battery heating wires. All wires outside enclosures are protected by an extra layer of tubing.

The best solution is to wire the voltmeter so that it is always on, giving you a continuous readout of battery status. You don't have to worry about draining the battery because a modern voltmeter's internal resistance is high enough to cause only a miniscule

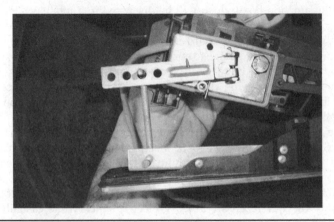

FIGURE 10-28   Electric Porsche throttle potentiometer. (*Paul Liddle.*)

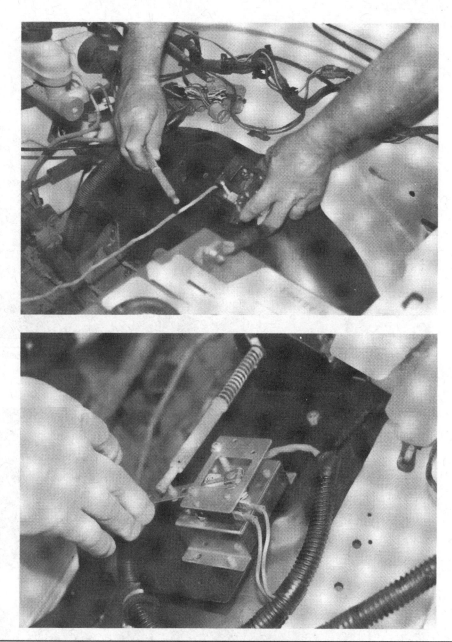

Figure 10-29   1987 Ford Ranger throttle potentiometer positioning (top) and mounted and wired throttle potentiometer (bottom).

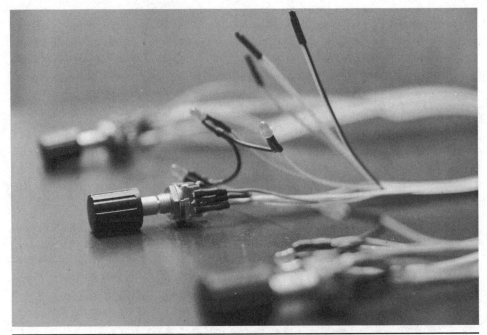

FIGURE 10-30   Potentiometer wires with light-emitting diodes (LEDs). (*Christian von Hösslin.*)

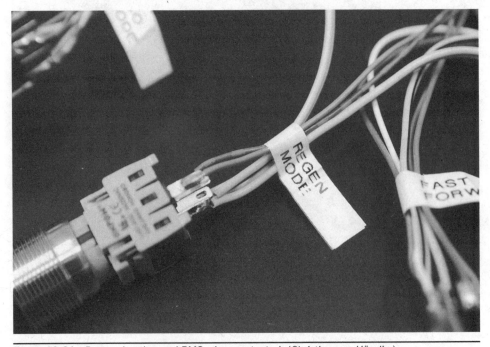

FIGURE 10-31   Battery heating and BMS wires protected. (*Christian von Hösslin.*)

current drain (an order of magnitude less than the battery's own internal self-discharge rate). The battery indicator or state-of-charge (SOC) meter also goes across the battery, but it should operate only when the key switch is on, so wire it to the on/off side of the main contactor (away from the battery). The temperature meter wiring is not particularly critical; just observe proper polarities and grounding, and make sure that the thermistors are securely attached to whatever you are measuring. Ford Motor Company does a quality test for everything. Besides all the instrumentation working, Ford ensures each battery pack that is built in the United States is ready to go. Figure 10-32 shows a Ford Motor Company engineer approving a battery pack before it goes in the car.

If you wish to use a more modern digital voltmeter readout in place of an analog meter, you need to adjust the sample-and-hold circuit (it memorizes the value at any instant) either to display the average of the last few moments' sample-and-hold values or to give a steady readout when a "Read" button is pressed. Otherwise, the rapidly changing voltage or current will be hard to interpret.

The subject of safety interlocks is an important one. Steve Clunn's Green Shed Conversions design uses three—all wired in series on the low-voltage 12-V key-switch line: a fuel injection impact switch, a main safety cutoff switch, and a charger cutoff switch (to be covered in the "Charger System" section). The main safety cutoff switch is a highly accessible dashboard-mounted switch wired in series with the key switch. Punching it immediately removes energizing voltage from the main high-current contactor. A few EV converters also use a seat interlock switch that latches closed when the driver's presence on the seat is detected. You might wish to consider this as an option.

Steve opted to use a battery as the source of the 12-V accessory system power. You can do the same or use the DC-to-DC converter shown in Chapter 9, which is driven from the main battery pack voltage. If you opt for the DC-to-DC converter, now is the time to install it and wire it in place; its input side goes directly across the main battery

FIGURE 10-32 Everything is wired up for this battery pack for the Ford Focus electric car. (*Ford Motor Company.*)

pack plus and minus terminals. Its output side provides +12 V at its positive terminal, and its negative terminal is wired to the chassis. If you elect to use a 12-V deep-cycle accessory battery, do the wiring for it now, but wait until the battery phase to purchase, install, and connect it up.

Try to use stranded insulated copper wire for the low-voltage system. The instrumentation gauges can be wired with AWG 16 or even AWG 18 wire. Safety fuses of the 1-A variety should be wired across the potentiometer and all delicate instrumentation meters. The key-switch circuit can use the original fuse panel, but don't use the original wiring for any loads greater than 20 A.

Unlike the high-current system, the low-voltage system is grounded to the frame; the negative terminal of the 12-V battery (or DC-to-DC converter) is wired directly to the frame or body. Most internal combustion engine chassis come this way. You eliminate rewiring, extra wiring, and potential ground loops by using the existing negative ground-to-frame convention.

Junction Box

A good junction-box design cleans up the hodgepodge of instrumentation wiring running every which way inside the engine compartment, enables you (on anyone else) to later retrace your wiring, and provides convenient mounting and tie-off points for various components. However, not all junction boxes are created equal. Paul Liddle's "magic box" in Figure 10-33 was more equal than most, Jim Harris did some great work for Bob Brant on his electric Ford Ranger—he combines simple design and layout with high utility. In the older editions of this book, *Build Your Own Electric Vehicle*, Jim Harris did a conversion for Bob Brant on his 1987 electric Ford Ranger. Figure 10-35 really explained it nicely. The high-current safety fuse (in the center, behind the power cable), the ammeter shunts (large one on center of back wall, small one in front of main contactor), and main contactor (on left side of box) form, along with the terminal strip, the "backbone" from which all interconnections are made.

Notice that all the safety fuses are located in one convenient area at the right rear of the box. Jim's objectives were to simplify in Figure 10-35. You'll see the continuous design evolution of the "magic box," as the conversion progressed. Figure 10-33 shows how Paul made it more advanced in his design. Figure 10-34 and the Ford Motor Company electric magic box for the Ford Focus electric. Figure 10-35 shows the original "magic box" in all of the editions of the book since it is so clean and descriptive.

Charger System

The benefits of the onboard charger are convenience and the ability to take advantage of on-the-road charging opportunities as they are presented. The dual objectives in wiring the onboard charging system are to prevent the charging routine from becoming a "shocking" experience (via proper grounding) and to prevent momentary distractions from causing you to drive away while the charging cable is still attached (via a charger safety interlock). There are four main components to the charging system:

- Compact onboard charger
- Lightweight line booster (optional)
- Safety charging interlock
- AC input system

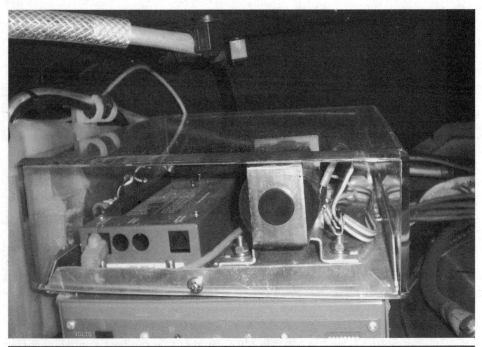

Figure 10-33   Paul Liddle's EV Porsche "magic box" for Porsche EV.

Figure 10-34   Ford Motor Company "magic box" for the Ford Focus EV in Detroit.

FIGURE 10-35   Jim Harris' prototype "magic box" for 1987 Ford Ranger.

Figure 10-36 shows a Manzita Micro charger that can be mounted by the driver's side of the aluminum heat sink plate, directly above the motor in the engine compartment. The line booster, required by the charger for 120-V operation, is mounted next to the throttle potentiometer on the driver's side fender well. The preferred location for most AC input charging connections in conversion vehicles is usually the position vacated by the gas tank filler neck opening.

Figure 10-37 really does a great job explaining the installation of the controller. This installation information comes from Café Electric LLC and is for a Zilla Controller.

FIGURE 10-36   Manzita Micro PFC 50M charger (50 A in and out).

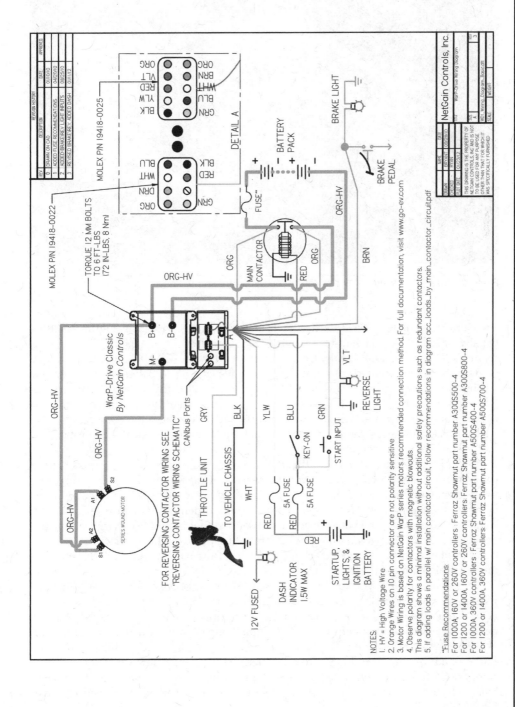

Figure 10-37 Controller installation manual from Café Electric and Manzita Micro.

315

Figure 10-38 Eltek Valere 3-kW charger for the EV Smart. (*TurnE Conversions.*)

The Eltek Valere 3-kW EV charger from Denmark is shown in Figure 10-38. It usually needs to be attached to a cooling plate, but it's only one-sixth the size of a regular EV charger. Yes, it is fully Controller Area Network (CAN) compatible.

Use stranded insulated copper wire for both the charger-to-battery and charger-to-AC-input receptacle connections. In order to prevent you (or anyone else) from casually driving away with the extension cord attached while charging, it's a good idea to implement a charger interlock system. When the 120-V AC line is plugged in, this relay latches open and keeps the main battery pack disconnected from the controller and motor—the vehicle is immobilized.

Other EV converters also have used the charger interlock feature to energize battery compartment fans (forced ventilation of the batteries) while charging. Additional interlock possibilities include sensors that inhibit drive away when the fault conditions of the engine compartment hood open, battery compartment hood open, or AC charging connector access door open are detected and sensors that inhibit the charging function during fault conditions such as engine or battery compartment hood open (because you don't want outsiders prying when charging currents and battery gases are present). You might consider any of these options.

In order to prevent you (or anyone else) from getting shocked when touching your EV's body while it's charging, the body and frame should be grounded to the AC neutral line (the third prong of the connector with the green wire leading to it). This neutral wire is connected between the AC input connector and the body or frame and is used only when charging. The batteries should be floating—no terminals touch the frame—and might even be further isolated by locating them inside their own compartment or battery box. Figure 10-39 shows an EV200 Kilova contactor for the Smart Electric Car, and Figure 10-40 shows the charger connector with security cover for the Smart EV.

Figure 10-39   EV200 Kilova contactor. (*TurnE Conversions.*)

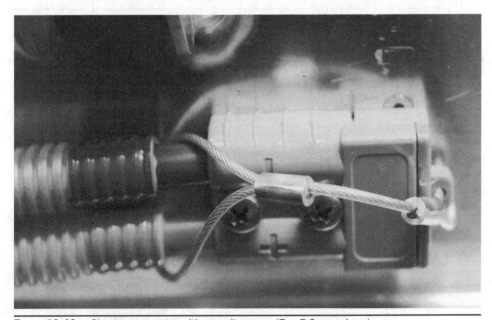

Figure 10-40   Charger connector with security cover. (*TurnE Conversions.*)

Figure 10-41 shows the Operation Z crew charging up their Nissan in a converted electric car. Figure 10-42 shows the charger port to a CODA electric vehicle. They built one and wanted to show their wares too.

Figure 10-41   Operation Z charging connector with plug installed. (*Operation Z.*)

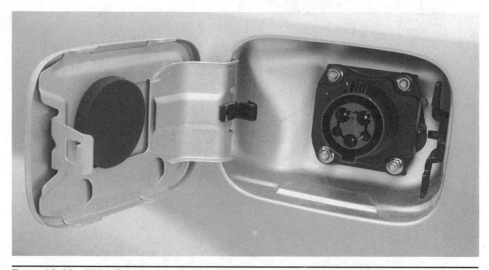

Figure 10-42   CODA Electric charging connector for the charging plug to be installed. (*CODA Automotive.*)

To guarantee no shocks, transformer-less chargers always should have a ground-fault interrupter installed, and transformer-based chargers should be of the isolation-type. If you prefer not to use an onboard charger, this changes your wiring and interlock design plans. In this case, you are going to be providing up to a 140-V DC input at currents of up to 30 A from a stationary charger. You also will need to charge the onboard 12-V accessory battery (unless you use a DC-to-DC converter). This means that your charging receptacle input needs at least two connectors—one for the high-current floating 140-V plus and minus input leads and the other for the low-voltage grounded-to-frame 12-V plus and minus input leads. The charger interlock design for the off-board charger is identical to the onboard case, except that you use a relay whose coil is energized by the presence of 120 V DC and whose contacts are in series with the 12-V key-switch line. When the 120-V DC or extension 120-V AC (depending on your charging preferences) extension cord from the charger is plugged in, this relay latches open, disconnects the main battery pack, and immobilizes the vehicle. Figure 10-43 shows the contact for Delphi manual service disconnects. The 560 V had to be divided in five smaller 105-V units by regulation. Figure 10-44 shows a close-up of the manual service disconnect. The socket is a plug with bridging wires for the communication wires of the BMS system.

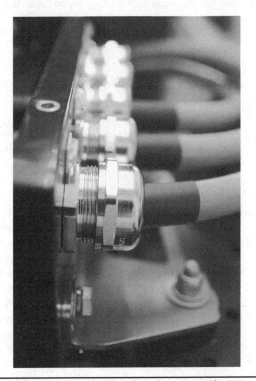

Figure 10-43   Contact for disconnects on the electric Smart car. (*Christian von Hösslin.*)

FIGURE 10-44   Manual service disconnect for the electric Smart car.

Batteries

Buying and installing your main propulsion battery pack is the last step in the EV conversion process. You buy your batteries last because you don't want to be tripping over them, nor do you want to keep charging them during the weeks (or months) of assembly. When your EV conversion is nearly complete, after all the battery-mounting frames are built and the wiring is done, you bring in the batteries. Let's take a closer look at the steps.

Battery Installation

The most important consideration in battery installation is to mount them in a location that will be easily accessible for servicing later. You've previously decided on your battery type and quantity, obtained their dimensions, allocated your front and rear mounting space, and designed and built your battery-mounting frames slightly oversized to accommodate battery expansion with use. All that remains is to install them.

With an engine power peak of 30 kW and 120 Nm of torque, the new Hotzenblitz has sporty handling and ensures astonishment of other road users at traffic lights. The motor is completely silent, and the transmission has a better connection to the engine and was optimized for low noise. The speed limit was designed to be 110 km/h. This was possible with the old Hotzenblitz, but the new, more efficient drive system provides higher sustained speeds (e.g., for cross-country trips). At 90 km/h, it cruises through the country quickly and easily. Thus, with a 2P32S configuration, the new Hotzenblitz battery pack delivers 107 V, 140 Ah, and 14,9 kWh. As shown in Figure 10-45, the battery

FIGURE 10-45   Battery pack encased in aluminum ready to be connected to the "magic box" and mounted to the TurnE Conversion of a Smart car.

pack is encased in aluminium and located in the bottom center of the vehicle. Figure 10-46 shows the freshly cut aluminum plates ready to be assembled into the charger bracket. Figure 10-47 shows all of Paul Liddle's batteries connected together and ready to be installed into his electric Porsche.

FIGURE 10-46   Battery mount for TurnE Conversions vehicle.

FIGURE 10-47 Paul Liddle's battery connectors ready to go.

Battery Wiring

The most important consideration in battery wiring is to make the connections clean and tight. Figure 10-48 shows two of the BMS units that measure the single-cell voltages connected to the BMW wires and battery heating wires.

Check that you haven't accidentally reversed the wiring to any battery in the string as you go. Double-check your work when you finish, and use a voltmeter to measure across the completed battery pack to see that it produces the nominal 120 V you expect. If not, measure across each battery separately to determine the problem. If reversed wiring is the culprit, the correctly installed and wired battery should fix it. Figure 10-48

FIGURE 10-48 Nice battery and BMS wiring of the A123 going into an all-electric Porsche. (*TurnE Conversions.*)

shows more BMS modules. This one is for 12 cells and two for 10 cells for a total of 32 cells in series as shown racked up.

Figure 10-49 shows BMS wires and battery heating wires. All wires outside of enclosures are protected by an extra layer of tubing in the Smart electric car from TurnE Conversions as seen in Figure 10-50. Figure 10-51 shows the proper ventilation needed for these energy-dense batteries. Figure 10-52 shows the build and double checking that occurs in building a battery pack from Ford Motor Company. Figure 10-53 shows the circuits being clamped to the box. The box on the other end is your dashboard. However, Figure 10-54, 10-55, and 10-56 really helps since it is a Toyota RAV4 EV to a dragster-style dashboard that really helps you realize you are almost there on your conversion/build. It helps to see what I'm talking about, so this picture from Toyota Motor Corporation really brings it home.

The Rema power connector shown in Figure 10-57 shows the connectors that are perfect for up to 160 V and 500 A. They are attached to the side of the battery case for easy connection. Figure 10-58 shows the Epic multipin connector to connect the signal wires of the car with counterparts of the Curtis controller that is used for this car. All wires except the power cables go through this connector.

If a badly discharged or defective battery is the culprit for any problems with the State of Charge, check to see that it comes up on charging and/or replace it with a good

Figure 10-49 BMS boxes, wires and battery heating wires.

FIGURE 10-50   Wired up and ready to go on battery cables.

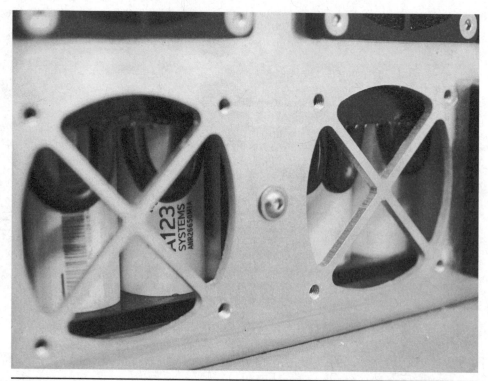

FIGURE 10-51   Proper ventilation for such energy-dense batteries.

FIGURE 10-52   Proper ventilation for Ken Koch Batteries. The undercarriage.

FIGURE 10-53   Circuit to switch to second, third, and fourth gears.

FIGURE 10-54   Circuits connected to the box for the electric RAV4.

FIGURE 10-55   Wires inevitably connect to a dashboard for an electric dragster from Ken Koch Conversion Company. (*Ken Koch.*)

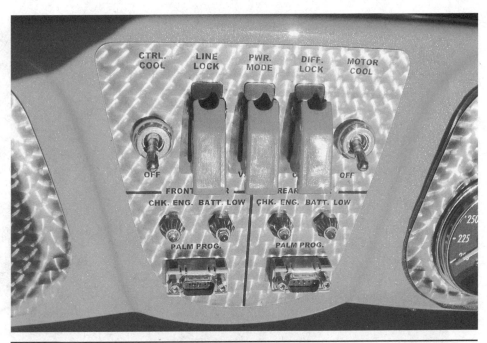

FIGURE 10-56   Don't forget an old fashioned dashboard too! (*Ken Koch.*)

FIGURE 10-57   Power Connectors for 160 V and 500 A.

FIGURE 10-58   Epic multipin contactor that connects the signal wires with the counterparts of the Curtis controller (the brains of the car).

battery from your dealer. A recharged "dead battery" will shorten the life for the entire battery pack. Please be careful to check all the batteries. *Important:* Make sure that the main circuit breaker is off before you connect the last power cable in the battery circuit. Better still, switch the main circuit breaker off, and wait until the system checkout phase before final battery connection.

12-V Accessory Battery

The accessory battery is another example of something to mount or test at this point. However, recycle the 12-V battery from the old car (if you can), and keep it in the same place. There is no need to reinvent where to put that 12-V battery.

After Conversion

This is the system checkout, trial run, and finishing-touches stage. First, make sure that everything works, then find out how well it works, and finally, try to make it work even better. When you're satisfied, you paint, polish, and sign your work. Let's look at the individual areas.

System Checkout on Blocks

This is the time to "jack up" the drive wheels of your conversion EV (or raise them up on work stands). The objective is to see that everything works right before you drive it out on the street. Figure 10-59 shows Tesla setting up the wiring prior to system check out on blocks. With your vehicle's drive wheels off the ground and the transmission in first gear, do the following:

- Before connecting the last battery cable, verify that the proper battery polarity connections have been made to the controller's B+ and B– terminals.

- Obtain a 100- to 200-Ω, 5- or 10-W resistor, and wire it in place across the main contactor's terminals. With the key switch off but the last battery cable connected and the main circuit breaker on, measure the voltage across the controller's B+ and B– terminals. It should measure approximately 90 percent of the main battery pack voltage with the correct polarity to match the terminals. If this does not happen, troubleshoot the wiring connections. If it does, you're ready to turn the key switch.

- Turn on the key switch with your foot off the accelerator pedal. If the motor runs without the accelerator pedal depressed, turn off the key switch and troubleshoot your wiring connections. If nothing happens when you turn on the key switch, go to the next step.

- With the transmission in first gear, slowly press the accelerator pedal and see if the wheels turn. If the wheels turn, good! Now look to see which way the wheels are turning. If the wheels are turning in the right direction, this is doubly good! If not, turn off the key switch and main breaker and interchange the DC series motor's field connections. If you are moving in the right direction, go to the next step.

FIGURE 10-59 Tesla Roadster raised up on blocks for the vehicle testing. (*Tesla Motors.*)

- If you have the high pedal disable option on the controller, turn off the key switch, depress the accelerator pedal, and turn on the key switch. The motor should not run. Now completely take your foot off the accelerator and slowly reapply it. The motor should run as before. If this does not work correctly, troubleshoot your wiring connections. If it works, you are ready for a road test! Turn off the key switch.

Further Improved Cooling

The eternal mechanic in Steve Clunn from Green Shed Conversions was still not satisfied—he wanted to have controller cooling independent of the vagaries of engine compartment airflow. So he added a fan and directional cooling shroud to the top of the heat sink. He also widened the mounting space between his "magic box" and the controller (everything is now mounted on the common-heat-sink aluminum plate) and relocated the battery charger to the passenger-side fender well—all to maximize cooling and airflow.

Improving Heating Too

The upper connector is 100 V for the heating element. The lower connector is for the charger. These connectors are made by a company named REMA. Figure 10-60 shows the heating element, which is 1,800 W. TurnE Conversions replaced the heat exchanger in the Smart car with this unit. It gives heat immediately. One is enough for a small car such as the Smart EV from TurnE Conversions. It literally takes all the five buttons to

FIGURE 10-60   Heating element for electric Smart car. (*TurnE Conversions.*)

FIGURE 10-61 Wiring of the switches. (*TurnE Conversions.*)

drive the car; and Figure 10-61 shows the wires for the switches. In the future, this will be assembled on a printed circuit board for easier assembly, according to Christian von Hösslin. Figure 10-62 shows a completed engine.

FIGURE 10-62 CODA Automotive electric vehicle wired up and ready to test. (*CODA Automotive.*)

Neighborhood Trial Run

- Check the SOC of your main battery pack. If it's fully charged or nearly full, you can proceed. If it's not charged, recharge it before taking the next step.
- After the batteries are fully charged, remove the jack and/or wheel stands from under your EV, open the garage door (believe me, it's a necessary step), check to see that all tools, parts, and electrical cords are out of the vehicle's path, and turn on the key switch. Put the car in gear, disengage the parking brake, step on the accelerator, and cruise off into the neighborhood.
- The vehicle should have smooth acceleration and a good top speed and should brake and handle normally. The overwhelming silence should enable you to hear anything out of the ordinary with the drive train, motor, or brake linings. Now you're ready to get fancy.

Paint, Polish, and Sign

After everything is running the way you like, it's time to "pimp your ride." Why? A car is an extension of oneself. The way the exterior of a vehicle looks reflects you; you are proud of your work and want to show it off in its best light. Without an outside "Electric Car" sign, the only way you can tell that a car is electrically powered is to look closely through its front or rear grille. Figure 10-63 shows TurnE Conversions designs. The company uses its own adhesive foils to brand the Smart car that it converts.

Onward and Upward

Notice the evolution of Paul Liddle's "magic box" for the Porsche electric car. Everything from the controller, junction box, heat sinks, and fans is packaged in a protective metal enclosure that is now centrally mounted above the motor in the engine compartment. With everything in one box, you just have to connect a few wires—the build or conversion is even easier to do and is much more reliable an operation.

FIGURE 10-63   You must have proper labeling to show off your car. (*TurnE Conversions.*)

Ken Koch is also proud of how his roadster looks, as shown in Figure 10-64. It has a gorgeous metallic orange paint job that just radiates out in the sunlight. The fiberglass body and metal hood are as straight and smooth as the best of show cars. The car's stance is perfect. With traditional stock-type chrome grille, headlights, taillights, bumpers, and standard windshield with wind wings, it's a mixture of the old and the new. The wheels make it look somewhat on the modern side, as do the seats. With the hood, side panels, and trunk lid closed up, nobody would ever suspect that it's a battery-powered car in Figure 10-65 with 800 hp lurking under the hood. The only clue that a casual observer might have is if he or she peeks through the louvered side panels. The peeker would see lots of chrome, polish, and hot-rod parts—but no engine. Whatta sleeper!

With the wheel wells into the trunk 12 inches on either side, there isn't a whole lot of room for batteries. You can see that the two battery boxes are a snug fit, as shown in Figure 10-65.

There is no room for a rumble seat and not even room for a bag of groceries. I guess that's okay—this car isn't a "grocery getter" anyway. On either side of the boxes there's a Manzita Micro PFC-20 battery charger, each mounted on its own shelf. The car can plug into either 120 or 240 V AC through two chrome recharge plugs mounted in the pan below. The left plug is in the same location as the original '34 gas filler. The battery boxes can be lifted into or out of the trunk with a hydraulic cherry picker. Balance is critical, and it's a delicate two-person job.

FIGURE 10-64 How Ken Koch's roadster looks today. (*Ken Koch, EV Consulting, Inc.*)

FIGURE 10-65   All wired up inside looking extremely impressive. (*Ken Koch, EV Consulting, Inc.*)

The roadster arrived home fully painted and with about 95 percent of the mechanical work completed. TurnE Conversions planned to do all the electrical work. First estimate was that I could design and install all the wiring in about two months. Boy, talk about another gross underestimation. Bob Brant didn't work on his electric vehicle conversion with the 1987 Ford Ranger each and every day. "I put in only four to five hours a day when I do work on it—so things are taking extra time. Actually, if the truth be known, I've taken my eye off of the finish line, but I'm enjoying the journey immensely."

The point-to-point wiring diagram from the car is 3 feet wide by 7½ feet long and continues to evolve as I go along. Not forgetting that the car should look consistently nice throughout, the wiring had to be top-notch as well. As we all know, top-notch takes a lot of planning and tons of extra time. Finally, all the wiring has been completed in the trunk, underneath the car, on the floorboard, and totally under the hood. The motor compartment itself took five weeks to complete, but I think that you'll agree that it looks neat and tidy. It's my philosophy to make all equipment serviceable. Everything in and on the roadster can be removed as a module or subassembly, thanks to a lot of connectors in the wiring. Even the 80-pound power electronic assembly under the hood can be disconnected and removed in minutes with lift rings. All 36 mating connector sets took a lot of extra time.

Idling is necessary because there's no torque converter on the automatic transmission.

There's a switch to turn on or off motor cooling. Another switch activates low or high controller cooling. And in the middle are covered switches used for racing only. The "LINE LOCK" on/off permits power burnouts. The "DIFF. LOCK" on/off electrically locks or unlocks the rear end. And the "PWR. MODE" selects either "VALET (300 hp)" or "RACING (800 hp)" modes.

Considering that the car weighs 2,450 pounds, 300 hp ought to keep up with street traffic with no problem. Actually, as of this writing, the wiring behind the dash is not yet completed. Also, not quite having 100 percent of the batteries and LED multiplex electronics in hand, the roadster has a little ways to go before it goes in for upholstery. The upholsterer will do the carpeting, seats, and trunk in a medium tan and will also need some time to build a custom removable Carson top.

Put Yourself in the Picture

Now that you've seen how it's done, you can do it yourself. It's a very simple project that virtually anyone can accomplish today just by asking for help in a few appropriate places and taking advantage of today's kit components and prebuilt conversion packages. Imagine yourself in Figures 10-66 through 10-72, and take the steps to make it happen. However, Figure 10-71 shows how Christian von Hösslin makes sure that the Porsche Roadster is also recognized as well. Finally, Figure 10-72 shows a Nissan Leaf with the charger port up in the front on National Plug-in Day in Long Island. The proudest time to show off your car!

FIGURE 10-66   Smart electric conversion by TurnE Conversions.

FIGURE 10-67 Put yourself in this picture—do an EV conversion of your own.

FIGURE 10-68 Joe Porcelli and Dave Kimmins of Operation Z.

Figure 10-69 First Tesla Model S off the block party! Now that's how I want my car delivered! (*Tesla Motors.*)

Figure 10-70 TurnE Conversions' Electric Porsche Roadster. (*Christian von Hösslin.*)

FIGURE 10-71   Make sure you just don't label the car electric. Also let them know you have your own Porsche Roadster. (*Christian von Hösslin.*)

Summary

One of the great supporters of this book and builds throughout the years has been Ken Koch, who in the first edition helped with the Ford EV conversion with KTA Services. "Having sold KTA Services, the EV parts supply business, I now am able to achieve some semblance of retirement. This has created extra time whereby I can work on a personal project I've wanted to do for a very long time: to build up a 1934 Ford roadster street rod. Back in high school, I used to race my car at a local drag strip on weekends, and this created a need for speed that's still in my blood today. When I was a teenager, everyone dreamt of having an early '30s street rod. I was no exception. Of course, at that time, not many could afford to fulfill this dream. For anyone who eventually could afford it, it simply had to wait until years later. Well, now's the time, before it's too late!

Probably most folks have figured out that KTA is still alive and well and accessible at the old website www.kta-ev.com. The new owner of KTA is Wistar Rhoads, as of April 2008, when he purchased the business. This shows that there are survivors such as ElectroAutomotive, EVA, Steve Clunn, and Ken Koch for starters in the conversion world who are amazing resources for your build. Ask for help. They are here on this earth for that reason.

FIGURE 10-72   Nissan Leaf showing inside and charging port on National Plug-In Day. (*Seth Leitman.*)

CHAPTER 11

How We Can Maximize Our Electric Vehicle Enjoyment

Besides having more charging stations across the globe and easy processing of your paperwork, you'll be driving for only pennies a day.
—Bob Brant

O nce you've driven your electric vehicle (EV) conversion or build around the block for the first time, it's time to start planning for the future. You need to register and insure it so that you can drive it farther than just around the block. You also need to learn how to drive and care for it to maximize your driving pleasure and its economy and longevity.

Registration and Insurance Overview

No one anywhere would seriously question registering and insuring a Porsche 911 EV conversion or a Ford Focus build. Nowadays too the dealer at any dealership can and will provide you with the plates, registration and title!

If you are converting the car, though, its chassis is in compliance with Federal Motor Vehicle Safety Standards (FMVSS) and National Highway Transportation Safety Administration (NHTSA) safety standards. New York and California, as well as most other states, for that matter, enable you to change the title of the vehicle from gas to electric; it's just a checkbox electric! Yes, it is that simple.

On the other hand, many jurisdictions used to balk at licensing and insuring the home-built and/or commercially produced golf cart–style EVs of the 1970s that could barely reach 45 mi/h and did raise issues of safety. This is why the FMVSS 500 ruling that allows for low-speed vehicles to be driven came into being. The ruling addresses golf cart–style EVs, but it requires some basic safety features of a regular vehicle (e.g., turn signals, brake lights, taillights, bumpers, reverse lights, and sound). Such vehicles are regularly called either *low-speed vehicles* (LSVs) or *neighborhood electric vehicles* (NEVs) and cannot go more than 25 mi/h and can only be driven on public roads that are posted at no more than 35 mi/h. In addition, each state also must approve the use of this type of vehicle on the road. Most states do allow NEVs to be driven on public roads.

Which end of the spectrum your EV conversion or build of today resembles directly determines its ability to clear any registering and insuring hurdles. Let's take a closer look at each area.

Getting Registered

Any vehicle's registration falls under the jurisdiction of the state in which it resides. Each state's motor vehicle codes, although based on common federal standards, are just a little bit different. While your internal combustion engine vehicle conversion or build chassis may be fully compliant with federal FMVSS and NHTSA safety standards, you need to check out what your state's motor vehicle code says. If you're doing a from-the-ground-up EV, you'd be well advised to check out these rules and regulations in advance.

In general, few states have specific EV regulations. You'll find that states with larger vehicle populations, such as California, New York, and Florida, are on the leading edge in terms of establishing guidelines for EVs. Check with your own state's motor vehicle department to be sure.

As for the registration process, most of the people who work for the Department of Motor Vehicles and/or the Department of Environmental Protection for each state are far more involved with the smog certification or Department of Environmental Quality or Environmental Protection (DEQ) or (DEP). It is very important that you check the vehicle compliance rules and regulations in your state to see what the process is for allowing a converted EV to receive license plates. However, in most states, owning an EV conversion or build short-circuits this process. In fact, you can offer to buy the whole local DEQ inspection team lunch if their meters find any emissions coming out of your EV at all (hybrid EV owners, please don't make this offer).

In most states, you can receive a tax credit for an EV, and there is also a federal tax credit for EVs. In some areas, you are entitled to a reduction in your electric power rate. Check with your local utility and city and state governments to see if you and your EV are entitled to something similar in your area.

Getting Insured

Insurance is roughly the same as registration in today's world. Ironically, back in 2008, when the second edition came out of this book, it was a little more confusing, but just use the Vehicle Identification Number (VIN) and you should be good!

You're not likely to have any trouble with your EV if it's been converted from an internal combustion engine chassis. While it is a process to explain that the vehicle has been converted to electric with your carrier, having a VIN and approval from the State Department of Motor Vehicles will give most large national insurance companies ease in underwriting an EV for insurance coverage. Verify that your design meets insurance requirements in advance because the company might ask you for some backup documents. Be sure to ask your conversion or build company if you choose to get an EV that way to provide you with the necessary schematics and everything, as in a regular car manual.

Safety Footnote

My basic assumption in this section is that you would put safety high on your list of desirable characteristics for your converted EV. This line of reasoning assumes that you leave intact the safety systems of the original internal combustion engine chassis: lights, horn, steering, brakes, parking brake, seat belts, windshield wipers, etc. It also assumes that you are thinking "Safety first!" when installing your new EV components.

Driving and Maintenance Overview

An EV is easier to drive and requires less maintenance than its internal combustion engine counterpart. But because it's driving and maintenance requirements are different, you'll need to adjust your acquired internal combustion engine vehicle habits. The driving part is very similar to the experience of a lifelong stick-shift driver who drives an automatic transmission vehicle for the first time. As for the maintenance, there's a whole lot less to do, but it has to be done conscientiously. Let's take a closer look at each area.

Driving Your EV

Your EV conversion or build still may look like its internal combustion engine ancestor, but it drives very differently. Here's a short list of reminders.

Starting

Starting an EV conversion or build involves a little bit of a science. There's no need to use the clutch on starting because the motor's not turning when your foot is off the pedal. On the other hand, there is a very definite need to have the EV in gear because you always want to start your series DC motor with a load on it so that it doesn't run away to high rpms and hurt itself. If you forget and accidentally leave the clutch in (or the transmission in neutral), back off immediately when you hear the electric motor winding up. The best analysis I have heard is that if your EV conversion or build has a lighter package (not a performance-based controller and heavy-duty motor), you can start the car in second gear without a problem because the torque of an electric motor will push the rpms (such as the the Ford Ranger or Focus plus the converted Porsche in Chapter 10). However, higher-performance vehicles (such as the Porsche discussed in Chapter 10) can easily be started in fourth gear because the vehicle's motor, controller, and batteries will have the necessary voltage requirements to accelerate from 0 to 60 mi/h in four to six seconds.

Shifting

If you do city driving, you'll wind up mostly using the first two gears. The lower the gear, the better is your range, so use the lowest gear possible at any given speed. However, if you do highway driving, expect to shift gears higher.

Economical Driving

If you keep an eye on your ammeter while driving, you'll soon learn the most economical way to drive, shift gears, and brake. For maximum range, the objective is to use the least current at all times. You'll immediately notice the difference in drag racing and going up hills—either alter your driving habits or plan on recharging more frequently.

Coasting

If you don't have regeneration, coasting in an EV is unlike anything you've ever encountered in your internal combustion engine vehicle—there's no engine compression to slow you down. You need to learn how to correctly "pulse" and when. I have ridden with drivers who floor it for three seconds, then coast, then floor it, and then coast. Heavy pulsing is not good for the vehicle and wastes energy. In most driving, a steady foot is better. Light pulsing is only an advantage when little power is needed.

Regeneration

Regenerative braking is a mechanism that reduces vehicle speed by converting some of its kinetic energy into another useful form of energy. This captured energy is then stored for future use or fed back into a power system for use by other vehicles. For example, electrical regenerative brakes in an electric railway vehicle feed the generated electricity back into the supply system. In battery electric and hybrid EVs, the energy is stored in a battery or bank of capacitors for later use. Other forms of energy storage that may be used include compressed air and flywheels.

Many people now have experienced coasting in a hybrid electric car. As with a hybrid, by definition, an EV is designed to be as frictionless as possible, so take advantage of this great characteristic. Learn to pulse your accelerator and coast to the next light or to the vehicle ahead of you in traffic. Hybrid electric car owners (especially New York City taxi owners) understand this concept.

When you accelerate, you do not need to floor it and then coast. You can slowly step on the accelerator and then take your foot off to coast. When you coast, you use the regenerative braking. It is much smoother and a more efficient use of the vehicle because it is moving forward and charging at the same time. While there are plenty of people who own converted EVs who like to floor it and coast, regeneration can be a very big help in most stop-and-go driving.

Determining Range

Use the battery voltmeter as a fuel gauge in conjunction with the odometer to tell you how far your batteries can take you. Tape a note to your dashboard or use a notebook to keep track of the elapsed mileage between full charges and voltmeter reading (e.g., the voltmeter (not gas) reading was x after you drove y miles), and you'll quickly get an idea of the pattern. Keep in mind that your battery pack will not reach its peak range until you've deep-cycled it a few times.

Running Out of Power

Before you totally run out of power, *get off the road!* If you totally run out of power and cannot find an electrical outlet, turn off your key switch and allow your batteries to rest for 20 to 30 minutes (shut down everything else electrical at this time too). Amazingly, you'll find extra energy in the batteries that may be just enough to take you to the power outlet you need. The convenience of an onboard charger is welcomed most of all in this particular circumstance. It is not good to be stuck in the middle of a busy freeway without power. You should understand that there is a significant power loss just before the no power condition. You also should know that a full discharge greatly shortens battery life.

Regular Driving

(Please note that the equations are based on gasoline at $4.50 per gallon. See the Preface for more details.)

Drive your EV regularly—several times a week. Remember, the chemical clock inside your lead-acid batteries is ticking whether you use them or not, so use them.

Better yet, think of how much money you are saving by using your EV. Even if good old Jim Harris today got a 60-mile range out of one charge, his average energy use would be 6.6 cents per mile (0.44 kWh/mi × $0.15/kWh).

If you compare this with gasoline at $4.50 per gallon and a typical 25 mi/gal for an internal combustion engine pickup before conversion or build, this works out to 18 cents per mile (1/25 gal/mi × $4.50/gal). This is almost a three-to-one savings; I'd say that you might want to take advantage of it. (Please note that when using updated 2012 prices, the price ratio regarding a gas car versus electric car is *better* for EVs. The ratio is even better than the 1993 ratio that held steady for the past 50+ years!)

Caring for Your EV

Now that you are driving for only pennies a day, it takes little more to make your pleasure complete. Actually, a properly designed and built EV conversion or build requires surprisingly little attention compared with a gas-powered vehicle. It comes down to the care and feeding of your batteries, minimizing friction, and preventative maintenance.

Battery Care

Check the battery terminals for corrosion, and correct any deficiencies using a baking soda solution. [*Note:* All distilled water is condensed water vapor. These days it is usually done at reduced pressure (and temperature) to save energy. Reverse osmosis is also effective in removing minerals.]

Thus, since lead-acid batteries have been the talk of this section, the question remains: Are lithium-polymer batteries the next big thing for EVs? I've been hearing some interesting news lately about a company called DBM that's developing Kolibri lithium-polymer batteries. The latest lithium-polymer battery technology, called *AlphaPolymer*, claims

- 400 miles on a charge
- Recharges in six minutes
- 5,000 life cycles[1]
- A tenth the price of other lithium chemistries
- Twenty-year battery life
- Weight 30 percent lighter than a Tesla battery pack

CarCharging Group, Inc., Offers EV Charging Services at Rapid Park in Lower Manhattan and Many Other Places Such as the Astoria Tower in Chicago

CarCharging Group, Inc.,[2] announced the availability of its EV charging services at the Rapid Park Garage at 25 Beekman Street, near the financial district, and South Street Seaport in New York City. EV drivers in lower Manhattan can now quickly and easily recharge their EVs at the Rapid Park facility, which is in close proximity to Pace University and New York Downtown Hospital.

"With a growing number of people driving EVs, we wanted to provide our customers the convenience and reassurance of recharging while at work or enjoying the many restaurants at nearby South Street Seaport," said Kevin Wolf, Rapid Park regional director. "Working with CarCharging to provide EV charging services as a sharable amenity, we can now offer Rapid Park's EV customers the ability to park and rapidly recharge their EV in one location."[3]

CarCharging provides EV charging services using a Coulomb Technologies EV fast-charging station known as level 2, which provides 240 V with 32 A of power to quickly recharge an EV's battery. EV charging stations use the standard SAE J1772 connector widely adopted by nearly all automobile manufacturers.

EV drivers can easily register and create a CarCharging account online and attach a small card to their keychain to initiate use and payment at any intelligent CarCharging station. The CarCharging keychain card also allows drivers to use charging locations on the ChargePoint Network. CarCharging also accepts direct payment via credit card.

Users can pinpoint EV charging station locations using the CarCharging map at http://www.carcharging.com/. The ChargePoint mobile application also provides real-time charging station locations with turn-by-turn directions. Drivers soon will be able to reserve a time slot, guaranteeing access to EV charging stations.

EVs already on the market include the Nissan Leaf, Chevy Volt, Toyota Prius Plug-in Hybrid, BMW ActiveE, Ford Focus Electric, Fisker Karma, and Tesla Model S. Additional EV models are expected to be available in the latter part of 2012 and for years to come.[4]

However I did notice in this ad in Figure 11-1 how even BMW (like all the others) are advertising the BMW ActiveE and are making electric vehicle charging and infrastructure accessible. Therefore, the need to be concerned is lessened through the automotive manufacturer or even you.

Tire Care

Tire rolling resistance is a major energy loss, so switch to low-rolling-resistance tires (if possible), and frequently check that your tires are inflated properly using a more accurate meter-style tire gauge. Proper inflation means 32 lb/in^2 and up; EV tires should be inflated hard. Special low-rolling-resistance tires that can handle 50 lb/in^2 (or more) are available for most cars. Talk to your local tire specialist about inflation limits versus loading.

Also, listen to be sure that no brake shoes or pads are dragging. See how in Figure 11-2 the brake pads and calipers look amazing. Keep it like that! (*Note:* 80 lb/in^2 tires are available for many vehicles. Their designation starts with "LT.")

Lubricants

The weight of viscosity of your drive train fluids (transmission and rear axle lubricants) also contributes to losses on an ongoing basis, so experiment with lightweight lubricants in both these areas. The EV conversion or build puts a much smaller design load on the mechanical drive train, so you ought to be able to drop down to a 50-weight lubricant for the rear axle and a lower-loss transmission fluid grade. Consider low-loss synthetic lubricants.

FIGURE 11-1 Finding a charger never got easier with BMW ActiveE ad. (*Seth Leitman.*)

FIGURE 11-2 Brake pads and calipers looking amazing. (*TurnE Conversions and Christian von Hösslin.*)

Checking Connections

Preventative maintenance mostly involves periodically checking the high-current wiring connections for tightness like on the charger in Figure 11-3. Use your hands here. Warmth is bad—it means a loose connection—and anything that moves when you pull it is also bad. A few open-end and box wrenches ought to make quick work of your retightening preventive maintenance routine.

Here are a few simple steps you should take to make the EV give years of great performance[5]:

1. Protection from the elements is important. Use good design concepts and materials to protect the motor from rain, snow, and ice. You don't want your exterior to get like Figure 11-4 from TurnE Conversions' old Smart car before it got converted to all electric.

2. Design the motor-mounting area to allow for good air flow. The motor needs a continuous supply of clean fresh air to cool properly.

3. Protect the motor from "dirty air" that may be used to cool it. Most airborne grit will act as an abrasive, which eventually will cause harm to the internal parts of your motor.

4. Clean the brushes and connectors for the batteries under the hood if any regularly for the dust/dirt that occurs during normal operation.

5. Regularly check connections, voltages, tolerances, and alignment to ensure that they are within normal specifications.

6. If you suspect or question the motor's operation, immediately shut it down. Record any visual signs, audio sounds, or scents at the time, and ask an expert for an opinion prior to operating the motor again.

Figure 11-3 Charger can be checked for tight connections. Make sure it is accessible. (*TurnE Conversions.*)

Figure 11-4   Make sure your car does not get rust like this. (*TurnE Conversions.*)

7. Always operate the motor within the normal safety ranges for voltage, amperage, and rpms.

8. Follow all the safety rules available to you.

9. Remember that your motor will take care of you if you take care of it.

Safety Information

This is not an all-inclusive list. Use common sense and act responsibly. Electric motors are extremely powerful and could cause death, dismemberment, or other serious injury if misused or not handled safely. Use caution when operating any motor. If you're not sure what you're doing, find a knowledgeable person to advise you. Remove all metal jewelry and metal objects from your hands, wrists, fingers, etc.

Before Working on Any Electric Motor

If you are working on an EV, make certain that the vehicle is positioned securely with the drive wheels safely clear of the floor and blocked up so that they cannot make contact with the floor under any circumstances. Also:

- Block the nondrive wheels if they remain in contact with the floor so that the vehicle cannot roll in either direction.

- Before troubleshooting or working on any EV, disconnect the battery, and discharge all capacitors as shown in Figure 11-5.

- Reconnect the battery only as needed for specific checks or tests.

Figure 11-3 Disconnect wires to the batteries! (*TurnE Conversions.*)

Motors

Electric motors must be connected to a power source only by knowledgeable and experienced personnel.

- Motors *never* should be run without a load. Running a motor without a load could result in harm to people or the motor. Absence of a load is considered misuse and could prove dangerous to anyone in the vicinity and void the motor warranty.
- Portions of the motor may become hot, and proper precautions must be taken.
- Motors are heavy and are likely to become damaged if dropped or cause damage to anything they fall on (including people and body parts).
- Use extreme caution when working with motors.
- Make certain that the motor is disconnected from any power source before servicing.
- Motors contain moving parts that could cause severe injury if the proper precautions are not taken. Never touch an operating motor.
- Motors should never be operated beyond the limits established by the manufacturer.
- Motors must not be modified in any manner; doing so will void the motor warranty and could prove extremely dangerous.
- Wear protective or safety equipment such as safety shoes, safety glasses and gloves when working with motors.

FIGURE 11-6   Fuses and circuits. (*TurnE Conversions.*)

- Make sure that you know where the closest functioning eyewash station is before working on or testing batteries.
- Do not defeat any safety circuits or safety devices like in Figure 11-6.
- Under no circumstances should you push in any contactor of an EV while the drive wheels are in contact with the floor. Pushing in a contactor when the drive wheels are in contact with the floor can cause serious property damage, personal injury, or death.[6]

Emergency Kit

While carrying extra onboard weight is a no-no, carrying a small highway kit ought to be just enough to give you on-the-road peace of mind, knowing that you have planned for most contingencies. At a minimum, your kit should have a small fire extinguisher, a small bottle of baking soda solution, a small toolkit (e.g., wrenches, flat and wire-cutting pliers, screwdrivers, wire, and tape), and a heavy-duty charger extension cord with multiple adapter plugs (male and female) like the AeroVironment EVSE-RS Plug-In Charger. The EVSE-RS Plug-In easily and securely plugs into a dedicated 240 V outlet and slides in and out of a wall-mounted bracket. The moveable charger, shown in Figure 11-7, weighs about 20 lb, which is less than your average suitcase, so take it with you when you're visiting relatives or a vacation home.

Figure 11-7   You might just want to have a spare charger in need. (*Aerovironment.*)

Recommendations for EV Mass Deployment

1. Expand national, regional, and local efforts that help to attract greater concentrations of EVs in communities across the country.

2. Remove unnecessary bureaucratic and market obstacles to vehicle electrification nationwide through a variety of policies that:

   - Bolster nationwide installation of and access to basic charging infrastructure, both at people's homes and in public places.

   - Incentivize the purchase of EVs and EV charging equipment and streamline the permitting application process for EV charging equipment.

   - Educate the public about the benefits of EVs and the costs, opportunities, and logistical considerations involved with EV charging infrastructure.

   - Ensure appropriate training for workers installing EV charging equipment and for first responders.

   - Encourage utilities to provide attractive rates and programs for EV owners and to increase off-peak charging.

- Assist in deployment of clean energy, energy efficiency, and energy management technologies jointly with vehicle charging.
- Accelerate advanced battery cost reduction by boosting EV use in fleets, in second use, and in stationary applications.

3. Ensure U.S. leadership in manufacturing of EVs, batteries, and components.

EVs will help us to achieve freedom from oil, and they also will mean less air pollution. Emissions from EVs are at least 30 percent lower than those from traditional vehicles—and that's on today's electric grid. As we clean up our grid and rely more and more on renewable sources of power, EVs will get even cleaner over time.[7]

To convert a critical mass of drivers to EVs, we must first drive market penetration by establishing policies to reduce ownership cost of EVs and plug-in hybrid EVs (PHEVs). This is according to an update on the challenges to Smart Grid in a new issue of *Proceedings of the IEEE*.[8] EVs are essential to the smart grid's future and are considered to be one of the most promising ways to reduce fossil fuel dependence.

As I stated before, charging at people's homes is important. That is why Leviton is working with Toyota and many other car companies to develop each charging with vehicles. The charging station companies have come up with an easy to plug in and hang on the wall a charger as seen in Figure 11-8. The evr-green™ 320 electric vehicle charging station for level 2 EV charging has 32 amp, 7.7kW output.

This smart grid special issue, "The Electric Energy System of the Future," published in the *Proceedings of the IEEE*, which is the world's most highly cited general-interest journal in electrical engineering and computer science since 1913, is one of the

FIGURE 11-8   evr-green™ 320 electric vehicle charging station for level 2 EV charging station. (*Leviton.*)

most comprehensive assessments of this topic to date. According to authors of this 13-article report, the overall driving forces for the smart grid, a fundamentally different energy system from the present one, are the new needs of more energy-knowledgeable, computer-savvy, and environmentally conscious customers combined with regulatory changes, availability of more intelligent technologies, and ever-greater demands for enough energy to drive the global economy.

The report focuses on technological progress and acknowledges that while much has been achieved to move Smart Grid forward, there is still a lot to be accomplished in three dominant areas: governmental policies at both the federal and state levels, customer efficiency needs, and new intelligent computer software and hardware technologies.

EV Update

In an article on EVs entitled, "Vehicle Electrification: Status and Issues," authors Albert G. Boulanger, Andrew C. Chu, Suzanne Maxx, and David L. Waltz concede that while automakers recognize that EVs are critical to the future of the industry, widespread consumer adoption of EVs is inhibited not only by actual costs but also by perception of costs. For example, consumers must grasp that although the current initial price for an EV is higher than for an internal combustion engine vehicle, EV operating costs are lower.

Other EV issues that need to be addressed to jump-start a transition to EVs include best practices for extraction and mining of rare earths and lithium, development and deployment of EV technologies, standardization of industry protocols of plugs and chargers, deployment of charging infrastructure, and public education and national and global political will for the adoption of smart grid technology and renewable energy sources.[9]

Fast Charging Will Accelerate Electric Car Market

TH!NK electric vehicles announced that it is working with AeroVironment (AV) to develop high-end charging or fast charging for the TH!NK City EV. The TH!NK City is the first vehicle of its type to be granted pan-European regulatory safety approval and Conformity in Europe (CE) certification and has over 30 million miles of on-the-road, real-world customer use and testing. TH!NK is also a leader in electric drive-system technology and was the first to market a "plug and play" EV mobility solution, as seen in the electric cars in Figure 11-9 at the NYPA/THINK Clean Commute Program I ran. We made sure to include charging stations not just at customers homes but at the train stations so that the plug and play option was available.

AV is one of the leaders in fast charging worldwide for electric cars. As you can imagine, this is intended to jump-start the fast-charge infrastructure in the United States. This is a huge development for the company because it will currently provide EV owners with the opportunity to fast charge their cars like they would refuel a gas-powered vehicle. It might take slightly longer, but it's cheaper with zero tailpipe emissions, and the TH!NK City is powered by an EnerDel battery pack in the United States.

A fleet of 60 TH!NK City EVs was available to rent in the central region of the Swiss Alps at a special rate of approximately $57 per day (60 Swiss francs) through 30 hotels, resorts, and other points of interest in the area. Alpmobil marketed the service and took reservations

FIGURE 11-9 Look how close the electric cars are to the train station platform. Talk about incentive! (*Town of New Castle.*)

through its website (www.alpmobile.com). EV charging was made available around the region through more than 20 battery-charging points that have been installed in the area—powered by renewable hydroelectric-generated electricity derived from local mountain waters, reservoirs, and dams.[10]

There is no question that for mass acceptance of the EV, more EVs need to be introduced into the marketplace. In addition, charging has to be as easy as fueling a car so that the transition is easier.

As AV recently announced that they were providing lower cost home chargers as seen in Figure 11-7. The more charging is accessible, the better of the entire electric vehicle market will become.

One of the big opportunities where this can be changed is in the commercial vehicle sector. There are over 230 million commercial cars and light-duty trucks on U.S. roads—a huge potential source for widespread EV adoption. But large companies, federal, state, and local governments, and other fleet operators have hesitated to make the switch. GE just started on a project that drives at changing that.

Researchers at GE's Global Research Center in Niskayuna, New York, will design and build a better commercial fleet charging station that will incorporate GE's existing smart grid technology and help to dramatically cut installation costs in the depots and garages housing bus, vehicle delivery, and other commercial and government fleets. The project also received support from the U.S. Department of Energy.[11]

ABB Global

ABB, the leading power and automation technology group, announced in September 2011 that it bought Epyon B.V., an early leader in EV charging infrastructure solutions focusing on direct-current (DC) fast-charging stations and network charger software. The acquisition is in line with ABB's strategy to expand its global offering of EV infrastructure solutions. "ABB's brand recognition and strong global presence will accelerate the growth of a combined Epyon-ABB offering and provide access to key customers and partners," said Hans Streng, Epyon's CEO, who led the company over the last year and who will stay as an experienced industrial leader of the combined business. "Epyon's existing business is complemented by ABB's strong power electronics platform and global manufacturing footprint, as well as its supply, marketing, and service network."[12]

The U.S. Ready Index by TH!NK Electric Cars

I wrote back in 2010 on PlanetGreen.com that the growing number of EVs is driving a global market opportunity for charging solutions, including supporting technologies to equip the electrical grid with more sophisticated monitoring systems and software. EV charging station unit sales are expected to multiply rapidly over the next five years and reach 1.6 million units globally by 2015, according to Pike Research.

Back in January, TH!NK Electric Car Company rated and created a "U.S. Ready Index" for cities across the nation. The purpose of TH!NK developing the EV-Ready Cities Index was to show which cities would start to see EVs first. The index was compiled for TH!NK by ASG Renaissance, a market research and business services firm located in Dearborn, Michigan.

What Does the EV-Ready Cities Index Take into Account?

According to TH!NK North America, the EV-Ready Cities Index takes into account purchase and use incentives, such as:

- High-occupancy vehicle (HOV) lane access and infrastructure support
- Hybrid EV sales
- Traffic congestion
- Environmental Protection Agency non–attainment zone status (air quality)
- Potential lower-carbon energy sources for vehicle recharging

Now Government Doesn't Need to Spend Money on a Report—It Can Just Do It

This is pretty much what government agencies do when they create reports like this for EV. When I worked for the State of New York, my colleagues and I took all those types of issues into account when implementing projects statewide. However, I would like to believe that TH!NK did all of us a service by indicating which cities will be ready for EVs. This will hopefully give the government more time to focus on getting the cars on the streets and ensuring that the battery technology is proven and "Made in the United States."

Which Cities First?
So here is the bottom line about this report: These are the cities that get "first dibs" on EVs. Every other city should take notice, read the report, and get ready to bring in EVs. The cities are

1. Los Angeles

2. San Francisco

3. Chicago and New York

4. San Diego

5. Portland, Oregon

6. Sacramento, CA

Rounding out the top 10 were Newark, Seattle, and Atlanta. The remaining cities include Denver, Boston, Washington, DC, Philadelphia, and Phoenix.

According to TH!NK North America, "We expect that the roll-out of EVs to the U.S. market will be quite focused in the early stages. Some cities are more likely to be early adopters of EV technology, and the EV-Ready Cities Index will be a helpful tool to guide and prioritize the development of those markets. It reflects the available government support, consumer acceptance, and opportunity for EVs to provide the maximum benefits possible from electric drive. EVs are a unique solution for congested urban environments, and we are taking a city-by-city approach rather than a national or state-by-state approach."[13]

Simply Put, Electric Cars Can Be Plugged In on a Plug
This is one of the greatest parts about EV. EVs *can* have plugs and chargers up to 20 A so that your car *can* plug into your standard household 120-V AC outlet. Higher-current-capacity EV chargers require a dedicated 240-V AC circuit—the kind that powers your household electric range or clothes dryer.

You see, you *can* install an EV charger, and it is *easier* to install than almost any other alternative.[14]

Conclusion

I was recently asked my opinion on the state of electric cars in America. That is a very difficult and also a very easy answer (at the same time).

The state is currently, "Eh." Why? There has been not been an effective deployment strategy of EVs and the placement of EV chargers in America. More important, there are no connections between mass transit and EVs for a clean commute. There also have not been smart integrations between clean commutes, EVs, and the potential for vehicle-to-grid technologies.

All these concepts without question would spur the economy because when I worked for the New York Power Authority and ran its clean commute program, it took minimal investments because the gross domestic product of that program was in the millions (with only 100 TH!NK City electric cars) after 9/11.

So what *is* a clean commute program, you ask? I'll tell you. The NYPA/TH!NK Clean Commute Program allowed interconnectivity between EVs and transit locations in New York.

People leased the TH!NK City for $199 per month and were able to get guaranteed up-front parking at train stations and free EV charging at train stations and at home. We even were able to get transit checks to customers. This reduced the EV driver's costs for transit.

Based on the success of that program, we were able to have a phase 2 that would have included solar charging, vehicle-to-grid demonstrations, and 400 members. By now, it would have been millions, I tell you. Millions!

This program was created during the time frame of the movie *Who Killed the Electric Car?*, which was set in the early 2000s, when car companies tried to kill the TH!NK City. So, no more program, no more coordination between agencies for this basic program that should be continued and enlarged by mass amounts. Nationwide even! Look, I am not looking for a job here (even though I am the typical starving author), but even Harvard and Yale evaluated the program and thought it to be the best thing since sliced bread.

Now, if people have problems with me, then create the program despite me. If not, please recognize that I am here. The economic impact of proper coordination for effective EV deployment in America would grow this country by leaps and bounds. There were people who were marketing the cars at train stations funded by Ford (jobs). Electricians and contractors were building EV charging stations and interconnecting with Con Edison or LIPA. Jobs. Engineers and designers were involved in NYPA. It was great to give back to NYPA for even allowing me to develop for NYPA this program.

So what is stopping us from unleashing the American spirit? The employment needed and the ability to reduce oil and help the grid? Me? Do I smell? Not today, but that's beside the point.

We need an American and EV deployment strategy for plug-in hybrid electric cars. Here's a hint people: If some people in California can work with the University of California and the federal government to create Prius/Ford Escape hybrid electric cars, why not create jobs and help these hybrids become Plug-In Hybrid Electric Vehicle (PHEV) cars. Then you can get the 1 million EV cars you want on roads.

A recent article noted that "Bright Automotive is an Indiana and Michigan-based company developing the Bright IDEA vehicle. The vehicle is a plug-in hybrid electric commercial van that delivers unprecedented fuel economy and utility while significantly lowering the total cost of ownership. Bright's technology is on the road and is being tested by the Department of Defense and other fleets. The company reports [that] it has anxious customers ready to buy its vehicles, with half its customer base small businesses—the growth engine of jobs in America—which will save thousands of dollars per vehicle every year." (Source: Bright Automotive)

While this is a plug-in hybrid electric car and EVs are better, we need to slowly reduce our reliance on oil. It's going to be a slow process to get us off our addiction. One day at a time.

Also to the car companies that are creating EVs, let's invigorate the American spirit of effective programs and spur this economy. Public-private partnerships, such as the NYPA/TH!NK Clean Commute Program, work and can continue to do so. As Tim Gunn says, "Make it work!" (My wife loves that guy.)

In addition, deploy all clean commute programs across the country where there is mass transit. I guarantee you that the numbers will jump. Washington, DC, people are working and commuting. Along the BART mass transit line—San Francisco, Berkeley, and Danville—people commute. Chicago, Boston, and all the cities I've reported on are ready for EV cars.

After you do that or get to a point, the free market will have developed batteries that are cheap to make and allow each car to have a range of 300 or 400 miles on a charge. If this can't happen today, then maybe the globe can take a hint. I'm available and ready to work.

Automotive manufacturers are looking to move plug-in electric vehicles (PEVs) from expensive early-adopter vehicles to more of a mass-market mode of personal transportation. Following several years of significant investment in engineering development and marketing resources, the PEV market was launched in 2011 with production volume of more than 20,000 PEVs, an all-time record for such vehicles. Yet, with a better understanding of the practical challenges of launching new models and expanding PEV sales to mainstream consumers unfamiliar with the technology, especially coming off of a global economic recession, optimism waned.

Now there are more inexpensive electric cars being sold, such as the CODA electric vehicle seen in Figure 11-10 charging away.

As battery electric vehicles (BEVs) and plug-in hybrid electric vehicles (PHEVs) are purchased in greater numbers over the next few years, Pike Research's analysis indicates that significant regional differences in adoption patterns will emerge. For example, while PHEVs will outsell BEVs in North America, the converse is true in Europe and Asia. Meanwhile, auto manufacturer strategies continue to evolve. The lion's share of original equipment manufacturers (OEMs) making plug-in EVs (PEVs) is doing so at a very measured pace, carefully vetting new technologies and consumer preferences with

Figure 11-10   CODA electric vehicle charging up. (*CODA Automotive.*)

a sharp focus on regional differences in demand. All are optimistic about the long-term prospects for the PEV sector, but the industry remains cautiously focused on growing at the pace of market demand.

So as I have always stated, build your own electric vehicle out of an existing vehicle. That is why Toyota recently announced their Scion EV, as seen in Figure 11-11. This vehicle is going to be demonstrated at college campuses in Colorado for starters, but watch out. Then make sure you check out the new charging door for the Ford Focus EV. No difference in the vehicle except it is electric as shown in Figure 11-12.

Meanwhile, Tesla Motors builds their own electric vehicle, shown being built throughout this book and also in Figure 11-13. Just remember we are building our own electric vehicles that inevitably you will sit in. They are being built for you just like gas cars are being built. Now just shift your thinking to all electric.

Finally, I reiterate that you must have an Electric Vehicle logo on your car to show everyone that you built your own electric vehicle. The Toyota Motor Corporation Scion IQ EV does that in Figure 11-14. So let us build our electric vehicles, regardless of conversion or car company. I want to see more images like the one in Figure 11-15 where a Ford Motor Company worker is installing the battery pack into an electric Ford Focus. We need more jobs like this one and this book helps not just the consumer but that guy to build electric vehicles. Electric vehicles will always remain an exciting opportunity for the world; to transform our economy and environment. So get an electric car and make sure everyone knows it!

FIGURE 11-11   Scion IQ from Toyota gets built to be an electric vehicle. (*Toyota Motor Corporation.*)

FIGURE 11-12 Ford built their own electric vehicle—the Ford Focus EV. Check out the charging door!

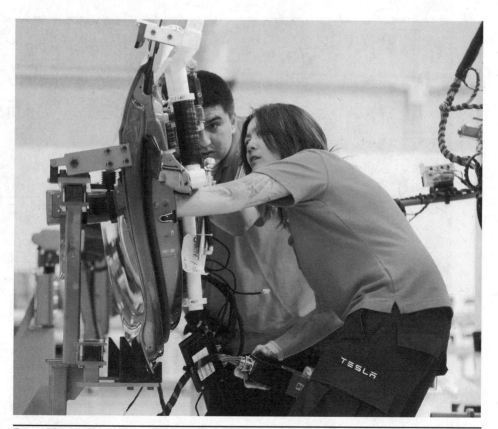

FIGURE 11-13 Toyota Motor Corporation working on building their own electric vehicle. Check out the door! (*Toyota Motor Corporation.*)

FIGURE 11-14 Scion IQ EV logo. Make sure your car has an EV logo. Everyone must know. (*Toyota Motor Company.*)

FIGURE 11-15 A Ford Motor Company worker installing the battery pack into an electric Ford Focus. (*Ford Motor Company.*)

Notes

Chapter 1

1. www.fueleconomy.gov/feg/oildep.shtml.
2. www.alt-e.blogspot.com/2005/01/alternative-fuel-cars-plug-in-hybrids.html.
3. www.greenconsumerguide.com/governmentll.php?CLASSIFICATION=114& PARENT=110.
4. www.drivingthefuture.com/97pct.html.
5. www.worldclassexotics.com/Electriccaronv.htm.
6. Michael d'Estries, *Groovy Green*. www.groovygreen.com/groove/?p=1962.
7. www.kiplinger.com/tools/hybrid_calculator/index.html.
8. www.kiplinger.com/columns/car/archive/green-cars-make-cents.html.
9. Ibid.
10. "The Green Vehicle Trend: Electric, Plug-in hybrid or Hydrogen Fuel Cell?," Fangzhu Zhang & Philip Cooke, Centre for Advanced Studies, Cardiff University, UK.
11. www.nhtsa.gov/cars/rules/rulings/cafe/alternativefuels/background.htm.
12. www.avinc.com/resources/press_release/washington_selects_aerovironment_to_light_up_nations_first_electric_highway.
13. http://businesscenter.jdpower.com/news/pressrelease.aspx?ID=2011039.

Chapter 2

1. www.treehugger.com/corporate-responsibility/electric-vehicles-bring-us-closer -to-freedom-from-oil.html.
2. Bob Batson, Electric Vehicles—The Clean Alternative, 1991. Electric Vehicles of America.
3. www.cbo.gov/sites/default/files/cbofiles/attachments/10-06-ClimateChange_ Brief.pdf.
4. Energy Convenient Solutions: How Americans Can Solve the Energy Crisis in Just Ten Years, www.senesisword.com.
5. www.cbo.gov/sites/default/files/cbofiles/attachments/10-06-ClimateChange_ Brief.pdf.
6. www.cbo.gov/publication/15353.
7. http://knowledgeproblem.com/2011/08/09/gasoline-taxes-and-cafe-regulations/.
8. www.washingtonpost.com/local/congress-heads-for-another-showdown-over -transportation-funding/2011/08/23/gIQAUoYEaJ_story.html.
9. www.infrastructurist.com/2011/08/01/are-consumers-getting-smarter-about -hybrid-cars/.
10. *Emission Impacts of Electric Vehicles.* Wang, Quanlu, University of California Transportation Center; DeLuchi, Mark A., University of California Transportation Center; Sperling Daniel, University of California Transportation Center. Publication Date: 09-01-1990.
11. www.ntnu.edu and EVWorldWire from www.greenlivingguy.com/greenliving/ 2011/5/5/study-finds-lithium-ion-batteries-cause-less-environmental-d.html.

12. www.huffingtonpost.com/rep-ed-markey/driven-to-change_b_913770.html and Office of Congressman Edward Markey.
13. www.eei.org/whatwedo/DataAnalysis/IndustryData/Pages/default.aspx.
14. www.eia.gov/electricity/monthly.
15. Jessica L. Anderson, Associate Editor, *Kiplinger's Personal Finance*; www.kiplinger .com/columns/car/archive/green-cars-make-cents.html#ixzz1X7DoAoEA; accessed September 5, 2011.

Chapter 3

1. Jonathan Kwitny, "The Great Transportation Conspiracy," *Harper's*, February 1981.
2. www.nhraracer.com/content/general.asp?articleid=46995&zoneid=175.
3. Courtesy of Electric Automotive Association President Ron Freund.
4. Chelsea Sexton, seen in the movie *Who Killed the Electric Car?* and producer of *Revenge of the Electric Car!*
5. White House press release, http://www.whitehouse.gov/the_press_office/president -obama-announces-2-point-4-billion-in-funding-to-support-next-generation-electric -vehicles.
6. David Welsh, "GM: Live Green or Die," *BusinessWeek*, May 26, 2008, pp. 38–39.
7. www.pbs.org/newshour/bb/transportation/jan-june08/electriccars_06-25.html.
8. http://media.ford.com/article_display.cfm?article_id=32959.
9. www.fordvehicles.com/technology/electric/.
10. Ibid.
11. Ibid.
12. Ibid.
13. http://money.cnn.com/2011/08/17/autos/cadillac_electric/index.htm.
14. www.daimler.com/dccom/0-5-633234-1-1499723-1-0-0-0-0-0-16696-0-0-0-0-0-0-0-0 .html.
15. http://content.usatoday.com/communities/driveon/post/2011/11/honda-prices -its-all-electric-fit-but-pushes-leases-/1.
16. Liberty Electric Cars, www.liberty-ecars.com/home/press; accessed April 2012.
17. www.stoffstrom.org/institut/arbeitsbereiche/elektromobilitaet/.
18. www.teslamotors.com/about/press/releases/tesla-expands-retail-network -model-s-deliveries.

Chapter 4

1. www.basf.com/group/corporate/site-ludwigshafen/en/news-and-media -relations/news-releases/P-11-396.
2. www.athloncarlease.com/athlon-com/About-us/Athlon-Car-Lease -International/Athlon-Car-Lease-in-the-news/?id_module_component=4043.
3. http://pressroom.toyota.com/toyota/toyota+rav4+ev+coming+to+market+in +2012.print.
4. www.nissan-global.com/EN/NEWS/2011/_STORY/110422-01-e.html.
5. Ibid.
6. www.treehugger.com/cars/new-reva-electric-car-boosts-range-and-speed.html.
7. Cool GreenSheds.
8. Interview with Lynne Mason of Electric-cars-are-for-girls.com.
9. Ibid.
10. www.evamerica.com/eveconomics.pdf.

11. www.nissanusa.com/leaf-electric-car/tags/show/performance#/leaf-electric
-car/range-disclaimer/index.

12. www.electric-cars-are-for-girls.com/ac-conversion.html.

Chapter 5

1. Volvo Cars press information provided at New York Auto Show 2011.
2. Ibid.
3. www.bmw-i-usa.com/en_us/bmw-i3/.
4. www.bmwusa.com/ActiveE.
5. www.nissan-global.com/EN/NEWS/2012/_STORY/120523-02-e.html.
6. www.bmwusa.com/ActiveE.
7. Ibid.
8. www.teslamotors.com/models/innovations.
9. General Motors Chief Executive Richard Wagoner.
10. TurnE Conversions.
11. Smith Electric Vehicles Corp., www.smithelectric.com and FedEx, news.fedex.com.
12. www.chevrolet.com/volt-electric-car/.

Chapter 6

1. auto.howstuffworks.com.
2. www.go-ev.com/TransWarP.html.
3. Ibid.
4. How Does a Shunt Motor Work? Available at: eHow.com http://www.ehow.com/
how-does_5020041_shunt-motor-work.html#ixzz1yALDQ41I.
5. www.micromotcontrols.com/htmls/Motor%20characteristics.html.
6. www.electric-cars-are-for-girls.com/electric-car-motors.html.
7. www.teslamotors.com/about/press/releases/tesla-motors-and-toyota-motor
-corporation-intend-work-jointly-ev-development-tm.
8. www.greenlivingguy.com/greenliving/2007/8/17/plug-in-hybrids-can-have
-dramatic-impact-on-emissions-oil-de.html.
9. http://blogs.wsj.com/drivers-seat/2011/10/02/tesla-model-s-quicker-than-a
-porsche-911-slideshow/?mod=google_news_blog.
10. BMW Motors Press Page Active E.
11. http://electronics.howstuffworks.com/motor7.htm.

Chapter 7

1. HowStuffWorks.com.
2. http://auto.howstuffworks.com/electric-car2.htm.
3. HowStuffWorks.com.
4. Christiansen, Donald and Charles K. Alexander. (2005); *Standard Handbook of
Electrical Engineering* (5th ed.). McGraw-Hill.
5. Zero-Emission Electric Vehicles Australia, www.zeva.com.au.
6. Café Electric LLC.
7. www.A123Systems.com.
8. www.ngcontrols.com/wd1.php.
9. www.zapiweb.com/main.htm or www.zapiinc.com.
10. Sevcon.
11. http://acpropulsion.com/products-drivesystem.html.
11. Ibid.

12. Ibid.
13. Ibid
14. Davide Andrea, Elithion, February 16, 2010; http://liionbms.com/php/wp_balance_current.php.

Chapter 8

1. Right now it's about $3,000 for lead batteries and $8,000 for LiFePO$_4$ batteries—but the LiFePO$_4$ batteries will easily last 2,000 to 3,000 cycles. Best-case scenario for lead batteries is 750 cycles (500 is more likely)—and only if you take *extremely* good care of them.
2. When I say OEMs, I'm referring to the makers of those EVs you don't build yourself, such as Chevy, Nissan, and Tesla.
3. www.npr.org/2011/06/08/136856479/leading-the-charge-to-make-better-electric-cars.
4. From Rod Wilde's TV show about EV racing on the Discovery Channel; www.suckamps.com/.
5. There's some debate among us who have cut LiFePO$_4$ batteries open about whether the "goop" is white foamy stuff or black goopy stuff. Maybe the difference is due to the electrolyte choice and/or whether the battery is charged or discharged when you cut it open.
6. Seth Fletcher, *Bottled Lightning: Superbatteries, Electric Cars, and the New Lithium Economy*, Kindle edition, loc. 2268. New York: Hill and Wang, 2011.
7. Carcinogenic, I'm sorry to say.
8. Chiang's 2011 patent application, entitled "Lithium Secondary Cell with High Charge and Discharge Rate Capability"; www.faqs.org/patents/app/20110081577.
9. www-nrd.nhtsa.dot.gov/pdf/esv/esv22/22ESV-000107.pdf.
10. Seth Fletcher, *Bottled Lightning: Superbatteries, Electric Cars, and the New Lithium Economy*, Kindle edition, loc. 2336-44. New York: Hill and Wang, 2011.
11. Okay, some of the prismatic cells do interlock a little, such as GBS. Still, they're not quite Legos.
12. Seth Fletcher, *Bottled Lightning: Superbatteries, Electric Cars, and the New Lithium Economy*, Kindle edition, loc. 2344-51. New York: Hill and Wang, 2011.
13. See the "Best Practices" section at the end of this chapter on the subject of battery cooling.
14. http://batteryuniversity.com/learn/article/pouch_cell_small_but_not_trouble_free.
15. http://jackrickard.blogspot.com/2011/11/elescalade-liftoff-and-a123-lifepo4.html.
16. http://batteryuniversity.com/learn/article/pouch_cell_small_but_not_trouble_free.
17. www.a123systems.com/Collateral/Documents/EnglishUS/A123%20Systems%20AMP20%20Data%20Sheet.pdf.
18. http://jackrickard.blogspot.com/2012/01/international-module.html.
19. http://batteryuniversity.com/learn/article/pouch_cell_small_but_not_trouble_free.
20. Seth Fletcher, *Bottled Lightning: Superbatteries, Electric Cars, and the New Lithium Economy*, Kindle edition, loc. 1620-28. New York: Hill and Wang, 2011.
21. S.Y. Chung, et al., "Electronically Conductive Phospho-Olivines as Lithium Storage Electrodes," *Nature Materials* 1, October 2002.

22. B. Kang and G. Ceder, "Battery Materials for Ultrafast Charging and Discharging," *Nature* 2009;458 (7235):190–193.
23. Chiang's 2011 patent application, entitled, "Lithium Secondary Cell with High Charge and Discharge Rate Capability"; www.faqs.org/patents/app/20110081577.
24. The LiFePO$_4$ *chemistry* should have no trouble with 10 years of service, but manufacturers don't like to warrant them for that long. Will the battery components, such as the separator and the case, actually last 10 years? This is a good question and one that I don't have an answer for.
25. http://batteryuniversity.com/learn/article/lithium_based_batteries.
26. Seth Fletcher, *Bottled Lightning: Superbatteries, Electric Cars, and the New Lithium Economy*, Kindle edition, loc. 1580-88. New York: Hill and Wang, 2011.
27. Y. H. Rho, et al., "Surface Chemistry of LiFePO$_4$ Studied by Mossbauer and X-ray Photoelectron Spectroscopy and Its Effect on Electrochemical Properties," *Journal of the Electrochemical Society* 2007;154(4): A283–A289.
28. S. Y. Chung, et al., "Electronically Conductive Phospho-Olivines as Lithium Storage Electrodes," *Nature Materials* 1 (October 2002).
29. J. Hu, et al., "Doping Effects on Electronic Conductivity and Electrochemical Performance of LiFePO$_4$," *J Mater. Sci. Technol.* 2009;25(3).
30. G. K. P. Dathar, et al., "Calculations of Li-Ion Diffusion in Olivine Phosphates," *Chemistry of Materials* 2011;23: 4032–4037.
31. Chiang's 2011 patent application, entitled "Lithium Secondary Cell with High Charge and Discharge Rate Capability"; www.faqs.org/patents/app/20110081577.
32. *Particle* in this case has a different meaning than the *charged particle* I spoke of when I was discussing chemistry. Ions are charged atoms. These particles are microscopic chunks of crystal made up of many gazillions of atoms each.
33. B. Kang and G. Ceder, "Battery Materials for Ultrafast Charging and Discharging," *Nature* 2009;458(7235): 190–193.
34. Chiang's 2011 patent application, entitled, "Lithium Secondary Cell with High Charge and Discharge Rate Capability." www.faqs.org/patents/app/20110081577.
35. Ibid.
36. Ibid.
37. http://hipowergroup.com/list.php?catid=36&page=3.
38. www-scf.usc.edu/~rzhao/LFP_study.pdf.
39. www.electric-cars-are-for-girls.com/lithium-battery-management-systems.html.
40. http://jackrickard.blogspot.com/2010/11/itsy-bitsy-spyder-550-and-battery.html.
41. http://batteryuniversity.com/learn/article/safety_concerns_with_li_ion.
42. http://jackrickard.blogspot.com/2012/01/struggle-continues.html.
43. http://jackrickard.blogspot.com/2010/04/april-23-friday-show.html.

Chapter 9

1. www.avinc.com/resources/press_release/zero_to_80_percent_in_15_minutes_new_benchmark_for_ev_fast-charging.
2. Delphi Automotive Wireless Charging Division. http://delphi.com/manufacturers/auto/hevevproducts/charging-cordsets/wrls-chrg/.
3. www.ecotality.com/media/press-releases/ecotality-partners-with-regency-centers-at-19-locations-nationwide/.
4. Miami Beach, FL, and San Francisco, CA, May 30, 2012.
5. www.carcharging.com/about/news/all/carcharging-and-ace-parking-management-offer-electric-vehicle-charging-services-in-the-san-francisco-bay-area/.

6. Ibid.
7. *Build Your Own Electric Vehicle*, Second Edition
8. Courtesy of Project Better Place.
9. www.betterplace.com/the-company-pressroom-pressreleases-detail/index/id/CSG-and-Better-Place-Open-Switchable-Electric-Car-Experience-Center-in-Guangzhou-China.
10. www.nissan-global.com/EN/NEWS/2011/_STORY/111110-03-e.html.
11. www.nissan-global.com/EN/NEWS/2011/_STORY/110321-01-e.html.
12. www.grassrootsev.com/contactors.htm.
13. Leviton and Toyota.
14. www.evnetics.com/evnetics-products/throttle-assembly.
15. www.grassrootsev.com/cpacks.htm.

Chapter 10

1. Interview with Ken Koch, EV Consulting, Inc. and information provided.
2. www.xaviertechnologies.com/ev_conversion.htm.
3. Interview with Ken Koch, EV Consulting, Inc. and information provided.
4. Ibid.
5. www.plasmaboyracing.com/whitezombie.php.
6. John Wyland, head of Plasma Boy Racing with approvals to use website for Plasma Boy Racing.

Chapter 11

1. http://green.autoblog.com/2011/04/05/dbm-energy-record-breaking-kolibri-battery-passes-government-tests/.
2. Fast EV charging service now available at 25 Beekman Street in Manhattan, New York. CarCharging Group.
3. Miami Beach, FL,. and New York, May 31, 2012 (Globe Newswire).
4. www.go-ev.com/end-users/005_005_SU060215_001_Motor_Care.pdf.
5. www.go-ev.com/end-users/005_003_Motor_Safety.pdf.
6. www.treehugger.com/corporate-responsibility/electric-vehicles-bring-us-closer-to-freedom-from-oil.html.
7. www.ieee.org/publications_standards/publications/proceedings/index.html.
8. Courtesy of Institute of Electrical and Electronics Engineers (IEEE).
9. http://thinkev.leftbankcompanies.com/buy-think-city/fleet/.
10. http://planetgreen.discovery.com/tech-transport/thinkcity-evready-index-electriccars.html.
11. General Electric.
12. ABB Global.
13. http://auto.howstuffworks.com/install-electric-car-charger.htm.
14. Courtesy of and information provided by Pike Research.

Index

5-Series (BMW), 94
102EX Phantom Experimental Electric
    (Rolls-Royce), 101–102
280Z (Nissan), 99–100
300ZX (Nissan), 59
911 Carrera (Porsche), 82, 103–104, 115, 190,
    196, 341

▬ A ▬

A8 (Audi), 94, 190
A123 Systems, 15, 106, 201, 219, 222,
    223–225, 228, 231–234
ABB, 17–18, 61, 356
AC controllers, 196, 198, 203–208, 214–215
AC drive systems, 106, 107, 110, 111–113,
    115–116
AC electric motors, 178–188, 204–208
    characteristics, 178–179
    future of, 185–187
    ratings, 187–188
    transformers versus, 179–180
    types, 180–185
AC induction motors, 180–185, 204, 208
    polyphase, 181–182, 183
    single-phase, 180
    split-phase, 181–182
    wound-rotor, 182–185
AC Propulsion, Inc., 9, 15, 190, 203, 204–208,
    247
    AC-150 Gen 2 system, 204, 205–206

AC-150 Gen 3 system, 205–208
    Autoport, Inc. partnership, 208
    eBox, 78, 79
    Tesla controllers with, 214–215
acceleration force, 125–126
accessory system power, 311–312, 328
Ace Parking, 252–253
ActiveE (BMW), 117, 120–121, 142, 143, 147,
    160–161, 190, 254, 346, 347
Acura, 49, 50
Advance Corp., 207
Advance DC motors, 184, 185
    Model FB1-4001, 149
aerodynamic drag force, 119–120, 130–135,
    157
    front airflow, 135
    lowest coefficient of drag, 130–133
    rear airflow, 134
    relative wind factor, 133
    speed of vehicle and, 134
    underbody airflow, 134–135
    wheel well airflow, 134–135
AeroVironment (AV), 21–23, 245, 354–355
    EVSE-RS Plug-In, 351–352
Agassi, Shai, 79, 258
air cooling systems, 115, 209, 273, 274, 324,
    325
air pollution, 2–3, 6, 24, 36, 45–51, 69
    emissions standards, 3, 13, 24, 31,
        34–35

airport tugs, 1
All Harley Drag Racing Association
    (AHDRA), 60
Alpmobil, 355
Altairnano, 77
Alternative Fuels Act of 1988, 65
AM General, 74–75
American Drag Racing League (ADRL), 60
American Recovery and Reinvestment Act
    of 2009, 69
American Wire Gauge (AWG), 275, 303,
    306–307, 312
ammeters, 272, 343
AMP Electric Vehicles, 79, 282
Ampera (Vauxhall), 87–89
ampere-hours (Ah), 225, 243
amperes, 221–222
Ampere's law, 166, 167
Anderson, Jessica L., 19
Andrea, David, 208–214
Android, 254
anions, 219
anodes, 220, 222
antisite defects, 230
area-specific impedance, 231–232
Arizona, 49, 60, 251, 258
Armand, Michel, 227–229
armature, 169–177, 198
Asea Brown Boveri (ABB), 17–18, 61, 356
ASG Renaissance, 356
Association of International Auto
    Manufacturers, Inc., 15
associations, 59–63
Athlon Car Lease, 92
AT&T, 74
AUCMA Electric Vehicle Co., Ltd., 203–204
Audi AG
    A8, 94, 190
    Q7, 82
    R8 e-tron, 77–78
    R8 LMS, 78
Australia, 83, 117, 183, 259
Autolite, 59
automatic transmission, 145, 291–292
AutoPort, Inc., 208
auxiliary relays, 270–271

AVEOX, 177
AVERE (European Association for Battery,
    Hybrid and Fuel Cell Electric Vehicles),
    62, 63
Azure Dynamics
    AC24 motor, 188, 189
    LEAD customer program, 74–75

▬▬ B ▬▬

Back to the Future (movie), 81
backup power supply, 206, 213, 255, 257
Baker Electric automobile, 59
Baker Institute for Public Policy (Rice
    University), 38
BART mass transit line, 359
BASF SE, 90, 91
Basseng, Marc, 78
BAT International, 61
Batson, Bob, 31, 99, 113, 114, 298
batteries, 6, 63, 109–110, 217–240, 320–328.
    See also charger systems; charging
    infrastructure
        30 percent or greater weight rule,
            129–130
        accessory system power, 311–312, 328
        ampere-hours, 225, 243
        anatomy, 222
        battery electric vehicles (BEVs), 54, 64,
            66, 359
        best practices, 235–239
        bottom balancing, 236
        building, 220–221
        C-rate, 226, 231
        cathode research, 228, 230–231
        charging infrastructure, 17–18, 20,
            21–23, 24, 30
        circuits, 221–222
        conversion process, 286, 325–328
        cooling, 237, 274, 324, 325
        costs of, 15, 30
        diffusion-limited lithiation, 230, 231
        discharging and charging cycle, 228,
            243, 244
        disconnecting from chassis, 291
        energy/power density, 228, 232–233
        Exide, 54

fast charging, 231–232, 254–255, 262–263, 354–355

flat discharge curve, 226–227

formats, 223–225

impedance, 231–232

installation, 320–321

installing battery mounts, 300–302

lead-acid, 9, 13, 37, 39, 61, 217, 218, 220, 225, 233, 242, 245

lithium-ion. *See* lithium-ion batteries

maintaining, 345

nanophosphate lithium-air, 9, 77, 106

nickel-metal hydride (NiMH), 9, 13, 37

onboard charger, 344, 351–352

range of EVs, 16–17, 20, 21, 24, 76, 77, 106, 128, 344

recycling infrastructure, 13–14, 37

replacement battery packs, 258–259, 261

running out of power, 344, 351–352

safety issues, 13–14, 222–223, 228, 233, 234–236, 239

state of charge (SOC), 209, 211, 221, 243, 246, 272, 311

sulfur, 61

testing, 242–243, 272–273

thermal runaway, 233, 234–236

voltage, 225, 243

wiring, 225–226, 238–239, 322–323

Battery Box, 59, 237–238

battery electric vehicles (BEVs), 54, 64, 66, 359

battery management system (BMS), 201–202, 209–214, 233–234, 236–237, 239, 308, 310, 322–323

battery mounts, 300–302

Battery Switch networks, 258–259

Battery University, 225

BBC, 73, 87, 102

Beal, Dennis, 159

Beiqi Foton Motor Co. Midi-E, 207

Benfield Motor Group, 263

Berube, Dennis, 60

Best Buy, 20

BGS Electric, 61

bikes, electric, 78

biofuels, 39, 43, 155

Bissonette, Mike, 23

Blankenship, George, 92

Blink network, 251–252, 254, 255, 264

BMS (battery management system), 201–202, 209–214, 233–234, 236–237, 239, 308, 310, 322–323

BMW Group, 63, 86

5-Series, 94

ActiveE, 117, 120–121, 142, 143, 147, 160–161, 190, 254, 346, 347

i3 Concept, 120, 190

Mini-E, 204, 208

Boniface, Bob, 160

Bonneville Salt Flats, 59

Bosch, 121

EM-motive, 76

Bottled Lightning (Fletcher), 220

bottom balancing, 236

Boulanger, Albert G., 354

Boxwell, Michael, 88, 90

Boyle, Matt, 204

BP *Deepwater Horizon* offshore drilling disaster, 29, 38, 42, 57

brakes

brake drag, 137

regenerative braking, 111, 142, 173, 175, 176

Brant, Bob, 14, 34, 130, 204, 258, 282, 289, 308, 312, 334, 341

Bright Automotive, 358

Bright IDEA vehicle, 358

British Petroleum (BP) oil spill disaster, 29, 38, 42, 57

brushes, motor, 111, 170

brushless DC motors, 177, 196

Build Your Own Electric Vehicle (Harris), 312

building electric vehicles, 20–25, 70–76. *See also* conversions and conversion process steps in process, 286

bulk diffusion, 231

Bush, George W., 39, 69

Butler, Don, 75

buying ready-to-run EVs, 86–98

━━ **C** ━━

C-rate, 226, 231

C-Zero (Citroen), 88

Cadillac. *See also* General Motors (GM)
    Converj Concept, 75
    ELR, 75
CAFE (corporate average fuel economy), 21,
    32, 34–36, 38–39, 77
Café Electric, 196, 199, 201–202, 314, 315
CALB (China Aviation), 224–226, 228, 236, 237
California
    California Air Resources Board
        (CARB), 10, 48, 64–66
    EV Project, 23, 251, 252–253
    guidelines for EVs, 341, 342
    Los Angeles Department of Water
        and Power, 48
    oil fields, 55
    Southern California Edison, 51, 74
    zero-emission vehicle (ZEV) mandate,
        54, 64–66, 71
Camissa, Mike, 15
Camry (Toyota), 114
Canada, 21–23, 60, 74, 227–228, 235–239
Canadian Electric Vehicles, 235–239
CANbus connectivity, 202
Canny, Richard, 245
CARB (California Air Resources Board), 10,
    48, 64–66
carbon dioxide emissions, 47, 49, 80
carbon monoxide emissions, 46–49, 49
CarCharging Group, Inc., 252–253, 345–346
Carter, Jimmy, 59
cathodes, 220, 222, 228, 230–231
cations, 218–219
Ceder, Gerb, 230, 231
cell shunts, 236
cell voltage, 243
Centurion (Rolls-Royce), 101
Chang, William, 207
charge-collecting coating, 231
charge-collecting layer, 230
Charge de Move (CHAdeMO) Protocol, 256
ChargePoint America program, 247, 254
ChargePoint Network, 247, 253–254, 346
charger systems, 6, 110, 241–257. *See also*
    batteries; charging infrastructure
        battery discharging and charging
            cycle, 243, 244

conversion process, 312–319
cooling battery for charging, 237
examples of, 246–250
first startup charge, 239
ideal, 244–246
induction charging, 255–257
onboard chargers, 344, 351–352
overview, 241
rapid charging, 231–232, 254–255,
    262–263, 354–355
solar energy, 48
warming battery for charging, 237
charging cycle, battery, 243, 244
charging infrastructure, 17–18, 21–23, 24, 30,
    250–254. *See also* batteries; charger
systems
    Best Buy charging stations, 20
    best practices, 236, 237–238, 239
    Blink network, 251–252, 254, 255, 264
    CarCharging Group, Inc., 252–253,
        345–346
    Cracker Barrel charging stations, 20,
        263–265
    Delphi Wireless Charging System,
        250–251, 259–260
    EV Power Station, 255–257
    EV Project, 22–23, 251, 254, 255
    first startup charge, 239
    home charging stations, 74, 76, 266
    Leviton charging station, 265–266, 353
chassis, 61, 108–109, 110, 117–161
    aerodynamic drag, 119–120, 130–135,
        157
    calculation overview, 148–149
    cleaning and preparing, 286, 290–291
    conversion process, 154–159, 286,
        288–292
    current, 147
    design process, 146–154
    drive trains, 138–140, 145–146, 157
    engine/electric motor, 139–144
    gears, 141, 144–146, 343
    horsepower, 147–148
    mounting, 295–297
    optimizing EV, 121–123
    purchasing other components, 289

removing chassis parts, 291–292
selecting, 118–121, 288–289
testing on component level, 118
tires and rolling resistance, 135–138, 157, 346
torque, 148, 149–154
used chassis, 12, 120–121, 155–156
weight issues, 119, 123, 124–130, 157
Chevrolet. *See also* General Motors (GM)
ChargePoint America program, 247, 254
Cruz LTZ, 19, 45
S-10 pickup truck conversions, 104, 105, 282
Turbo 400 transmission, 172
Volt, 3, 13, 16–17, 19, 21–23, 44–45, 68–69, 88, 89, 114, 115, 160, 254, 346
Chiang, Yet-Ming, 227–231
China, 69, 117, 203–204, 207–208, 217, 261
China Aviation (CALB), 224–226, 228, 236, 237
China Southern Power Grid (CSG), 261
Chrysler, 63, 64
Dodge Dakota conversion, 282
Dodge Ram, 282
Epic minivan, 66
Global Electric Motorcars, 3, 248–250
Chu, Andrew C., 354
circuit breakers, 267, 268
circuits, 221–222
Citroen C-Zero, 88
CityEl, 77
Civic/Civic hybrid (Honda), 64, 136
Clean Air Act of 1968, 57
Clean Air Act of 1990, 38
Clean Liner, 59
Clinton, Bill, 64
Clio (Renault), 88–89
Clooney, George, 184
Clunn, Steve, 69, 281, 294, 298, 307, 311–312, 330, 338
clutch, 139, 142, 145. *See also* transmission
coal, 2–3, 33, 36, 40, 47–51
coasting, 343–344
CODA Automotive, 29, 76, 102, 254, 317, 318, 331, 359
Collier, Mike, 235–239

Colorado, 360
Commonwealth Club of California, 155
commutator, DC electric motor, 169
Commuter Cars Corporation, Tango, 1, 70, 184, 201
Compact Power, Inc. (CPI), 72–73
compound DC motors, 175–176
computer-aided design (CAD), 237, 238
computers, 63, 95, 237, 238
conductive foil tabs, 224
Conformity in Europe (CE), 354
connections/connectors, 275–276, 348–349
contactors
main, 267, 268
reversing, 267, 270
continuous acceleration, 108
Controller Area Network (CAN), 316
controllers, 9–10, 110, 193–215
AC, 196, 198, 203–208, 214–215
battery management systems (BMS), 201–202, 209–214
DC, 196–203, 215
electronic, 195
failure, 112, 203
overview, 193–195
role of, 193–195
selecting, 196–215
solid-state, 195
vehicle management systems (VMS), 208–209
Converj Concept (Cadillac), 75
conversion companies/shops, 20, 69, 79–81, 97–98, 102–104, 281–282, 284, 286–289
conversions and conversion process, 4–8, 70, 98–114
advantages of, 4
batteries. *See* batteries
battery mounts, 300–302
charger systems. *See* charger systems
chassis and design. *See* chassis
checking stage, 276–277, 328–332
controllers. *See* controllers
costs of, 85–86, 98, 112–116, 155–156
decisions in, 105–110
doing stage, 286–328
electric motors. *See* electric motors

conversions and conversion process (*cont.*)
  electrical systems. *See* electrical
    systems
  examples, 282–284, 332–335
  of existing vehicles, 99–101
  help in, 284, 291
  high-end, 101–105, 115–116
  kits, 289
  manuals, 157
  motors. *See* electric motors
  needs list, 156–157
  overview, 281–284, 287
  painting and polishing, 332
  planning stage, 146–154, 284–286
  pros and cons, 154–155
  purchases and deliveries for, 285
  selling engine parts, 157
  space for, 285
  system checkout on blocks, 329–330
  systems needed for, 4–6
  tools for, 285, 291
  weight during and after conversion,
    124–125, 157
converters, 275, 311–312, 319
Cool GreenSheds, 203, 294
CoolGreenCar, 103
cooling
  air cooling systems, 115, 209, 273, 274,
    324, 325
  battery, 237, 274, 324, 325
  liquid cooling systems, 160–161, 186,
    190, 195, 201–203, 275
Corbin Motors Sparrow, 70, 86, 248
corporate average fuel economy (CAFE), 21,
  32, 34–36, 38–39, 77
Corvette, 190
costs of EVs, 18–20
  battery prices, 15, 30
  conversions, 85–86, 98, 112–116,
    155–156
  economical driving, 343
  financial incentives, 19, 21, 35–36,
    65–66, 72, 79, 87–88, 257
  gasoline prices and, 19, 31, 35–36,
    44–45, 54–59, 64, 66, 114, 345
  oil costs and, 34

  oil industry subsidies and, 15
  operating, 12, 19–20, 24, 44–45, 114
  purchase price, 12, 19–20, 24, 31, 44–45,
    71, 72, 78, 95, 98, 106
Coulomb Technologies, 253
Cracker Barrel charging stations, 20,
  263–265
crash, 235
Cruz LTZ (Chevrolet), 19, 45
CRX (Honda), 204
Cugnot steam tractor, 55
current, 147, 233, 242–243, 266, 269, 275,
  303–312
  high-current electrical systems,
    303–307
Curtis battery tester, 272, 273
Curtis Instruments, Inc., 106, 196–198, 269
Curtis PWM DC motor controller, 110,
  196–198
Curtis throttle potentiometer, 269
cycle life, battery, 228, 243, 244
cylindrical cells, 223–224

D

Daily Green, 13–14, 96, 263–264
Daimler AG, 86, 91
Davenport, Chris, 88
Days of Thunder (movie), 135
DBM, 345
DC controllers, 196–203, 215
DC drive systems, 106, 107, 110, 111–113
DC electric motors, 165–178
  characteristics, 168
  components of, 169–170
  magnetism and, 164, 166–170, 176–177
  selecting, 170–171
  types, 171–178, 196
DC-to-DC converters, 275, 311–312, 319
de la Guardia, Manuel, 95
dealers, 20
DELIVER project, 79–80
delivery vehicles, 31–32, 79–80, 158–159,
  208. *See also* trucks; vans
DeLorean electric car, 11, 81, 83
Delphi Automotive
  DC-to-DC converter, 275

Delphi Wireless Charging System, 250–251, 259–260
DeLuchi, Mark A., 36
Denmark, 258–259, 316
Department of Motor Vehicles (DMV), VINs and, 3, 342
depth of discharge (DOD), 209, 226–227
design process, 146–154
    calculation overview, 148–149
    current, 147
    horsepower, 147–148
    torque, 148, 149–154
Detroit Edison, 31
Detroit Electric Car, 61, 62, 68
Deutsche Accumotive batteries, 75–76
Diablo conversions (Lamborghini), 103, 104
DiCaprio, Leonardo, 184
Dick, Michael, 77, 78
diesel engines, 23, 24, 44–45, 49, 156, 164
differential, 139, 144
diffusion-limited lithiation, 230, 231
discharge cycle, battery, 243, 244
Dodge
    Dakota conversion, 282
    Ram, 282
dolly, engine-lifting, 291
Dominique, Larry, 13–14
doping, 229–230
Dow Kokam, 305–306
drag force. See aerodynamic drag force
drag racing, 59–62, 108, 112, 305–306
drive axles, 139, 144
drive systems
    AC, 106, 107, 110, 111–113, 115–116
    DC, 106, 107, 110, 111–113
drive trains, 138–140, 145–146, 157
driveshaft, 139, 144, 172
driving EVs, 10–12, 74, 343–345
    coasting, 343–344
    economical, 343
    range determination, 344
    regenerative braking, 111, 142, 173, 175, 176, 344
    regular, 344–345
    running out of power, 344, 351–352
    safety issues, 120, 344
    shifting, 343
    starting, 343
Dube, Bill, 60

E

E-Car-Tech, 81
E-laad.nl Foundation, 95–96
E-Mobility research institute (Germany), 80
e-NV200 electric van (Nissan), 132–133
E-Smart prototype, 80–81
E Source, 17
Eagle Pitcher, 59
East Coast Electric Drag Racing Association, 62
eBox (AC Propulsion), 78, 79
eCars-Now!, 63, 81
ECOtality, Inc.
    Blink network, 251–252, 254, 255, 264
    EV Project, 22–23, 251, 254, 255
    Regency Centers partnership for charging, 251–252
eDrive system, 120
efficiency
    electric vehicle (EV), 49, 50, 91, 164
    internal combustion engine (ICE), 39–40, 50
Eisenhower, Dwight D., 56
Electric Automobile Association (EAA), 10, 20, 60, 62, 111, 146–147
electric bikes, 78
Electric Cars are for Girls, 103–104, 106–108, 176–178, 217–218
Electric DeLorean, 11, 81, 83
Electric Drive Transportation Association (EDTA), 61, 64
electric motorcycle, 78
electric motors, 8–9, 140–144, 163–192
    AC, 178–188, 204–208
    advantages of, 164
    common uses of, 191–192
    conversion process, 286, 292–300
    critical distance, 294–295, 297
    DC, 165–178, 196
    engines versus, 6–8, 14, 27–32, 36, 46–47, 50, 139–144

electric motors (*cont.*)
    flywheel-to-motor-shaft connection,
      297–298
    future of, 185–190
    horsepower, 164–165
    magnetism and, 164, 166–170, 176–177
    mounting, 292–294, 295–300
    ratings, 187–188
    rear support, 297
    safety issues, 349–351
    testing, 298–300
    torque, 164, 167, 170
electric-only vehicles, 17. *See also* Leaf
    (Nissan)
electric power plants
    coal-burning, 2–3, 33, 36, 40, 47–51
    hydroelectric power, 33, 40, 227–228
    natural gas, 2, 33, 40, 47–48, 57
    nuclear, 33, 40, 48, 57, 58
Electric Power Resource Institute (EPRI), 51
electric trains, 1, 53, 59–60
electric utility companies, 30. *See also*
    electric power plants
    air pollution issues, 45–51
    load leveling, 40, 51, 257
Electric Vehicle Association of Asia Pacific
    (EVAAP), 61
Electric Vehicle Technology Competitions
    (EVTC), 60
electric vehicles (EVs)
    advantages of, 1–3, 4, 15, 20–25,
      27–30
    alternatives, 3
    availability of, 29–30
    block diagram, 9
    chassis and design. *See* chassis
    components of, 6, 8–18
    controllers. *See* controllers
    conversions. *See* conversions and
      conversion process
    costs of. *See* costs of EVs
    customizing, 12–13
    dealers, 20
    disadvantages of, 20
    driving, 10–12, 74, 120, 343–345
    efficiency, 49, 50, 91, 164

films about, 10, 64–66, 68, 71, 79, 81,
    358
financial incentives for, 19, 21, 35–36,
    65–66, 72, 79, 87–88, 257
history of. *See* history of EVs
insurance, 3, 341, 342
internal combustion engines versus,
    6–8, 14, 27–32, 36, 46–47, 50
list of types, 1
maintaining, 345–349
mass deployment recommendations,
    352–354
motors. *See* electric motors
myths concerning. *See* myths
    concerning EVs
perceptions of, 23–25
production targets, 65–66
recall and destruction, 66, 68
registration, 341–343
resale, 20
safety issues. *See* safety issues
time to purchase/build, 20–25
update on, 354–357
Electric Vehicles International, 102
Electric Vehicles of America, Inc. (EVA), 31,
    97–98, 104, 113, 114, 203, 282, 289
electrical systems, 157, 266–275
    ammeter, 272, 343
    auxiliary relays, 270–271
    battery indicator, 209, 211, 221, 243,
      246, 272, 311
    charger systems. *See* charger systems
    conversion process, 286, 302–320
    DC-to-DC converter, 275, 311–312, 319
    fans, 273, 274
    fuses, 239, 267, 268, 273, 351
    high-current systems, 303–307
    high-voltage, high-current power, 266
    interlocks, 275, 311
    low-voltage, low-current power, 269,
      275, 308–312
    main circuit breaker, 267, 268
    main contactor, 267, 268
    reversing contactor, 267, 270
    rotary switch, 273
    safety issues, 267–269, 273–275, 276

shunts, 272
summary of parts, 277–279
terminal strip, 271–272
throttle potentiometer, 269–270,
    308–310
voltmeter, 272, 311
wiring, 267, 275–277
Electrical Vehicle Charge Systems (Esarj), 63
ElectroAutomotive, 185–187, 282, 338
electrodes, 220
electrolyte solutions, 219, 222
electronic conductivity, 221
electronic controllers, 195
electrons, 219, 221
ELR (Cadillac), 75
Eltek Valere chargers, 248, 249, 316
emergency kit, 351
emissions standards, 3, 13, 24, 31, 34–35
end-of-charge current (EOCC), 242–243
Energy, Convenient Solutions (Johnson),
    32–33
energy density, battery, 232–233
Energy Independence and Security Act of
    2007, 39
Energy Policy Act of 2005, 24
engine-lifting dolly, 291
engines. *See also* conversions and
    conversion process; internal combustion
    engine (ICE)
        electric motors versus, 139–144
environmental issues. *See also* air pollution;
    water pollution
        advantages of EVs, 1–3, 4, 15, 21, 22,
        23–25, 27–30, 31–32, 38, 63
        for internal combustion engines (ICE),
        6, 14, 21, 27–32, 36, 38, 41
        oil use and, 56–57
Epic minivan (Chrysler), 66
Epyon BV, 17–18, 356
ethanol, 43, 155
European Association for Battery, Hybrid
    and Fuel Cell Electric Vehicles (AVERE),
    62, 63
European Community (EC). *See also names
    of specific counties*
        DELIVER project, 79–80

EV Convert, 121–123
EV Cup, 63
EV Photo Album, 103–104
EV Porsche, 282
EV Power Station, 255–257
EV Project, 22–23, 251, 254, 255
EV Propulsion, 196
EV supply equipment (EVSE), 254, 266,
    351–352
EV1 (General Motors), 3, 10, 49, 50, 66,
    68–69, 256
EV2 (General Motors), 66
EV200 Kilova contactor, 316, 317
EVA (Electric Vehicles of America, Inc.), 31,
    97–98, 104, 113, 114, 203, 282, 289
EVE (Italy), 81
EVermont Project, 49
Evonik (Deutsche Accumotive), 76
EVPlus (Honda), 66
EVTV Motor Verks, 106–108, 235–239
Exide batteries, 54
Exxon Valdez oil spill disaster, 38

F

fans, 273, 274
Faraday, Michael, 166, 218
Farkas, Michael D., 253
fast charging, 231–232, 254–255, 262–263,
    354–355
FB1-4001A motor, 164–165
Federal Motor Vehicle Safety Standards
    (FMVSS), 341–342
Federation of International Automobiles
    (FIA), 60
FedEx
        DELIVER project, 79–80
        zero-emission step vans, 158–159
Feynman, Richard, 217
Fiat, 79
field poles, DC electric motor, 169
field weakening
        series DC motor, 172
        shunt DC motor, 174
Fiesta (Ford), 201
films about EVs, 10, 64–66, 68, 71, 79, 81, 358
Fischer, Christian, 91

Fisker, 1, 10
    Karma, 254, 346
Fit EV (Honda), 76
Fitzpatrick, Kevin, 263
flat discharge curve, 226–227
Fletcher, Seth, 220
Flex (Ford), 100–101
floating propulsion system ground, 276,
    307
floor jacks, 298
Florida, guidelines for EVs, 342
Fluence ZE (Renault), 96, 258
Flux Power, 97–98
Focus Electric (Ford), 1, 3, 10, 15, 27, 72–75,
    86, 87, 92, 117, 254, 290, 292, 293, 311–313,
    341, 346, 361
Ford, Henry, 55
Ford, Mrs. Henry, 61, 62
Ford Motor Company, 32, 63, 64, 78–79, 83,
    111, 203. See also TH!NK City
    battery pack, 323, 325
    ChargePoint America program, 247
    F-150 trucks, 61
    Fiesta, 201
    Flex, 100–101
    Focus Electric, 1, 3, 10, 15, 27, 72–75,
        86, 87, 92, 117, 254, 290, 292, 293,
        311–313, 341, 346, 361
    Model T, 54, 55
    Prius/Ford Escape hybrid, 358
    quality testing battery packs, 311
    Ranger pickup conversion, 10, 66,
        110–111, 126, 128–129, 134, 136, 137,
        144–145, 147, 149–152, 282, 294, 308,
        309, 312, 314, 334, 343
    Taurus, 131, 144, 145
    TH!NK City and, 16. See also TH!NK
        City
    Transit Connect van, 73–75
forklifts, 1, 54
frame, DC electric motor, 170
France, 47, 48, 56, 59–63, 66, 76, 92, 117
Freakonomics, 34–35
front airflow, 134
fuel economy, 21, 32, 34–36, 38–39, 77
Fuel Economy Guide, 96

Fuji Electric, 59
fuses, 239, 267, 268, 273, 351

G

G-Van (Vehma International), 31
G-Wiz/REVA Electric Car, 20, 27, 77, 97
Gage, Tom, 208
gas tax, 35–36
gasoline prices, 19, 31, 35–36, 44–45, 54–59,
    64, 66, 114, 345
gear ratios, 141, 144, 145, 147
gears, 141, 144–146, 343. See also transmission
GEM car (Global Electric Motorcars), 3,
    248–250
General Electric (GE), 195
    smart grid technology, 355
General Motors (GM), 17, 32, 55–56, 59, 63,
    79, 86, 111, 155. See also Chevrolet
    EV1, 3, 10, 49, 50, 66, 68–69, 256
    EV2, 66
    Hughes Division, 256
Geo Metro, 61
Germany, 48, 56, 77–78, 80, 83, 92, 237, 248,
    282
Ghosn, Carlos, 94
Giberson, Michael, 35
Gioia, Nancy, 74
Global Electric Motorcars, GEM car, 3,
    248–250
golf carts, 64, 185–186, 341
Goodenough, John, 227–228
Goodier, Mark, 263
Goodyear Tires, 136
Grannes, Dean, 60
graphite anodes, 220
Grass Roots Electric Vehicles, 69, 270
Green Genuity program, 252
Green Guru Guide Book, 292
Green Shed Conversions, 281, 282, 289, 292,
    297, 311–312, 330
Greenpeace, 66
Gribben, Chip, 45–51
grid-connected hybrids, 3
Gross, Darwin, 53
gross axle weight rating (GAWR), 129
gross vehicle weight rating (GVWR), 21, 119

grounding, 276, 307
Gumpert Apollo Sport, 77
Gutzwiller, Frank W., 195

━━ **H** ━━

Haase, Christopher, 78
Hall, Robert N., 195
Halstead, Jerry, 123
Hammond, Paula, 21–22
Hansel, Bryan, 159
Harris, Jim, 312, 314, 345
Harrison, Scott, 74–75
Haug, Birger N., 72
Haugerud, Nils, 72
heating element, 330–331
Hedlund, Roger, 59
Henney Coachworks, 54
Henney Kilowatt, 54
Hi Performance Electric Vehicle Systems
    (HPGC), 106
high-current electrical systems, 303–307
high-end conversions, 101–105, 115–116. See
    also Porsche
High Occupancy Vehicle HOV lane, 66
High Performance Electric Vehicle Systems
    (HPEVS), 106, 185–188
high-voltage electrical systems, 266
Hild, Ruediger, 223
hill-climbing force, 126, 127
HiPer Drive motors, 185–187
HiPower batteries, 233
history of EVs, 53–83
        associations in, 59–63
        movies in, 10, 64–66, 68, 71, 79, 81, 358
        return of interest in EVs, 59–61
        timeline of vehicle history, 54
        timelines of electric cars, 54–56
Holmquist, Randy, 235–239
home charging stations, 74, 76, 266
HomePlug, 87
Honda, 32
        Civic/Civic hybrid, 64, 136
        CRX, 204
        EVPlus, 66
        Fit EV, 76
        Petrolia, 89

Horlacher Sport, 61
horsepower, 107–108, 123, 126–128, 147–148,
    164–165, 233
Hösslin, Christian von, 81, 296–297, 300,
    303, 308, 319, 331, 335
Howard, Brian Clark, 13–14
HPEVs (High Performance Electric Vehicle
    Systems), 106, 185–188
HPGC (Hi Performance Electric Vehicle
    Systems), 106
HPL, 79
Hurricane Katrina, 69
hybrid vehicles, 64, 76–77. See also plug-in
    hybrid electric vehicles (PHEVs); Prius
    hybrid (Toyota)
        cost of, 24, 44–45
        minimum noise requirement, 14–15
Hydro-Quebec (H-Q), 227–228
hydrocarbon emissions, 47, 48
hydroelectric power, 33, 40, 227–228
hydrogen fuel cells, 64
Hypermini (Nissan), 66
Hyundai Sonata, 13

━━ **I** ━━

i-MiEV (Mitsubishi), 88, 96
i3 Concept (BMW), 120, 190
IKA, 79
impedance, 231–232
Implementing Agreement for Cooperation
    on Hybrid and Electric Vehicle
    Technologies and Programs (IA-HEV),
    62
ImPulse, 203
Inconvenient Truth, An (documentary
    movie), 64
India, 20, 69, 77, 97, 117
Indica Vista (Tata), 90
induction charging, 255–257
induction motors. See AC induction motors
information and communication technology
    (ICT) system, 95
insulated-gate bipolar transistor (IGBT), 195
insurance, 3, 341, 342
Interbrand, 88
intercalation, 220

internal combustion engine (ICE), 55–59. *See also* conversions and conversion process
  components, 6, 7
  conversion to EV. *See* conversion companies/shops; conversions and conversion process
  efficiency, 39–40, 50
  electric motors versus, 6–8, 14, 27–32, 36, 46–47, 50, 139–144
  environmental issues for, 6, 14, 21, 27–32, 36, 38, 41
  evolution of, 6, 21
  "golden years," 54, 56, 64
internal resistance, 232
International Energy Agency (IEA), 62
International Maritime Organization, 42–43
Interstate Highway Bill of 1956, 56–57
ionic conductivity, 221
ions, 218–220, 221, 229–230
Ishitani, Hisashi, 61–62
isovalent doping, 230
Israel, 79, 258–259
Italy, 81, 92
Ivarsson, Jan, 117, 118

J

J. D. Power and Associates, U.S. Green Automotive Study, 23–25
Japan, 48, 56, 57, 64, 66
  Nissan Leaf in, 71–72, 95
  rapid charging, 254–255, 262–263
Japan Automotive Research Institute (JARI), 256
jeeps, 31
Jinyu, Wang, 207
Johnson, Howard, 32–33
Johnson Controls, Inc., 74, 106, 225
JTB Okinawa, 95
junction box, 312

K

Kangoo ZE van (Renault), 96
Karma (Fisker), 254, 346
Kenwood, 59
kerosene, 55, 56
Khomeini, Ayatollah, 58

kill switches, 269, 275, 307
Killacycle, 224
Kimbell, Mike, 130
Kimmins, Dave, 99–100, 282, 283, 336
kinetic energy, 221
King, Julia, 97
Kirchner, Roy, 208
kits
  emergency, 351
  list of conversion vendors, 289
Knowledge Problem, 35
Koch, Ken, 286, 295–298, 325, 326, 327, 333, 334, 338
Kodjak, Drew, 51
KTA Services, Inc., 286, 288, 289, 292, 338
Kwong, Bill, 15

L

Lamborghini Diablo conversions, 103, 104
Le Mans, 77, 78
lead-acid batteries, 9, 13, 37, 39, 61, 217, 218, 220, 225, 233, 242, 245
Leaf (Nissan), 1, 3, 10, 13, 15, 17, 18–19, 21–23, 27, 66, 67, 71–72, 85, 86, 87–90, 94–95, 114, 115, 117, 132, 335, 339, 346
  costs, 44–45
  EV Power Station, 255–257
  EV Project, 22–23, 254, 255
  Leaf to Home power-supply system, 255–257
  tax credit, 19
  United Kingdom deliveries, 263
Lear steam cars, 55
leasing space, 285
leasing vehicles, 76, 86–98, 189, 357–358
"lemons," 155
Leviton, 265–266, 353
LG Chemical, 72–73
Liberty Electric Cars, 79–80
Liddle, Paul, 101, 102–104, 282, 283, 289, 295, 300–302, 305, 306, 308, 312, 313, 321, 322, 332
LifeDrive concept, 120
lift trucks, 31
liquid cooling systems, 160–161, 186, 190, 195, 201–203, 275

lithium-ion batteries, 4, 9, 13, 15, 21, 63, 72–76, 78, 106, 109, 211–213, 217–240. *See also* lithium-iron phosphate (LFP) batteries
 ampere-hours, 225
 C-rate, 226, 231
 constructing, 220–221
 flat discharge curve, 226–227
 formats, 223–225
 lithium-cobalt-oxide, 220, 231, 235
 lithium-manganese-oxide, 220
 lithium-polymer, 104, 222, 225, 345
 nickel-cobalt-manganese (NCM), 37
 Peukert effect, 217, 227, 233
 voltage, 225
 wiring, 225–226, 238–239
lithium-iron phosphate (LFP) batteries, 37, 217–240
 best practices, 235–239
 development, 227–231
Lithium Storage, 235–239
load leveling, 40, 51, 257
loaded phase, 230
LoFaso, Rick, 266
Los Angeles Department of Water and Power, 48
low-speed vehicles (LSVs), 3, 64–66, 77, 92, 341–342
low-voltage electrical systems, 269, 275, 308–312
LPG, 263
lubricants, 346
Lutz, Bob, 68–69

M

Mabus, Roy, 29
Madslien, Jorn, 102
magnetism, DC electric motor, 164, 166–170, 176–177
maintaining EVs, 345–349
manual transmission, 139, 142, 145, 291–292
manuals, 157
Manzita Micro, 201–202, 289, 314, 315
 PFC-20 charger, 247–248, 333
Marakby, Sherif, 74
Maritz Research, 19, 45
Markey, Ed, 39

Maryland, 14–15, 47, 51
Mason, Lynne, 100, 103–104, 106–108, 115–116
mass factor, 126
Massachusetts v. the U.S. Environmental Protection Agency, 39
Matsumora, Stephanie, 60
Mattei, Michael, 266
Maurer, Marc, 14–15
Mavis Tire Stores, 291
Maxwell, James Clerk, 166
Maxx, Suzanne, 354
Mayer, Jeffrey, 16–17
Mayer, Kathryn, 13
Mazda Miata, 59, 86, 108
McMinn, Nigel, 263
Megane (Renault), 88
metal oxide semiconductor field-effect transistor (MOSFET), 195, 203
Metric Mind, 111–112, 179, 181
MG convertible, 115–116
Miata (Mazda), 59, 86, 108
Michelin, 79
Michigan Economic Development Corporation, 75
Midi-E (Beiqi Foton), 207
Mini E (BMW), 204, 208
minivans, 66
Mitsubishi Motors i-MiEV, 88, 96
Modec, 77, 79–80
Model S Electric Car (Tesla), 1, 3, 10, 77, 82, 92, 93, 117, 154, 158, 159, 190, 289, 290, 301–302, 337, 346
Model T (Ford), 54, 55
Model X Electric Car (Tesla), 1–3, 4, 10, 81–82, 85, 117
Morris, Eric, 34–35
motor case, DC electric motor, 170
motor ratings, 187–188
motorcycles, 60, 78
motors. *See* electric motors
mounting electric motors, 292–294, 295–300
movies about EVs, 10, 64–66, 68, 71, 79, 81, 358
Multilateral Cooperation of Advance Electric Vehicles, 62
Murano (Nissan), 88
Musk, Elon, 1, 4, 189, 190

MXenergy, 16–17
Myers Motors, 86
Mynott, Casey, 61
myths concerning EVs, 15–19
    battery charging, 17–18
    controllers and, 203–204
    costs, 18–20
    range limitations, 16–17, 20, 21, 24, 61,
        76, 77, 106, 128, 344
    smokestack pollution, 45–51
    speed limitations. *See* speed of EVs

━━ N ━━

nanophosphate, 231
nanophosphate lithium-air batteries, 9, 77, 106
NanoSafe batteries (Altairnano), 77
National Ambient Air Quality Standard, 2
National Electric Drag Racing Association
    (NEDRA), 59, 60–62, 306
National Energy Policy Act of 1992, 66
National Federation of the Blind, 14–15
National Hot Rod Association (NHRA), 60,
    306
National Plug-In Day, 292, 293, 335, 339
National Renewable Energy Laboratory, 42
National Traffic and Highway
    Administration (NHTSA), 3, 222, 341–342
National Union Electric Company, 54
natural gas, 2, 33, 40, 47–48, 57
neighborhood electric vehicles (NEVs), 3,
    64–66, 77, 92, 341–342
NESCAUM (Northeast States for
    Coordinated Air Use Management), 49, 51
NetGain Controls, Inc., 202–203, 289
NetGain Technologies, 165–166, 172
Netherlands, 66, 92, 95–96
New York Metropolitan Transportation
    Authority (MTA), 53
New York Power Authority (NYPA), 10, 53,
    64, 74
        NYPA/TH!NK Clean Commute
            Program, 92, 354, 357–358
New York State
    Energy Research and Development
        Authority (NYSERDA), 255
    guidelines for EVs, 341, 342

Newton, Isaac, 125
Newton step van (Smith), 158–159
NHRA Alternative Sanctioning
    Organization (ASO), 60
Nichicon Corporation, 255–257
nickel-cobalt-manganese lithium-ion
    (NCM) batteries, 37
nickel-metal hydride batteries (NiMH), 9,
    13, 37
Nielsen Energy Survey, 16
Nippon Telephone and Telegraph (NTT), 227
Nissan, 5, 13–14, 203
        280Z rear battery compartment, 99–100
        300ZX, 59
        e-NV200 electric van, 132–133
        EV-4H truck, 61
        Hypermini, 66
        Leaf. *See* Leaf (Nissan)
        Murano, 88
        Qashqai, 88
        Renault-Nissan Alliance, 79, 95–96
        Versa S hatchback, 19, 45
Nissan North America Co., DC quick
    charger, 262–263
nitrogen oxide/dioxide emissions, 46–48
nonstoichiometry, 230
Noppert, Brian, 286
Northeast States for Coordinated Air Use
    Management (NESCAUM), 49, 51
Norway, 37, 72
nuclear electric power, 33, 40, 48, 57, 58
Nürburgring Nordschleife, 77–78

━━ O ━━

Obama, Barack, 34, 38–39, 54, 59, 69
Oceana, 42–44
Ofer, Idan, 259
Ohm's law, 166
oil industry
    costs of maintaining oil transit routes,
        29
    gasoline prices, 19, 31, 35–36, 44–45,
        54–59, 64, 66, 114, 345
    oil costs, 34
    oil discoveries in U.S., 55
    oil shocks, 34, 57–59

oil spill disasters, 29, 38, 42, 57
OPEC oil cartel, 1–2, 34, 39, 44, 57, 59
price subsidies, 15
reducing dependence on, 41–44
supply limitations, 30, 32–33, 34, 57–59
U.S. dependence on oil, 1–2, 29, 30,
    33–37, 41–44, 56–57
Oklahoma, 55
OLEDs (organic light-emitting diodes), 91
onboard chargers, 344, 351–352
OPEC (Organization of Petroleum Exporting
    Countries), 1–2, 34, 39, 44, 57, 59
Operation Z, 282, 283, 305, 317–318, 336
optimization, 121–123
Oregon, 23, 251
Organization of Petroleum Exporting
    Countries (OPEC), 1–2, 34, 39, 44, 57, 59
overcharging, 235
ozone, 47

━━ **P** ━━

Paine, Chris, 64, 65, 66
painting stage of conversion, 332
Palm Pilot handheld terminal, 201
Palmer, Andy, 132–133
Parks, Wally, 60
Partnership for a New Generation of
    Vehicles program, 64
Patnoe, Brian, 29
Paturet, Olivier, 72
PBS, 69
peak discharge current, 233
pedestrians, minimum noise requirement
    for hybrids, 14–15
Pelosi, Nancy, 39
permanent-magnet DC motors, 176–177
Peterson, Josh, 60
Petrolia (Honda), 89
Peugeot iOn, 91
Peukert effect, 217, 227, 233
Phoenix Motorcars, 70, 77, 201
Phoenix Solar & Electric 500, 258
pickup trucks, 31, 131
    chassis for conversions, 108–109
    Chevrolet S-10 conversions, 104, 105,
        282

conversion cost estimates, 113
    Ranger conversions (Ford), 10, 66,
        110–111, 126, 128–129, 134, 136, 137,
        144–145, 147, 149–152, 282, 294, 308,
        309, 312, 314, 334, 343
pigtails (motor brushes), 111, 170
Pike Research, 18, 356, 359
PlanetGreen.com, 356
planning stage of conversion, 146–154,
    284–286
Plug In America (formerly dontcrush.com),
    62, 66
Plug-In Day, 292, 293, 335, 339
plug-in hybrid electric vehicles (PHEVs),
    3, 15–17, 64–65, 74, 76–77, 189, 351–352,
    353, 358, 359–360. *See also* Volt
    (Chevrolet)
Poland, 92
polarity, reversing, 194
POLIS, 79
polishing stage of conversion, 332
polyphase AC induction motors, 181–182,
    183
Porcelli, Joe, 99–100, 282, 283, 336
Porsche, 3, 4, 5, 10, 12, 86
    911 Carrera, 82, 103–104, 115, 190, 196,
        341
    conversions, 52, 102–104, 106–108,
        115–116, 190, 196, 274, 281, 298–302,
        305–308, 311, 312, 313, 322, 332, 333,
        335, 337, 338, 341
    Roadster, 323, 333, 335, 337, 338
    Speedster, 274, 282, 289
Portugal, 92
Posawetz, Tony, 13
potential energy, 221
potentiometers, 193, 269–270, 308–310
pouch cells, 224–225
power, 123
power density, battery, 228, 232–233
power electronics unit (PEU), 207
power transistors, 194
prebuilt EVs, 20–25, 86–98
Prigge, Scott, 252
prismatic batteries, 224–225
prismatic pouch cells, 224–225

Prius hybrid (Toyota), 15, 19, 45, 64, 76–77, 114, 189, 201–202, 254, 346, 358
Proceedings of the IEEE, 353–354
programmability, 111–112
Project Better Place, 258–259, 261
Project i, 63
public transportation, 1, 53, 56, 359
pulse-width modulation (PWM), 195–198

Q

Q7 (Audi), 82
Qashqai (Nissan), 88
Quadricycle, 55

R

R8 e-tron (Audi), 77–78
R8 LMS (Audi), 78
radial tires, 137
railroads, 56–57, 59–60
Rainforest Action Network, 66
Ram (Dodge), 282
range of EVs, 16–17, 20, 21, 24, 61, 76, 77, 106, 128, 344
Ranger pickup (Ford), 10, 66, 110–111, 126, 128–129, 134, 136, 137, 144–145, 147, 149–152, 282, 294, 308, 309, 312, 314, 334, 343
rapid charging, 231–232, 254–255, 262–263, 354–355
Raptor controller, 201
RAV4 (Toyota), 1, 3, 10–11, 20, 27, 31, 66, 67, 71, 88, 93, 117, 207, 256, 265–266, 283, 284, 323
Read, Colin, 251
Read, Jonathan, 264
ready-to-run EVs, 20–25, 86–98
rear airflow, 134
recall of EVs, 66, 67
rectifiers, 195
recycling, battery, 13–14, 37
Regency Centers, 251–252
regenerative braking, 111, 142, 344
    compound DC motor, 176
    series DC motor, 173
    shunt DC motor, 175
registration, 341–343

relative wind factor, 133
relays, auxiliary, 270–271
REMA, 323, 330
Renault
    Clio, 88–89
    Fluence ZE, 96, 258
    Kangoo ZE van, 96
    Megane, 88
    Renault-Nissan Alliance, 79, 95–96
    Twizy, 96
    ZOE, 96
replacement battery packs, 258–259, 261
resale of EVs, 20
resistance, 222, 232
REVA Electric Car/G-Wiz, 20, 27, 77, 97
Revenge of the Electric Car (documentary movie), 10, 64, 65
reverse, electric, 112
reversing compound DC motor, 176
reversing contactors, 267, 270
reversing polarity, 194
reversing series DC motor, 172
reversing shunt DC motor, 175
Rhoads, Wistar, 338
Rice University, 38
Rickard, Jack, 106–108, 235–239
Roadster (Porsche), 323, 333, 335, 337, 338
Roadster/Roadster Sport (Tesla Motors), 1, 10, 16, 70–71, 77, 78, 83, 92, 117, 189, 190, 196, 214–215, 301, 329
Rohrer, Finlo, 87–91
rolling resistance, 135–138, 157, 346
Rolls-Royce, 101–102
        102EX Phantom Experimental Electric, 101–102
        Centurion conversion, 101
rotary switch, 273
rotor, 164
routing, 276
Rudiger, 80
Rudman, Rich, 201–202
running out of power, 344, 351–352
rust, 156

S

Saab, 78–79

safety issues, 341–343
    battery, 13–14, 222–223, 228, 233, 234–236, 239
    controller failure, 112, 203
    driving EVs, 120, 344
    electric motors, 349–351
    electrical system, 267–269, 273–275, 276
    emergency kit, 351
    kill switches, 269, 275, 307
    safety standards, 321–322
    thermal runaway, 233, 234–236
salts, 219–220
Savitz, Jacqueline, 42
Science Daily, 38
Scion iQ (Toyota), 20, 71, 93, 360–361, 362
SCR (silicon-controlled rectifiers), 195
September 11, 2001 terrorist attacks, 66, 69, 357–358
series DC motors, 169, 171, 172–173
Sevcon
    controllers, 203–204
    DC-to-DC converters, 275
Sherman Antitrust Act, 56
short circuits, 235
shunt DC motors, 169, 171, 173–175
shunts, 272
silicon-controlled rectifiers (SCR), 195
Silverman, Beth, 188
single-phase AC induction motors, 180
Sinopoly, 224, 237
Six-Day War (1967), 57
Smalley, Joe, 201
Smart Electric Car, 237, 323
    battery case, 238
    ChargePoint America program, 247, 254
    conversions, 298–300
    Smart fortwo, 75–76, 81, 90–91
    Smart forvision, 90–91
smart grid, 353–354
Smith, Colin, 80
Smith Electric Vehicles Corp., Newton step van, 158–159
Society of Automotive Engineers (SAE), 15, 250
Solar & Electric 500 (Phoenix), 258

solar energy, 2, 40, 48, 49, 91, 258
Solectria Force, 10
Solhjell, Bård Vegar, 72
solid-state controllers, 195
solid waste pollution, 13–14, 37
Sonata (Hyundai), 13
Southern California Edison (SCE), 51, 74
SP, 79
Spain, 92, 95, 132–133
Sparrow (Corbin Motors), 70, 86, 248
specific energy, 232
specific gravity, 243
speed of EVs, 16, 59–60, 61, 70, 73, 82, 107–108, 123
    aerodynamic drag and, 134
    compound DC motor, 175–176
    DC motor speed, 172–173, 174
    weight and, 126–128
Speedster (Porsche), 274, 282, 289
Sperling, Dan, 36
Spiesshofer, Ulrich, 18
spinel, 220
split-phase AC induction motors, 181–182
sport utility vehicles (SUVs), 1, 3, 51, 64, 77, 100–101, 120, 281–282, 290
sports utility truck (SUT), 77
Stanley Steamer automobiles, 55
starting EVs, 343
state of charge (SOC), 209, 211, 221, 243, 246, 272, 311
state of health (SOH), 210
stator, 164
steam-powered vehicles, 54, 55
steering scuff, 137
Stippler, Frank, 78
Streng, Hans, 18, 356
Subaru, 115–116
subways, 1, 53
Sugiyama, Satoru, 59
sulfur batteries, 61
sulfur oxide/dioxide emissions, 46–48
Sumitomo Corporation, DC quick charger, 262–263
Sumner, Randy, 250–251
super ultralow-emissions vehicles (SULEVs), 66

supervalent ions, 229–230
SUVs (sport utility vehicles), 1, 3, 51, 64, 77, 100–101, 120, 281–282, 290
Sweden, 92
Switchable Electric Car Experience Center, 261
Switzerland, 83

━━ **T** ━━

T-Rex controller, 201
Tabe, Masahiko, 72
Tango (Commuter Cars Corporation), 1, 70, 184, 201
Tata Indica Vista, 90
Tauber, Lou, 60
Taurus (Ford), 131, 144, 145
taxes
    gas taxes, 35–36
    tax credits for EVs, 19, 21, 66
telematics communication unit (TCU), 95
Tennessee, 251, 263–265
terminal strips, 271–272
Terrorist attacks (9/11/2001), 66, 69, 357–358
Tesco, 79–80
Tesla, Nikola, 180
Tesla Motors, 15, 27, 30–31, 106, 203, 207, 224, 265–266, 360
    Model S Electric Car, 1, 3, 10, 77, 82, 92, 93, 117, 154, 158, 159, 190, 289, 290, 301–302, 337, 346
    Model X Electric Car, 1–3, 4, 10, 81–82, 85, 117
    Roadster/Roadster Sport, 1, 10, 16, 70–71, 77, 78, 83, 92, 117, 189, 190, 196, 214–215, 301, 329
    Toyota Motor Corporation joint venture, 188–189
testing
    battery, 242–243, 272–273
    checking stage of conversion process, 276–277, 328–332
    on component level, 118
    electric motor, 298–300
Texas, 51, 55, 227–228, 251
TGV electric train, 59–60
TheDailyGreen.com, 13–14, 96, 263–264

thermal runaway, 233, 234–236
TH!NK City, 1, 3, 10, 16, 27, 78–79, 83, 245, 250, 354–355
    NYPA/TH!NK Clean Commute Program, 92, 354, 357–358
TH!NK Electric Car Company, U.S. Ready Cities Index, 356–357
TH!NK North America, 78–79, 357
Three Mile Island nuclear accident (1979), 58
throttle-position sensor (TPS), 270
throttle potentiometer, 269–270, 308–310
Thundersky batteries, 236
Tire and Rim Association, 136
tires, 135–138, 157, 346
Tischer, Eric, 115–116
Todd, Richard, 263
tools for conversions, 285
Toronto Atmospheric Fund EV300, 74
torque, 108, 111, 123, 126–128, 141–142, 145, 148
    available, 149–154
    compound DC motor, 175
    electric motor, 164, 167, 170
    required, 149–154
    shunt DC motor, 173
toxic input fluids pollution, 38–39
Toyoda, Akio, 188
Toyota Motor Corporation, 32, 69, 78–79, 111, 353
    Camry, 114
    Prius hybrid, 15, 19, 45, 64, 76–77, 114, 189, 201–202, 254, 346, 358
    RAV4, 1, 3, 10–11, 20, 27, 31, 66, 67, 71, 88, 93, 117, 207, 256, 265–266, 283, 284, 323
    Scion iQ, 20, 71, 93, 360–361, 362
    Tesla Motors joint venture, 188–189
Toyota Motor Sales USA, 15
trains, 1, 53, 56–57, 59–60
transformers, 179–180
transistors, 54, 194, 195
Transit Connect van (Ford), 73–75
transmission, 139, 142, 144–146, 145, 172, 291–292, 343
TransWarP, 172, 203
Treehugger.com, 38

Triton Trikes, 98

Trojan batteries, 217, 218, 225, 242

trucks, 1, 10, 61, 64, 66, 79–80, 102, 281–282, 290. *See also* pickup trucks
  delivery vehicles, 31–32, 79–80
  electric, 31–32, 203–204
  lift trucks, 31
  sports utility truck (SUT), 77

tuning, motor, 188

TurnE Conversions, 80–81, 119, 145, 161, 223, 224, 237, 238, 239, 248, 249, 272, 274, 282, 289, 292–303, 348

Twizy (Renault), 96

▬ U ▬

ULEVs (ultra-low emission vehicles), 45–46, 49

Ultimate Customs, 286, 288

ultra-low emission vehicles (ULEVs), 45–46, 49

underbody airflow, 134–135

Union of Concerned Scientists, 49

United Kingdom, 20, 31, 47, 48, 77, 87–88, 97, 117, 263

U.S. Armed Forces, 29

U.S. Congressional Budget Office, 34–35

U.S. Department of Defense, 358

U.S. Department of Energy (DOE), 1–2, 22–23, 69, 254, 355
  Pacific Northwest National Laboratory, 33–34

U.S. Energy Information Administration, 42–44

U.S. Environmental Protection Agency (EPA), 21, 32, 39, 51, 76, 96

U.S. Federal Aviation Administration (FAA), 35–36

U.S. Green Automotive Study, J. D. Power and Associates, 23–25

U.S. Postal Service vehicles, 31, 208

U.S. Supreme Court, *Massachusetts v. the U.S. Environmental Protection Agency*, 39

universal DC motors, 177, 178

University of California, 358

University of Texas, 227–228

unloaded phase, 230

UPS, 79–80

UQM, 102, 177

USA Today, 76

used chassis, 12, 120–121, 155–156

used engine parts, 157

utilities. *See* electric utility companies

▬ V ▬

VanNieuwkuyk, Mike, 24

vans, 1, 31, 66, 74–75, 96, 100–101, 102, 124, 131, 132–133, 158–159

Vauxhall Ampera, 87–89

Vectrix, 78, 224

vehicle identification numbers (VINs), 3, 342

vehicle management system (VMS), 208–209

Vehicle Motor Safety Standards Rule No. 500, 92

Vehicle to Grid (V2G), 255

Vehma International G-Van, 31

Venturi Fetish, 207

Veolia, 79–80

Vermont, 49

Versa S hatchback (Nissan), 19, 45

Vezzi, Roberto, 81

Vincentric, 19, 45

VINs (vehicle identification numbers), 3, 342

Vogel, Carl, 292

Volkswagen conversions, 79, 86, 136, 161, 297, 298–300

Volt (Chevrolet), 3, 13, 16–17, 21–23, 68–69, 88, 89, 114, 115, 160, 254, 346
  costs, 44–45
  tax credit, 19

voltage, 166
  cell, 225, 243
  high-voltage electrical systems, 266
  low-voltage electrical systems, 269, 275, 308–312

voltmeters, 272, 311

volts, 221–222

Volvo Car Corporation, 75, 78–79, 117, 118

▬ W ▬

Wagoner, Rick, 69, 155

Waltz, David L., 354

Wang, Quanlu, 36
warming
      battery, 237
      heating element, 330–331
WarP-Drive controllers, 202–203
WarP motor, 61, 165–166, 306
warranties, 13
Washington, D. C., 251, 359
Washington State Department of
      Transportation (WSDOT), 21–23
water pollution
      battery, 13
      oil spill disasters, 29, 38, 42, 57
Wayland, John, 60
weight issues
      acceleration and, 125–126
      battery weight, 129–130
      chassis and, 119, 123, 124–130, 157
      during and after conversion, 124–125,
            157
      gross vehicle weight rating (GVWR),
            21, 119
      hill-climbing force and, 126, 127
      range and, 128
      removing unessential weight, 124
      removing weight to keep balance,
            128–129
      speed and, 126–128
Welburn, Ed, 75
Westin Hotel example, 220, 231
Westinghouse, George, 163
wheel well airflow, 134
Whittaker, Tony, 263
Who Killed the Electric Car? (documentary
      movie), 10, 64, 65, 66, 68, 71, 79, 358
Wicks, Jonathan, 252
Wilde, Roderick, 60
Wilde Evolutions, 60
wind energy, 40, 43, 48
wind tunnel test, 134
Winkelhock, Markus, 77–78
Winkler, Annette, 91

wiring
      battery, 225–226, 238–239, 322–323
      electrical system, 267, 275–277
WiTricity Corporation, 250–251
Wolf, Kevin, 346
Woodbury, Rick, 184
Woodhouse, Michael A., 265
Woodyard, Chris, 76
World Class Exotics, 103–104
World Electric Vehicles Association
      (WEVA), 61–62
World Resources Institute, 47
World War I, 56–57
World War II, 56–57
World's Columbian Exposition (Chicago,
      1893), 55
wound cells, 224
wound-rotor AC induction motors, 182–185
Wrightspeed, 207
Wyland, John, 305–306

X

X2000 electric train, 59–60
Xcel Energy, 74

Y

yoke, DC electric motor, 170
Yom Kippur War (1973), 34, 57
YouTube, 233
Yulon MPV, 204

Z

ZAPI, 203
zero emission vehicles (ZEVs), 2, 13, 27–28,
      29, 36, 45–46, 54, 64–66, 71
Zhu, Raymond, 203–204
Zilla controllers, 61, 104, 196, 199–202, 288,
      314
      Hairball 2 interface box, 200–201, 296
ZipCars, 354, 355
Zivan NG3 charger, 248–250
ZOE (Renault), 96

Electric Car Conversion
and Electric Transport News

http://buildyourownelectricvehicle.wordpress.com/

Please follow us on Twitter: @BuildYourOwnEV

Electric Cars are for Girls

Mostly Painless Guide to Electric Vehicles and Conversions

Converting your gas-guzzler to electric?

Ever ask yourself...

- Can an alternator power an electric car?
- Does an electric vehicle need a transmission?
- Can I convert a front wheel drive to electric?
- Can I keep the power steering and brakes in my conversion?

You're not alone!

Come see our huge library of conversion questions and answers at
Electric-Cars-are-for-Girls.com.

National Electric Drag Racing Association

"World Class Electric Drag Racing"

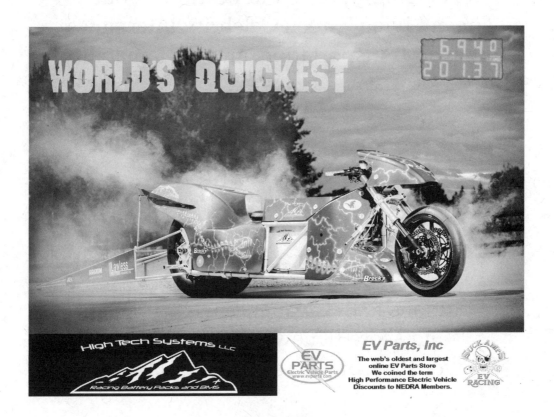